Date Due

PURE AND APPLIED MATHEMATICS

A Program of Monographs, Textbooks, and Lecture Notes

MONOGRAPHS AND TEXTBOOKS IN PURE AND APPLIED MATHEMATICS

1. K. Yano. Integral Formulas in Riemannian Geometry (1970)
2. S. Kobayashi. Hyperbolic Manifolds and Holomorphic Mappings (1970)
3. V. S. Vladimirov. Equations of Mathematical Physics (A. Jeffrey, editor; A. Littlewood, translator) (1970)
4. B. N. Pshenichnyi. Necessary Conditions for an Extremum (L. Neustadt, translation editor; K. Makowski, translator) (1971)
5. L. Narici, E. Beckenstein, and G. Bachman. Functional Analysis and Valuation Theory (1971)
6. D. S. Passman. Infinite Group Rings (1971)
7. L. Dornhoff. Group Representation Theory (in two parts). Part A: Ordinary Representation Theory. Part B: Modular Representation Theory (1971, 1972)
8. W. Boothby and G. L. Weiss (eds.). Symmetric Spaces: Short Courses Presented at Washington University (1972)
9. Y. Matsushima. Differentiable Manifolds (E. T. Kobayashi, translator) (1972)
10. L. E. Ward, Jr. Topology: An Outline for a First Course (1972)
11. A. Babakhanian. Cohomological Methods in Group Theory (1972)
12. R. Gilmer. Multiplicative Ideal Theory (1972)
13. J. Yeh. Stochastic Processes and the Wiener Integral (1973)
14. J. Barros-Neto. Introduction to the Theory of Distributions (1973)
15. R. Larsen. Functional Analysis: An Introduction (1973)
16. K. Yano and S. Ishihara. Tangent and Cotangent Bundles: Differential Geometry (1973)
17. C. Procesi. Rings with Polynomial Identities (1973)
18. R. Hermann. Geometry, Physics, and Systems (1973)
19. N. R. Wallach. Harmonic Analysis on Homogeneous Spaces (1973)
20. J. Dieudonné. Introduction to the Theory of Formal Groups (1973)
21. I. Vaisman. Cohomology and Differential Forms (1973)
22. B.-Y. Chen. Geometry of Submanifolds (1973)
23. M. Marcus. Finite Dimensional Multilinear Algebra (in two parts) (1973, 1975)
24. R. Larsen. Banach Algebras: An Introduction (1973)
25. R. O. Kujala and A. L. Vitter (eds). Value Distribution Theory: Part A; Part B. Deficit and Bezout Estimates by Wilhelm Stoll (1973)
26. K. B. Stolarsky. Algebraic Numbers and Diophantine Approximation (1974)
27. A. R. Magid. The Separable Galois Theory of Commutative Rings (1974)
28. B. R. McDonald. Finite Rings with Identity (1974)
29. I. Satake. Linear Algebra (S. Koh, T. Akiba, and S. Ihara, translators) (1975)
30. J. S. Golan. Localization of Noncommutative Rings (1975)
31. G. Klambauer. Mathematical Analysis (1975)

MATHEMATICAL ANALYSIS

GABRIEL KLAMBAUER

Department of Mathematics
University of Ottawa
Ottawa, Ontario, Canada

MARCEL DEKKER, INC. New York

MARCEL DEKKER, INC.
270 Madison Avenue, New York, New York 10016

LIBRARY OF CONGRES CATALOG CARD NUMBER: 75-11419
ISBN: 0-8247-6329-7

Current printing (last digit):
10 9 8 7 6 5 4 3 2 1

PRINTED IN THE UNITED STATES OF AMERICA

TO THE MEMORY OF MY MOTHER

CONTENTS

PREFACE

This book treats basic material concerning the real number system, sequences and series, continuity, differential and integral calculus of functions of one variable, uniform convergence, and the elements of the theory of metric spaces. It is designed for undergraduate students who have completed the ordinary courses in elementary calculus and are about to face a shift of emphasis from computational "know-how" to conceptual "know-why" in their further mathematical training.

The style of presentation is concise and the approach is to proceed from the more familiar situations to the more abstract theory. The proofs are presented with care and important results are often proved in several ways to aid the understanding of the student. The text is supplemented by stimulating exercises; many of these exercises are provided with solutions. The exercises serve the purpose of extending the results discussed in the text and to illustrate their usefulness.

In writing a textbook, the matter of what to put into it is certainly important; yet I feel that the question of what to leave out is even more important because sometimes 'less is more' in human affairs. I have therefore concentrated attention on topics whose importance is beyond question and tried to be reasonably comprehensive only in terms of the true objectives of the book. For more advanced topics and further study I refer the reader to my graduate text *Real Analysis* (American Elsevier Publishing Company, New York, 1973).

Finally, I am pleased to express my sense of gratitude to Dr. Antoine D'Iorio, Dean, Faculty of Science and Engineering, University of Ottawa; without the benefit of a sabbatical leave the progress in writing this book would not have been as rapid. I am deeply indebted to my colleague Dr. E. L. Cohen, my former student Dr. G. K. R. Rao, and my son John for their valuable assistance, and to Mrs. Solange Zayat who typed Chapter 5. I am particularly pleased to record my thanks to the publishers Dr. Maurits and Marcel Dekker and to their distinguished group of editors.

Gabriel Klambauer

Mathematical Analysis

Chapter 1

THE REAL NUMBER SYSTEM

In this chapter we consider properties of the real number system that are essential for a proper understanding of the basic concepts of calculus (e.g., convergence, continuity, differentiation, and integration). However, the details of proof in the first two sections of this chapter may be omitted on the first reading by the beginner.

1. On Integers, Rational Numbers, and Irrational Numbers

The set N of positive integers (natural numbers) satisfies two important properties; we shall state these properties and establish their equivalence.

Well-Ordering Principle Any nonempty set of positive integers has a smallest member.

Principle of Mathematical Induction If P is a set of positive integers with the properties:

(i) P contains the number 1, and

(ii) whenever P contains the positive integer n it also contains the positive integer $n + 1$,

then P contains all positive integers, that is, $P = N$.

Proposition 1 The number 1 is the smallest positive integer.

Proof 1 (Based on the Principle of Mathematical Induction) Let S be the set of all positive integers ≥ 1. Clearly, 1 is in S. If the integer n is in S, then $n \geq 1$. Thus, $n + 1 > n \geq 1$ and so the integer $n + 1$ is in S. It follows from the Principle of Mathematical Induction that $S = N$. Therefore each positive integer is greater than or equal to 1. □

Proof 2 *(Based on the Well-Ordering Principle)* It follows from the Well-Ordering Principle that there is a smallest positive integer, say s. Assume that $s < 1$. On multiplying the inequality $0 < s < 1$ by s, we obtain $0 < s^2 < s$ which implies that s is not the smallest positive integer. Since our assumption led to a contradiction, it is false. Therefore, 1 is the smallest integer. □

Corollary If n is an integer, there is no integer between n and $n + 1$.

Proof Suppose that there was an integer k such that $n < k < n + 1$. Then $0 < k - n < 1$ would follow, contradicting that 1 is the smallest integer. □

Remark Observe that the false claim "There is a largest positive integer n" implies that $n^2 = n$, which in turn would mean that n would have to be equal to 1.

Proposition 2 The Principle of Mathematical Induction and the

Well-Ordering Principle are equivalent in the sense that we can deduce one from the other, provided that the usual arithmetic properties of the integers are assumed.

Proof (Part I) Assume the Well-Ordering Principle to hold and let S be the set of positive integers with the properties:

(i) 1 is in S, and

(ii) if n is in S, n + 1 is also in S.

We must show that S is the set N of all positive integers.

Let T be the set of all positive integers not in S. If T is nonempty, it follows from the Well-Ordering Principle that T has a smallest member, say t. Since 1 is in S and 1 is the smallest positive integer (established in Proposition 1 using the Well-Ordering Principle), then $t > 1$. Thus $t - 1$ is a positive integer that must be in S because $t - 1 < t$. The second property of S insures that t, equal to $(t - 1) + 1$, is also in S, which is a contradiction since S and T are disjoint sets. Because our assumption that T is nonempty leads to a contradiction, it is false. The set T is therefore empty and $S = N$.

(Part II) Assume that the Principle of Mathematical Induction holds. Suppose, moreover, that there is a nonempty set of positive integers, say S, that does not have a smallest member. Since 1 is the smallest positive integer (established in Proposition 1 using the Principle of Mathematical Induction), then 1 is not in S and is smaller than all of the elements of S.

Let T be the set of all positive integers that are less than all of the elements of S. We have established that 1 is in T. Suppose the integer n is in T. If n + 1 is in S, then, because there is no integer between n and n + 1 (by the corollary of Proposition 1), n + 1 is the smallest element of S; but this is a contradiction to our assumption about S. Therefore, if n is in T, n + 1 must also be in T. It follows from the Principle of Mathematical Induction that T contains all the positive integers, and so S is empty. This, however, is contrary to our original assumption that S is nonempty. Therefore, if S is a nonempty set of positive integers, then S has a smallest element and the proof is complete. □

A modified, but equivalent, form of the Principle of Mathematical Induction reads:

A statement about positive integers that has the following two properties is true for all integers:

(i) The statement is true for the integer 1.

(ii) If the statement is true for the positive integer n, it must also be true for the integer n + 1.

Definition An integer a is said to be *divisible by* an integer b $\neq 0$ if there is a third integer c such that bc = a. We also say that b *divides* a, or a is a *multiple of* b, and we denote this fact by $b|a$. If $b \neq 0$ and a is not divisible by b, we write $b \nmid a$.

It is easy to verify the following properties:

$b|a$ and $a > 0$, $b > 0$ implies $1 \leq b \leq a$;

$c|b$ and $b|a$ implies $c|a$;

$c|a$ and $c|b$ implies $c|ma + nb$ for all integers m, n.

Definition An integer $p > 1$, not divisible by any smaller positive integer except 1, is called a *prime number*. A positive integer greater than 1 that is not a prime number is called a *composite number*.

Proposition 3 An integer greater than 1 is either a prime number or it is a product of prime numbers. Moreover, the factorization into primes is unique except for the order of the factors.

Proof We first verify the fact that a composite number can be factored as a product of primes.

Suppose the claim is false. Then there is a composite number that cannot be written as a product of primes. Let n be the smallest such number (such n exists by the Well-Ordering Principle). Since n is a composite, we can write

$$n = ab, \quad 1 < a < n, \quad 1 < b < n.$$

As a is smaller than n, a is a prime or a product of primes; a similar statement holds for b. But then n = ab is a product of primes, which is a contradiction. Thus, each composite number can be factored as a product of prime numbers.

We can now prove that the factorization is unique. Suppose that a composite number exists for which the factorization is not unique. Let (and here we invoke the Well-Ordering Principle)

$$Q = p_1 p_2 \cdots p_k = q_1 q_2 \cdots q_j$$

be the smallest such number, where $p_1, p_2, \ldots, p_k$ and $q_1, q_2, \ldots, q_j$ are prime numbers. Evidently, all the p's are different from all the q's, for otherwise we could cancel some p and some q and have a smaller integer with two distinct factorizations. We may assume that

$$p_1 \le p_2 \le \cdots \le p_k \quad \text{and} \quad q_1 \le q_2 \le \cdots \le q_j.$$

Since $p_1 \ne q_1$, we may suppose that $p_1 < q_1$ without any restriction. We now consider

$$P = p_1 q_2 \cdots q_j.$$

Clearly $p_1 | P$ and $p_1 | Q$, so that $p_1 | (Q - P)$. But $Q - P$ is positive since

$$Q - P = (q_1 - p_1) q_2 q_3 \cdots q_j.$$

We write $q_1 - p_1$ as a product of primes, say

$$q_1 - p_1 = r_1 \cdots r_s.$$

Then

$$Q - P = r_1 \cdots r_s q_2 \cdots q_j.$$

We have seen already that p_1 is none of the q_i. Since $p_1 \nmid (q_1 - p_1)$, it follows that $p_1 \nmid r_i$ for all i. Thus all the q's and p's are distinct from p_1. On the other hand, we have seen that $Q - P$ is divisible by p_1 and, therefore,

$$Q - P = p_1 t_1 \cdots t_h,$$

where the t's are primes. Thus we have two distinct factorizations of $Q - P$, in one of which p_1 appears and in the other not. This contradicts the minimum property of Q because $0 < Q - P < Q$. □

Proposition 4 Outside of any given finite set of prime numbers there is another one, i.e., there is no largest prime number.

Proof Let $q = 2\cdot3\cdot5 \cdots p$ be the product of all prime numbers $\le p$. The number

$$q + 1 = (2\cdot3\cdot5 \cdots p) + 1$$

is not divisible by any of these prime numbers. But $q + 1 > 1$, and so, either $q + 1$ is a prime number greater than p, or $q + 1$ is divisible by prime numbers greater than p. Hence the number of primes must be infinite. □

Remark Let p_n denote the n-th prime number; from the proof of the foregoing proposition we see that

$$p_m | p_1p_2 \cdots p_n + 1$$

for some $m > n$, and so $p_{n+1} \le p_m < p_n^n + 1$.

We can now easily verify the estimate

$$p_n < 2^{2^n}$$

by the Principle of Mathematical Induction and suppose that

$$p_1 < 2^2, \quad p_2 < 2^{2^2}, \quad \ldots, \quad p_n < 2^{2^n}.$$

Then

$$p_{n+1} \le p_1p_2 \cdots p_n + 1 < 2^{2+4+\cdots+2^n} + 1 < 2^{2^{n+1}}$$

holds and the claimed estimate is established.

Proposition 5 If p is a prime, then $\sqrt{p}$ is not a rational, that is, a ratio of two integers, but an irrational number.

Proof The claim is that

$$p = \left(\frac{m}{n}\right)^2 = \frac{m^2}{n^2}$$

is impossible in natural numbers m, n. Suppose it were possible. Then we would have the equation

$$pn^2 = m^2.$$

But a square number always has an even number of prime factors. The preceding equation is therefore impossible, since on the left-hand side we have a number with an odd number of prime factors, and on the right-hand side the same number with an even number of prime factors, contrary to the uniqueness of prime factors. □

Remark From the proof of the foregoing proposition we can easily see that if k (> 1) is not a square integer, then $\sqrt{k}$ must be an irrational number. Indeed, let P be a prime factor of k with an odd exponent. Then $kn^2 = m^2$ would have on the left the prime factor P with an odd exponent, but not on the right, violating Proposition 3.

Proposition 6 Let k be a natural number, but not a square integer. Let U denote the set of all positive rational numbers whose square is larger than k, and let L denote all rational numbers not belonging to the set U. Then U has no smallest member, and L has no largest member.

Proof Suppose that L contained a largest member, say a, and U contained a smallest member, say b. Thus, by assumption, $a^2 < k$ and $b^2 > k$. We now select two positive integers m and n such that

$$m^2 > kn^2$$

and put

$$\frac{ma + nk}{na + m} = a' \quad \text{and} \quad \frac{mb + nk}{nb + m} = b'.$$

Then

$$a' - a = \frac{n(k - a^2)}{na + m} > 0 \quad \text{and} \quad b' - b = \frac{m(k - b^2)}{nb + m} < 0,$$

$$(a')^2 - k = \frac{(ma + nk)^2 - k(na + m)^2}{(na + m)^2} = \frac{(m^2 - kn^2)(a^2 - k)}{(na + m)^2} < 0,$$

$$(b')^2 - k = \frac{(mb + nk)^2 - k(nb + m)^2}{(nb + m)^2} = \frac{(m^2 - kn^2)(b^2 - k)}{(nb + m)^2} > 0.$$

Thus $a' > a$, $(a')^2 < k$, and $0 < b' < b$, $(b')^2 > k$; this means that a' is in L and b' is in U, in violation of the assumption that a is the largest element of L and that b is the smallest element of U. □

Remarks Proposition 6 shows that not every nonempty set of positive rational numbers has a smallest member. Proposition 5 and the remark that follows it show that there are infinitely many irrational numbers. To appreciate the overwhelming preponderance of irrational numbers over the rational numbers we note that we can form a set of open intervals which cover all the points on the number line, that is, every rational point of the number line belongs to some member of this set of open intervals, and yet the sum of the lengths of the intervals is arbitrarily small; this is true even though the set of rational numbers is dense in the sense that between any two rational numbers there is another rational, e.g., the midpoint. Indeed, since the rationals can be put in one-to-one correspondence with the set N of positive integers (a fact that we shall show presently), we can consider an enumeration

$$r_1, r_2, r_3, \ldots$$

for the set of all rational numbers and construct around the rational point r_k of the number line an open interval I_k with r_k as midpoint and I_k having length $2^{-k}c$. The intervals so constructed cover the set of all rational points, and we have that the sum of their lengths is

$$\frac{c}{2} + \frac{c}{2^2} + \frac{c}{2^3} + \cdots = c.$$

It remains to verify that the set of rational numbers can be put in one-to-one correspondence with the set N of positive integers. We do this now.

We first consider a rule for ordering the positive rational numbers n/m, assuming that the largest natural number which divides both n and m is 1 in order to avoid duplicates. We say that n_1/m_1 *precedes* n_2/m_2 in case either of the following conditions (i) or (ii) holds

(i) $n_1 + m_1 < n_2 + m_2$;

(ii) $n_1 + m_1 = n_2 + m_2$ and $n_1 < n_2$.

This procedure gives us an ordering of the positive rational numbers beginning thus:

$$\frac{1}{1}, \frac{1}{2}, \frac{2}{1}, \frac{1}{3}, \frac{3}{1}, \frac{1}{4}, \frac{2}{3}, \frac{3}{2}, \frac{4}{1}, \frac{1}{5}, \frac{5}{1}, \ldots.$$

This enumeration $t_1, t_2, t_3, \ldots$, which includes every positive rational number, can be modified to become an enumeration of all rational numbers by inserting the negatives and zero in some fashion as

$$0, t_1, -t_1, t_2, -t_2, t_3, -t_3, \ldots$$

and we have what we set out to do.

N.B. The reader may be interested to look up Exercise 74 at the end of this chapter in connection with the foregoing remarks.

2. *Dedekind Cuts*

In this section we study a method due to Richard Dedekind for the construction of the real numbers as cuts in the set of rational numbers. The chief aim of this study is to show that the real number system is order-complete (see Proposition 30).

Consider the set Q of all rational numbers (i.e., numbers of the form n/m, where n and m are integers, $m \neq 0$) sprinkled in their natural position on the number line. If A is a subset of Q, A' will denote the set of all rational numbers not belonging to A.

Basic Definition A set α of rational numbers is called a *cut* if

(I) α contains at least one rational, but not every rational;

(II) if $p \in \alpha$ and $q < p$ with q rational, then $q \in \alpha$;

(III) α contains no largest rational.

Notational Convention In this section, we shall always use $p, q, r, s, \ldots$ to denote rational numbers, while cuts will be denoted by α, β, γ, δ, ..., except where otherwise specified.

Proposition 7 If $p \in \alpha$, q is rational, and $q \notin \alpha$, then $q > p$.

Proof The assumption $q \le p$ leads, via (II), to the contradiction $q \in \alpha$. □

In view of the foregoing proposition, the members of α are sometimes referred to as the *lower numbers of* α, whereas the rational numbers that are not in α are called the *upper numbers of* α. The set of all upper numbers of a cut α we shall denote by α'. A *proper upper number of* α is any upper number of α such that there is a smaller upper number of α.

Proposition 8 Let r be a rational number and α be a set consisting of all rational numbers p such that $p < r$. Then α is a cut, and r is the smallest, or least, upper number of α.

Proof Conditions (I) and (II) are trivially satisfied for α; condition (III) is satisfied too, for $p \in \alpha$ implies that $(p + r)/2$ belongs to α since $p < (p + r)/2 < r$. Thus α is a cut. Since $r < r$ is false, we have that $r \notin \alpha$. Since $p < r$ implies that $p \in \alpha$, r is the smallest upper number of α. □

The cut constructed in the foregoing proposition is called a *rational cut*. When we wish to indicate that a cut α is the rational cut related to r by the foregoing construction, we write $\alpha = r^*$.

Definition Let α and β be cuts. We say that α and β are *equal* and write $\alpha = \beta$ if $p \in \alpha$ implies $p \in \beta$, and $q \in \beta$ implies $q \in \alpha$, that is, if the two sets α and β coincide; otherwise we write $\alpha \ne \beta$. We say that the cut β is *greater than* the cut α and we write $\beta > \alpha$ (or $\alpha < \beta$) if there is a rational number p such that p belongs to β and p does not belong to α, that is, if α is a proper subset of β. The notation $\alpha \le \beta$ means that $\alpha = \beta$ or $\alpha < \beta$, and the notation $\alpha \ge \beta$ means that

$\beta \le \alpha$. If $\alpha \ge 0^*$, we say that α is *nonnegative*. Similarly, if $\alpha < 0^*$, α is said to be *negative*, and if $\alpha \le 0^*$, α is called *nonpositive*.

Observe that the symbol $<$ is used to denote inequality between two rational numbers as well as between two cuts; the context in which the symbol is used will make clear, however, which meaning is to be attached to it.

Proposition 9 Cuts satisfy the following laws:

(i) (Trichotomy) For any two cuts α and β, either $\alpha = \beta$ or $\alpha < \beta$ or $\beta < \alpha$ holds.

(ii) (Transitivity) For any three cuts α, β, and γ, if $\alpha < \beta$ and $\beta < \gamma$, then $\alpha < \gamma$.

Proof We consider the trichotomy law first. Clearly, if $\alpha = \beta$, then neither $\alpha < \beta$ nor $\beta < \alpha$ can hold. Moreover, $\alpha < \beta$ and $\beta < \alpha$ are mutually exclusive because $p < q$ and $q < p$ cannot hold simultaneously for rational numbers p and q. Thus at most one of the three relations can hold. If $\alpha \ne \beta$, then the two sets are not identical; that is, either there is a rational number p in α but not in β (meaning that $\beta < \alpha$), or there is a q in β but not in α (meaning that $\alpha < \beta$).

To prove the transitivity, we observe that $\alpha < \beta$ means that there is a rational number p such that $p \in \beta$ and $p \notin \alpha$; also, since $\beta < \gamma$ there is a rational number q such that $q \in \gamma$, and $q \notin \beta$. But $p \in \beta$ and $q \notin \beta$ implies that $p < q$; this together with $p \notin \alpha$ implies $q \notin \alpha$. Hence $q \in \gamma$ and $q \notin \alpha$, meaning that $\alpha < \gamma$. □

Remark Observe that the property $\alpha = \beta$, or $\alpha < \beta$, or $\beta < \alpha$ (and only one of these holds) implies that $\alpha < \alpha$ is false for any cut α.

Proposition 10 Let α and β be cuts and let γ be the set of all rational numbers r such that $r = p + q$, where $p \in \alpha$ and $q \in \beta$. Then γ is a cut. We call γ the *sum of* α *and* β and denote it by $\alpha + \beta$.

Proof We have to show that γ is a cut. Since α and β are nonempty, γ is nonempty; also, since α' and β' are nonempty sets (recall that the prime denotes the upper numbers of the cut), there is a p_0 and a q_0 such that $p_0 > p$, $q_0 > q$ for every p in α and every q in β. Thus, $p_0 + q_0 \notin \gamma$, and γ therefore does not contain every rational number.

Next, we have to show that if $r \in \gamma$ and $s < r$, with s rational, then $s \in \gamma$. Since $r \in \gamma$, $r = p + q$ for some $p \in \alpha$ and $q \in \beta$. But $s = [p + (s - r)] + q$ with $p + (s - r) \in \alpha$ (since α is a cut) and $q \in \beta$; so $s \in \gamma$.

Finally, let $r \in \gamma$ and $r = p + q$, $p \in \alpha$, $q \in \beta$. Then there is an s such that $s > p$, $s \in \alpha$. Hence $s + q > p + q$ and $s + q \in \gamma$; this means that r is not the largest rational in γ. □

Proposition 11 If α, β, γ are cuts, then

(i) $(\alpha + \beta) + \gamma = \alpha + (\beta + \gamma)$;

(ii) $\alpha + \beta = \beta + \alpha$;

(iii) $\alpha + 0^* = \alpha$.

Proof Properties (i) and (ii) follow easily from the associative and commutative laws of rationals. For (iii) we have: if r belongs to $\alpha + 0^*$, then $r = p + q$ for some $p \in \alpha$, $q \in 0^*$ (i.e., $q < 0$), and hence $r = p + q < p$, giving $r \in \alpha$; if $r \in \alpha$, then there is an s such that $s \in \alpha$, $s > r$, and thus $r = s + (r - s)$ where $s \in \alpha$, $r - s < 0$ (i.e., $r - s \in 0^*$), giving $r \in \alpha + 0^*$. This, however, means that the two sets α and $\alpha + 0^*$ coincide. □

Proposition 12 Given any cut α and any positive rational number r. Then there exist p, q such that $p \in \alpha$, $q \in \alpha'$, $q - p = r$, and q is not the smallest member of α' (if such a smallest member exists), i.e., q is a proper upper number of α.

Proof We choose an s in α and a t in α'. Then $t - s > 0$ and $r > 0$, so that, by the Archimedean property of rational numbers, there is a positive integer n such that

$nr > t - s$; that is, $t < s + nr$.

Hence $s + nr \in \alpha'$ and there is a unique positive integer $m \leq n$ such that

$$s + (m - 1)r \in \alpha \quad \text{and} \quad s + mr \in \alpha'.$$

If $s + mr$ is not the smallest upper number of α, we take

$$p = s + (m - 1)r \quad \text{and} \quad q = s + mr,$$

but if $s + mr$ is the smallest upper number of α, we take

$$p = s + \left(m - \frac{1}{2}\right)r \quad \text{and} \quad q = s + \left(m + \frac{1}{2}\right)r.$$

In either case $p \in \alpha$, q is not the smallest upper number of α and $q - p = r$. □

Remark In the foregoing proof we used the *Archimedean property* of rational numbers, which reads: If a and b are positive rational numbers, then there exists a natural number n such that $na > b$. This property is of course equivalent to the property: If r is a positive rational number, then there exists a natural number n such that $n > r$. Since r can be viewed as a ratio of two positive integers, say p/q, the inequality $n > r$ amounts to $nq > p$. But the Archimedean property clearly holds for positive integers. In fact, if p and q are positive integers, then pq is greater than or equal to both p and q.

Proposition 13 For any cuts α, β, γ, if $\alpha < \beta$, then $\alpha + \gamma < \beta + \gamma$.

Proof Since $\alpha < \beta$, there exists p such that $p \in \beta$, $p \in \alpha'$, and hence there exists q such that $q \in \beta$, $q \in \alpha'$, $q > p$. Thus, by Proposition 12, there exist r and s such that $r \in \gamma$, $s \in \gamma'$ and $s - r = q - p$, i.e., $p + s = q + r$. But $q + r \in \beta + \gamma$, and (since p is larger than every v in α and s is larger than every w in γ) $p + s \notin \alpha + \gamma$. Thus $\alpha + \gamma < \beta + \gamma$. □

Corollary If $\alpha + \gamma = \beta + \gamma$, then $\alpha = \beta$. Also, if $\alpha + \gamma = \alpha$, then $\gamma = 0^*$.

Proposition 14 Let α be a cut. Then the set $(-\alpha)$, defined by the collection of all rational numbers $-p$ such that p is a proper upper number of α, is a cut and $\alpha + (-\alpha) = 0^*$. Also, if β is a cut and $\alpha + \beta = 0^*$, then $\beta = (-\alpha)$.

Proof Let q be an upper number of α and put $s = \max\{q,0\}$, where $\max\{q,0\}$ denotes the larger of the two numbers q and 0. Then $s + 1$ is a positive rational number larger than q, and $-(s + 1)$ belongs to $(-\alpha)$; so $(-\alpha)$ is nonempty. Moreover, any negative $q \in \alpha$ (and there are some) is such that $-q$ is rational but $-(-q)$ is not a proper upper number of α so that $-q \notin (-\alpha)$; hence $(-\alpha)$ does not contain all rational numbers.

To check condition (II) of the Basic Definition, take $r \in (-\alpha)$, s rational and $s < r$. Then $-r$ is a proper upper number of α. Also $-s > -r$ so that $-s$ is a proper upper number of α as well; that is, $s \in (-\alpha)$.

To check condition (III) of the Basic Definition, take $r \in (-\alpha)$. Then $-r$ is a proper upper number of α and hence there is a smaller upper number p of α; i.e., $p < -r$ and $p \notin \alpha$. Hence $r < -p$ and $s = \frac{1}{2}[r + (-p)]$ is a rational such that $r < s < -p$ and $p < -s < -r$. From $p < -s$ and $p \notin \alpha$ we obtain that $-s$ is a proper upper number of α and therefore $s = -(-s) \in (-\alpha)$. Then $r < s$ shows that $(-\alpha)$ has no largest element.

Having established that $(-\alpha)$ is a cut, we now verify that $\alpha + (-\alpha) = 0^*$.

Take any $r \in \alpha + (-\alpha)$ and write $r = s + (-p)$ with $s \in \alpha$ and $-p \in (-\alpha)$. Hence $p \notin \alpha$ (in fact p is a proper upper number of α) and therefore $s < p$ so that $r = s - p < 0$. Consequently $r \in 0^*$ and $\alpha + (-\alpha)$ is a subset of 0^*. On the other hand, take any $r \in 0^*$. Hence $-r > 0$. We use $-r$ in place of the r in Proposition 13 to obtain a $p \in \alpha$ and a proper upper number q of α such that $q - p = -r$. Hence $-q \in (-\alpha)$ and $r = p - q = p + (-q) \in \alpha + (-\alpha)$. Thus 0^* is a subset of $\alpha + (-\alpha)$. We therefore have that $\alpha + (-\alpha) = 0^*$.

Finally, let β be any cut such that $\alpha + \beta = 0^*$. Then

$$\beta = 0^* + \beta = [\alpha + (-\alpha)] + \beta = (-\alpha) + (\alpha + \beta) = (-\alpha) + 0^* = (-\alpha)$$

and the proof is complete. □

Proposition 15 If α and β are cuts, then there is a unique cut γ such that $\alpha + \gamma = \beta$.

Proof Set $\gamma = \beta + (-\alpha)$. Then

$$\alpha + \gamma = \alpha + [\beta + (-\alpha)] = [\alpha + (-\alpha)] + \beta = 0^* + \beta = \beta.$$

Let δ be any cut such that $\alpha + \delta = \beta$. Then

$$\delta = 0^* + \gamma = (-\alpha) + (\alpha + \delta) = (-\alpha) + \beta = \gamma.$$ □

Remark The cut γ constructed in Proposition 15 is denoted by $\beta - \alpha$. That is, we write $\beta - \alpha$ in place of $\beta + (-\alpha)$ and $-\alpha$ in place of $(-\alpha)$.

Proposition 16 For any cuts α and β, $-(-\alpha) = \alpha$ and $-(\alpha - \beta) = -\alpha + \beta$.

Proof Since $(-\alpha) + [-(-\alpha)] = 0^* = (-\alpha) + \alpha$ we see that $-(-\alpha) = \alpha$ by the corollary to Proposition 13. If $\alpha + \delta = \beta$, then δ is necessarily $\beta - \alpha$ by Proposition 15, and so $-(\beta - \alpha) = \alpha - \beta$. □

Remark It is easy to see that $\alpha > 0^*$, $< 0^*$, $= 0^*$ according as $-\alpha < 0^*$, $> 0^*$, $= 0^*$.

Proposition 17 Let α and β be cuts such that $\alpha \geq 0^*$ and $\beta \geq 0^*$. Let γ be the subset of Q (Q denoting the set of all rational numbers) consisting of all members $r < 0$ together with all members r such that $r = pq$, where $p \in \alpha$, $q \in \beta$, $p \geq 0$, $q \geq 0$. Then γ is a cut. The cut γ so constructed is denoted by $\alpha\beta$ and is called the *product of α and β*.

Proof It is clear that γ is a nonempty set. We also see that γ' is nonempty since for any $s \in \alpha'$, $t \in \beta'$, $s > 0$, $t > 0$, we must have $v = st \in \gamma'$ (because if not, then $v = pq$, where $p \in \alpha$, $q \in \beta$, $p > 0$, $q > 0$; but $0 < p < s$, $0 < q < t$, resulting in $pq < st$, which is impossible). We therefore see that condition (I) of the Basic

Definition is satisfied.

To see that condition (II) of the Basic Definition is fulfilled, we let $r \in \gamma$ and $t < r$. Then either $r < 0$ or $r = pq$, where $p \in \alpha$, $q \in \beta$, $p \geq 0$, $q \geq 0$. In the first case, $t < 0$ so that $t \in \gamma$. In the second case, either t is again less than 0 (in which case $t \in \gamma$) or $0 \leq t < pq$, where $p \in \alpha$, $q \in \beta$, $p > 0$, $q > 0$. Let $s = t/p$, so that $0 \leq s < q$. Then, since $q \in \beta$, we have $s \in \beta$, and thus $t = ps$, where $p \in \alpha$, $s \in \beta$, $p > 0$, $s \geq 0$, giving $t \in \gamma$.

Finally, we check out condition (III) of the Basic Definition. Let $r \in \gamma$. Then, either $r < 0$, in which case $r < r/2 < 0$ (so that $r/2 \in \gamma$), or $r = pq$, $p \in \alpha$, $q \in \beta$, $p \geq 0$, $q \geq 0$. In the latter case, there exists $s > p$, $t > q$ such that $s \in \alpha$, $t \in \beta$; and so $st \in \gamma$, $st > r$. □

Proposition 18 If $\alpha \geq 0^*$, $\beta \geq 0^*$, then $\gamma = \alpha\beta \geq 0^*$. If $\alpha > 0^*$, $\beta > 0^*$, then $\gamma = \alpha\beta > 0^*$.

Proof In the first case, γ contains all $r < 0$. Thus $\gamma \geq 0^*$. In the second case, we use the readily established fact that for any cut δ, $\delta > 0^*$ if and only if $0 \in \delta$ (i.e., $r \in \delta$ for some $r > 0$). Hence, $0 \in \alpha$, $0 \in \beta$ and so $0 = 0 \cdot 0 \in \gamma$, giving $\gamma > 0^*$. □

Remark From Proposition 18 we see that if $\alpha > 0^*$, $\beta > 0^*$, then $0 \in \gamma = \alpha\beta$; thus $r \in \gamma$ if and only if $r \leq 0$ or $r = pq$, where $p \in \alpha$, $q \in \beta$, $p > 0$, $q > 0$. This alternate way of expressing $\gamma = \alpha\beta$, in the case where $\alpha > 0^*$, $\beta > 0^*$, is sometimes useful.

Definition With every cut α we associate a cut $|\alpha|$, which we call the *absolute value of* α, as follows:

$$|\alpha| = \alpha \quad \text{if} \quad \alpha \geq 0^*,$$
$$= -\alpha \quad \text{if} \quad \alpha < 0^*.$$

Clearly, $|\alpha| \geq 0^*$ for all α, and $|\alpha| = 0^*$ only if $\alpha = 0^*$.

We can now complete the definition of multiplication of two cuts.

Definition For any two cuts α and β, we define the *product of* α *and* β (written $\alpha\beta$) as follows:

$$\begin{aligned} \alpha\beta &= -(|\alpha||\beta|) \quad \text{if} \quad \alpha < 0^*, \quad \beta \geq 0^*, \\ &= -(|\alpha||\beta|) \quad \text{if} \quad \alpha \geq 0^*, \quad \beta < 0^*, \\ &= |\alpha||\beta| \qquad\quad \text{if} \quad \alpha < 0^*, \quad \beta < 0^*. \end{aligned}$$

Note that the product $|\alpha||\beta|$ has already been defined in Proposition 17, since $|\alpha| \geq 0^*$, $|\beta| \geq 0^*$. It is also clear that $\alpha\beta$ is a cut in all cases.

The proof of the following proposition is left as an exercise to the reader.

Proposition 19 For any cuts α, β, γ we have

(i) $\alpha\beta = \beta\alpha$,

(ii) $(\alpha\beta)\gamma = \alpha(\beta\gamma)$,

(iii) $\alpha(\beta + \gamma) = \alpha\beta + \alpha\gamma$,

(iv) $\alpha 0^* = 0^*$,

(v) $\alpha\beta = 0^*$ only if $\alpha = 0^*$ or $\beta = 0^*$,

(vi) $\alpha 1^* = \alpha$,

(vii) if $0^* < \alpha < \beta$, and $\gamma > 0^*$, then $\alpha\gamma < \beta\gamma$.

Proposition 20 For a given cut $\alpha > 0^*$ and a given rational number r such that $r > 1$, there are rational numbers p and q such that $p > 0$, $q > 0$, $p \in \alpha$, $q \in \alpha'$ (nonleast), and $q/p = r$.

Proof Let $s \in \alpha$, $s > 0$. We put, for $n = 0, 1, 2, \ldots,$

$$t_n = sr^n,$$

where $r^0 = 1$, $r^{n+1} = r^n r$ for $n \geq 0$. Then, by reasoning similar to that used in the proof of Proposition 12, we shall see below that t_n belongs to α' for all large n, and thus that there is a unique m for which $t_m \in \alpha$, $t_{m+1} \in \alpha'$, $t_{m+1}/t_m = r$. If t_{m+1} is nonleast, we choose $p = t_m$, $q = t_{m+1}$. If t_{m+1} is the smallest upper number of α, we choose $p = t_m + v$ and $q = t_{m+1} + (t_{m+1}/t_m)v$, where v is positive and so small that $t_m + v$ belongs to α. In both cases $q/p = r$, and the result is established.

To see that sr^n is in α' for large n, suppose that there is a t in α' such that $sr^n < t$ for every $n \geq 1$, that is,

$$s(1 + d)^n < t \quad \text{for every} \quad n \geq 1$$

where $d = r - 1 > 0$. Since $(1 + d)^n \geq 1 + nd$ for every $n \geq 1$, by the binomial theorem (see Exercise 1 at the end of this chapter) or a simple induction proof, we have that $s(1 + nd) < t$ for $n \geq 1$, i.e., $n < (t - s)/(ds)$ for $n \geq 1$. But this leads to a contradiction. □

Proposition 21 For every cut $\alpha > 0^*$, the set γ satisfying

$$\gamma = \{r: r \leq 0 \text{ or } [r > 0 \text{ and } 1/r \in \alpha' \text{ (nonleast)}]\}$$

is a cut, and $\gamma > 0^*$.

Proof It will be enough to show that γ is a cut; $\gamma > 0^*$ because $0 \in \gamma$.

The set γ is nonempty since γ contains all $r \leq 0$; also, since $\alpha > 0^*$, there is an r such that $r > 0$, $r \in \alpha$, and hence $1/r \in \gamma'$; whence γ' is nonempty.

If $q < r$, $r \in \gamma$ then either $q \leq 0$ or $1/q > 1/r > 0$, so that in either case $q \in \gamma$.

Let $r \in \gamma$. If $r < 0$, then $r < r/2 < 0$ so that $r/2 \in \gamma$; if $r = 0$, then for any $1/s \in \alpha'$ (nonleast) we have $s \in \gamma$, $s > r$. If $r > 0$, then $1/r \in \alpha'$ (nonleast), so that there exists $1/t$ satisfying

$$0 < 1/t < 1/r, \quad 1/t \in \alpha' \text{ (nonleast)},$$

and hence $t \in \gamma$, $t > r$. □

Proposition 22 For any given cut $\alpha > 0^*$, there exists a unique cut β satisfying $\alpha\beta = 1^*$ $(= \beta\alpha)$.

Proof The uniqueness follows from the fact that if $\alpha\beta = 1^*$ and $\alpha\delta = 1^*$, then by Proposition 19,

$$\beta = \beta 1^* = \beta(\alpha\delta) = (\alpha\beta)\delta = 1^*\delta = \delta.$$

We now show that if $\alpha > 0^*$, then the cut γ defined in Proposition 21 satisfies $\alpha\gamma = 1^*$.

Suppose first that $r \in \alpha\gamma$. Then, since $\alpha > 0^*$, $r \le 0$ or $r = pq$, where $p \in \alpha$, $q \in \gamma$, $p > 0$, $q > 0$. In the first case, $r \in 1^*$; in the second case, $r = pq$, where $p \in \alpha$, $1/q \in \alpha'$ (nonleast), $p > 0$, $q > 0$. Since $1/q \in \alpha'$, $1/q > p > 0$ and hence $pq < 1$, so that $r \in 1^*$. Thus, $\alpha\gamma < 1^*$. Suppose next that $r \in 1^*$, i.e., $r < 1$. Then, either $r \le 0$ or $1/r > 1$. In the first case, $r \in \alpha\gamma$ since $\alpha\gamma > 0^*$; in the second case, by Proposition 20, there exist $p \in \alpha$, $q \in \alpha'$ (nonleast), $p > 0$, $q > 0$, such that $q/p = 1/r$. Thus, $r = p(1/q)$, where $p \in \alpha$, $1/q \in \gamma$, $p > 0$, $1/q > 0$, giving $r \in \alpha\gamma$. Thus $1^* \le \alpha\gamma$. □

Definition For any cut $\alpha \ne 0^*$, α^{-1} (or $1/\alpha$) is the cut γ defined in Proposition 21 if $\alpha > 0^*$ and $-(-\alpha)^{-1}$ if $\alpha < 0^*$. It is called the *multiplicative inverse of* α.

We leave the proof of the following proposition to the reader.

Proposition 23 Let $\alpha \ne 0^*$. Then

(i) $\alpha\alpha^{-1} = 1^* = \alpha^{-1}\alpha$;

(ii) if $\alpha\gamma = 1^*$, then $\gamma = \alpha^{-1}$;

(iii) $(\alpha^{-1})^{-1} = \alpha$.

Proposition 24 If $\alpha \ne 0^*$, then for any cut β there is a unique cut γ such that $\alpha\gamma = \beta$.

Proof Firstly, $\alpha(\alpha^{-1}\beta) = \beta$, so that $\alpha^{-1}\beta$ is a value for γ; and secondly, if $\alpha\gamma = \beta$, then

$$\gamma = 1^*\gamma = (\alpha^{-1}\alpha)\gamma = \alpha^{-1}(\alpha\gamma) = \alpha^{-1}\beta. \qquad \square$$

Definition If α and β are any given cuts, with $\alpha \ne 0^*$, we write β/α for $\alpha^{-1}\beta$. Then β/α satisfies $\alpha(\beta/\alpha) = \beta$ and, if $\alpha\gamma = \beta$, then necessarily $\gamma = \beta/\alpha$.

We next consider three useful results connecting cuts and rational cuts.

Proposition 25 If α and β are any cuts such that $\alpha < \beta$, then there is a rational cut r^* such that $\alpha < r^* < \beta$.

Proof Since $\alpha < \beta$, there is a p such that $p \in \beta$, $p \notin \alpha$. Choosing any fixed $r > p$, $r \in \beta$, we have $p \in r^*$, $p \notin \alpha$, and $r \in \beta$, $r \notin r^*$, whence $\alpha < r^*$ and $r^* < \beta$, i.e., $\alpha < r^* < \beta$. □

Proposition 26 For any cut α and any rational number p, $p \in \alpha$ if and only if $p^* < \alpha$.

Proof If $p \in \alpha$, then (since $p \notin p^*$), $p^* < \alpha$. If $p^* < \alpha$, then there is a q such that $q \in \alpha$, $q \notin p^*$, that is, $q \in \alpha$, $q \geq p$, giving $p \in \alpha$. □

Proposition 27 For a given cut α, the set α' of all upper numbers of α has a smallest member if and only if α is a rational cut r^* (say). In the latter case the least member of α' is r.

Proof If $\alpha = r^*$, then (since $r \notin r^*$) $r \in (r^*)'$; and if $p < r$, then $p \in r^*$. Hence, r is the least or smallest member of $(r^*)'$. If α' has a least member r (say), then $p \in \alpha$ if and only if $p \in r^*$. Hence, $\alpha = r^*$. □

Proposition 28 Let p and q be any given rational numbers. Then

(i) $p^* < q^*$, $p^* > q^*$, $p^* = q^*$ according as $p < q$, $p > q$, $p = q$;

(ii) $p^* + q^* = (p + q)^*$;

(iii) $p^*q^* = (pq)^*$.

Proof

(i) If $p^* < q^*$, then there is a rational number r such that $r \in p^*$, $r \notin q^*$, i.e., $p \leq r < q$ and hence $p < q$. If $p < q$, then

$p \in q^*$ so that, since $p \notin p^*$, $p^* < q^*$. The proof of the other cases is similar.

(ii) If $r \in p^* + q^*$, then $r = s + t$, where $s \in p^*$, $t \in q^*$, i.e., $s < p$, $t < q$, and hence $r < p + q$, i.e., $r \in (p + q)^*$. If r belongs to $(p + q)^*$, i.e., $r < p + q$, then writing $h = p + q - r > 0$ we have $r = (p - h/2) + (q - h/2)$. Since $p - h/2$ and $q - h/2$ are in p^* and q^*, respectively, this gives $r \in p^* + q^*$.

(iii) The case where p and q is 0 is easily handled. If p and q are both larger than 0, then by part (i) p^* and q^* are both $> 0^*$ and hence $p^*q^* > 0^*$. Suppose $r \in p^*q^*$. Then, either $r \le 0$ or $r = st$, $0 < s < p$, $0 < t < q$, giving $r < pq$, i.e., $r \in (pq)^*$, in either case. Suppose $r \in (pq)^*$. Then $r \le 0$ or $0 < r < pq$. In the first case, r belongs to p^*q^*, since $p^*q^* > 0^*$. In the second case, $0 < r/q < p$ so that there exists an s such that $0 < r/q < s < p$. This gives r equals $(r/s)s$, where $0 < r/s < q$ and $0 < s < p$. Hence $r \in p^*q^*$. If, finally, p or q is less than 0, we use the fact that $-(r^*) = (-r)^*$, which is easily verified of course. □

The following proposition states a very fundamental property of cuts.

Proposition 29 (Dedekind's Theorem) Let L and U be two sets of cuts satisfying

(i) neither L nor U is empty;

(ii) every cut is either in L or in U;

(iii) if α is in L and β is in U, then $\alpha < \beta$.

Then there is one and only one cut γ such that

$$\alpha \le \gamma \le \beta \quad \text{for all } \alpha \text{ in } L \text{ and all } \beta \text{ in } U. \tag{1}$$

Also, this cut γ is either the largest element of L or else the smallest element of U.

Proof An immediate consequence of condition (iii) is that the sets L and U are disjoint.

Let γ be the set of all rational numbers p such that $p \in \alpha$ for some $\alpha \in L$. We shall show first that γ is a cut.

Since L is not empty, neither is γ. If $\beta \in U$ and $q \notin \beta$, then q does not belong to α for any $\alpha \in L$ [since $\alpha < \beta$ by condition (iii)], hence $q \notin \gamma$.

If $p \in \gamma$ and $q < p$, then $p \in \alpha$ for some $\alpha \in L$; hence $q \in \alpha$; hence $q \in \gamma$.

If $p \in \gamma$, then $p \in \alpha$ for some $\alpha \in L$; hence there exists $q > p$ such that $q \in \alpha$ (because no cut has a largest element); hence $q \in \gamma$.

Thus γ is a cut and we next show that γ satisfies (1).

It is clear that $\alpha \leq \gamma$ for all $\alpha \in L$. If there were some $\beta \in U$ for which $\beta < \gamma$, then there would be a rational number p such that $p \in \gamma$, and $p \notin \beta$; but if $p \in \gamma$, then $p \in \alpha$ for some $\alpha \in L$, and this implies that $\beta < \alpha$, a contradiction to condition (iii). Thus $\gamma \leq \beta$ for all $\beta \in U$.

Thus the cut γ satisfies (1) and we next assume there are two cuts γ_1 and γ_2, so named that $\gamma_1 < \gamma_2$, both satisfying

$$\alpha \leq \gamma_1 \leq \beta \quad \text{and} \quad \alpha \leq \gamma_2 \leq \beta \quad \text{for all } \alpha \in L \text{ and all } \beta \in U.$$

There is then a rational number r such that the cut r^* satisfies $\gamma_1 < r^* < \gamma_2$, by Proposition 25. Then $\alpha \leq \gamma_2$ for all α in L puts r^* in L, whereas $\gamma_1 \leq \beta$ for all $\beta \in U$ puts r^* in U. This contradiction to L and U being disjoint establishes uniqueness.

Since every cut is either in L or in U and since L and U are disjoint, we know that γ is in one of L or U but not in both. Hence, from (1), if $\gamma \in L$, then γ is the largest member of L, but if $\gamma \in U$, then γ is the smallest member of U. $\square$

Definition Let A be a set of cuts. If there is a cut β such that $\alpha < \beta$ for all $\alpha \in A$, we say that A is *bounded above*, and call β an *upper bound of* A.

Lower bounds are defined similarly. If A is bounded above and below, then A is said to be *bounded*.

Definition Let A be a set of cuts bounded above. Suppose that the cut γ has the following properties:

(i) γ is an upper bound of A;

(ii) if $\beta < \gamma$, then β is not an upper bound of A.

Then γ is called the *least upper bound of* A [that there is at most one such γ is clear from (ii)], or the *supremum of* A.

The *greatest lower bound*, or *infimum*, of any set A of cuts which is bounded below is defined similarly.

Proposition 30 If A is a nonempty set of cuts which is bounded above, then A has a least upper bound, or supremum, in the set of all cuts.

Proof Let L be the following set of cuts: $\delta \in L$ if and only if there is some $\alpha \in A$ such that $\delta < \alpha$. Let U consist of all cuts not in L.

We check that L and U satisfy the hypothesis of Proposition 29.

We take any $\alpha \in A$ (possible since A is not empty), then pick any $r \in \alpha$. Hence $r^* < \alpha$, by Proposition 26. Therefore $r^* \in L$ and hence L is nonempty. Let β be a cut and be an upper bound of A, which is possible since A is bounded above. Take any proper upper number s of β and note that $\alpha \in A$ implies $\alpha \leq \beta < s^*$. Hence $s^* \notin L$, $s^* \in U$ and thus U is nonempty as well.

By the definition of U every cut is either in L or in U.

Again by the definition of U we have that L and U are disjoint sets of cuts. We thus assume that there is a δ in L and a β in U with $\beta \leq \delta$. Since δ is in L, pick an $\alpha \in A$ such that $\delta < \alpha$. Then $\beta \leq \delta < \alpha$, i.e., $\beta < \alpha$; hence $\beta \in L$, that is $\beta \notin U$.

L and U, therefore, satisfy the conditions of Proposition 29 and hence there is a unique cut γ such that $\delta \leq \gamma \leq \beta$ for all $\delta \in L$ and all $\beta \in U$.

Finally, assume $\gamma < \alpha$ for some $\alpha \in A$. Pick a rational r such that $\gamma < r^* < \alpha$. Then $r^* < \alpha$ implies $r^* \in L$ but $\gamma < r^*$ implies $r^* \in U$ which in turn implies that $r^* \notin L$. Thus γ is an upper bound of A. To see that γ is the least upper bound of A, we note that if a cut γ_0 is an upper bound of A such that $\gamma_0 < \gamma$, then $\gamma_0 \in L$ and hence there is an $\alpha \in A$ such that $\gamma_0 < \alpha$, contradicting γ_0 being an upper bound of A. □

Proposition 31 If $0^* < \alpha < \beta$, then there is a positive integer n such that $n^*\alpha > \beta$.

Proof If the proposition were false, the inequality $n^*\alpha \le \beta$ would hold for all positive integers n. That is, the set of cuts

$$\{\alpha, 2^*\alpha, 3^*\alpha, \ldots\}$$

would be bounded above. Let γ be the least upper bound of this set. Then $n^*\alpha \le \gamma$ for all n, and hence $(n + 1)^*\alpha \le \gamma$ for all n. Therefore $n^*\alpha + \alpha \le \gamma$, or $n^*\alpha \le \gamma - \alpha$, for all n. Thus $\gamma - \alpha$ is an upper bound that is less than the least upper bound γ. This is the desired contradiction. □

Remark An alternate way of proving Proposition 31 is as follows. For $0^* < \alpha < \beta$, let r and s be rational numbers such that

$$0^* < r^* < \alpha < \beta < s^* < \alpha + \beta.$$

By the Archimedean property for rationals, there is a positive integer n such that $n^*r^* = (nr)^* > s^*$. But then $n^*\alpha > n^*r^* > s^* > \beta$.

Definition A *field* is a nonempty set F, together with two binary operations on F into F, called *addition* and *multiplication*, such that:

(a) For addition:

 (i) The associative law holds: $x, y, z \in F$ implies $x + (y + z) = (x + y) + z$.

 (ii) There exists a member 0 of F such that $x \in F$ implies $x + 0 = x$.

 (iii) $x \in F$ implies that there exists a $(-x) \in F$ such that $x + (-x) = 0$.

 (iv) The commutative law holds: $x, y \in F$ implies $x + y = y + x$.

(b) For multiplication:

 (i) The associative law holds: $x, y, z \in F$ implies $x(yz) = (xy)z$.

 (ii) There exists a member 1 of F such that $1 \neq 0$ and $x \in F$ implies $x1 = x$.

 (iii) $x \in F$, $x \neq 0$ implies that there exists an $x^{-1} \in F$ such that $xx^{-1} = 1$.

(iv) The commutative law holds: $x\ y \in F$ implies $xy = yx$.

(c) For addition and multiplication:
The distributive law holds (more precisely, multiplication is distributive over addition): $x, y, z \in F$ implies $x(y + z) = xy + xz$.

Definition An *ordered field* is a field F that contains a subset P such that:

(i) P is closed with respect to addition; that is, $x \in P$, $y \in P$ implies $x + y \in P$.

(ii) P is closed with respect to multiplication; that is, $x \in P$, $y \in P$ implies $xy \in P$.

(iii) $x \in F$ implies exactly that one of the following statements is true:

$$x \in P, \quad x = 0, \quad -x \in P.$$

A member x of F is *positive* if and only if $x \in P$; x is *negative* if and only if $-x \in P$. We write $x < y$ if $y - x \in P$.

Definition A *complete ordered field* is an ordered field F in which a least upper bound, or supremum, exists for every nonempty set in F that is bounded above in F. An ordered field F is called *Archimedean* if for any two positive x and y in F there exists a positive integer n such that $nx - y$ is positive.

Remark It can be shown that any two complete ordered Archimedean fields F and F' are isomorphic in the sense that there exists a one-to-one correspondence

$$x \leftrightarrow x', \quad \text{where } x \in F \text{ and } x' \in F',$$

that preserves binary operations and order; that is,

$$(x + y)' = x' + y', \quad (xy)' = x'y', \quad x < y \text{ if and only if } x' < y'.$$

Remark We note that the set of all cuts forms a complete ordered Archimedean field. Thus there is one and (except for isomorphic fields) only one complete Archimedean field and the following Definition therefore introduces a mere change in terminology.

Definition Cuts will from now on be called *real numbers*. Rational cuts will be identified with rational numbers (and will be called *rational numbers*). All other cuts will be called *irrational numbers*. The set of all real numbers we denote by R^1.

Proposition 28 shows that the ordered field of rational numbers is isomorphic to the ordered field of rational cuts, and it enables us to identify the rational cut r^* with the rational number r. Proposition 25 shows that between any two real numbers there is a rational number; Proposition 26 shows that every real number α is the set of all rational numbers p such that $p < \alpha$. The Archimedean property of real numbers was established in Proposition 31.

Definition The *extended real number system* consists of the real number system R^1 to which two symbols, $+\infty$ and $-\infty$, have been adjoined, with the properties:

(i) If x is a real number, then $-\infty < x < +\infty$, and

$$x + \infty = +\infty, \quad x - \infty = -\infty, \quad \frac{x}{+\infty} = \frac{x}{-\infty} = 0.$$

(ii) If $x > 0$, then $x\cdot(+\infty) = +\infty$, $x\cdot(-\infty) = -\infty$.

(iii) If $x < 0$, then $x\cdot(+\infty) = -\infty$, $x\cdot(-\infty) = +\infty$.

The operation $\infty - \infty$ is left undefined. The set of all extended real numbers we denote by R_e^1. When it is desired to make the distinction between real numbers on the one hand and the symbols $+\infty$ and $-\infty$ on the other hand, the former are called *finite*.

Remark One use of the extended real numbers is in the expession "sup S". If S is a nonempty set of real numbers with an upper bound, we define sup S to be the least upper bound of S. If S has no upper bound, that is, if for every $x \in R^1$ there is an $s \in S$ such that $x < s$, we define sup S to be $+\infty$. Then sup S is defined for all nonempty sets S, and if we define the supremum of the empty set to be $-\infty$, then in all cases sup S is the smallest extended real number which is greater than or equal to each element of S. Similar conventions are adopted with regards to "inf S".

3. *Inequalities*

We recall that with every real number a we may associate a real number $|a|$, called the *absolute value of* a, as follows:

$$|a| = a \quad \text{if } a > 0,$$
$$= 0 \quad \text{if } a = 0,$$
$$= -a \quad \text{if } a < 0.$$

Geometrically, $|x|$ means the distance between the points x and 0 on the number line; $|x - a|$ means the distance between x and a. Let $\delta > 0$; then $|x| < \delta$ means that $-\delta < x < \delta$ and $|x - a| < \delta$ means that $a - \delta < x < a + \delta$. It is obvious that $|-x| = |x|$.

The proof of the following proposition is trivial.

Proposition 32 Let a be any real number. Then

$$\frac{a + |a|}{2} = a \quad \text{if } a \ge 0,$$
$$= 0 \quad \text{if } a \le 0,$$

and

$$\frac{a - |a|}{2} = 0 \quad \text{if } a \ge 0,$$
$$= a \quad \text{if } a \le 0.$$

Corollary Let a and b be any real numbers and let max{a,b} and min{a,b} denote the larger and the smaller of the two numbers a and b, respectively. Then

$$\frac{a + b + |b - a|}{2} = \max\{a,b\}$$

and

$$\frac{a + b - |b - a|}{2} = \min\{a,b\}.$$

Proof The claim follows readily when we replace in Proposition 32 the number a by the number b - a and then add a to both sides. □

N.B.: We will sometimes denote by $a \vee b$ the number $\max\{a,b\}$ and denote by $a \wedge b$ the number $\min\{a,b\}$.

Proposition 33 Let a and b be any real numbers. Then

$$|a + b| \le |a| + |b|.$$

Proof If we add the (trivial) inequalities

$$-|a| \le a \le |a| \quad \text{and} \quad -|b| \le b \le |b|,$$

we obtain

$$-(|a| + |b|) \le a + b \le |a| + |b|;$$

but the inequality $|A| \le B$ is clearly equivalent to the inequality $-B \le A \le B$ for any two real numbers A and B. □

Remark The inequality in Proposition 33 is called the *triangle inequality*. Using the Principle of Mathematical Induction, this inequality can be extended to an arbitrary finite number of real numbers: If a, b, ..., c are any finite collection of real numbers, then

$$|a + b + \cdots + c| \le |a| + |b| + \cdots + |c|.$$

Moreover, the inequalities

$$|a - b| \le |a| + |b|,$$

$$\big||a| - |b|\big| \le |a - b|,$$

and

$$\big||a| - |b|\big| \le |a + b|$$

are simple consequence of the triangle inequality and we leave their verification to the reader.

Proposition 34 We have the following inequalities:

(i) $2^n > n$ for $n = 0, 1, 2, \ldots;$

(ii) $2^n > n^2$ for $n = 5, 6, \ldots.$

Proof

(i) By the Binomial Theorem (see Exercise 1 at the end of this chapter)

$$2^n = (1 + 1)^n = 1 + \binom{n}{1} + \binom{n}{2} + \cdots + \binom{n}{n} > \binom{n}{1} = n.$$

(ii) This is clear for $n = 5$. Suppose $2^k > k^2$ for some integer k larger than 5. Multiplying with the inequality $2 > [(k + 1)/k]^2$ (which is certainly valid since

$$\left(\frac{k + 1}{k}\right)^2 = 1 + \frac{2}{k} + \frac{1}{k^2} \le 1 + \frac{2}{5} + \frac{1}{25} < 2 \quad \text{for} \quad k \ge 5)$$

we get

$$2 \cdot 2^k > \left(\frac{k + 1}{k}\right)^2 k^2, \quad \text{or} \quad 2^{k+1} > (k + 1)^2. \qquad \square$$

Proposition 35 For nonnegative a and b and any natural number n we have

$$a^n + b^n \le (a + b)^n.$$

Proof The claim follows from the Binomial Theorem. □

Proposition 36 If $a/b < c/d$ and b and d are positive, then

$$\frac{a}{b} < \frac{a + c}{b + d} < \frac{c}{d}.$$

Proof Since b and d are positive, $a/b < c/d$ means that $ad < bc$. But then $a(b + d) < b(a + c)$ and so $a/b < (a + c)/(b + d)$. Again, since $ad < bc$, we see that $ad + cd < bc + cd$, or

$$(a + c)/(b + d) < c/d. \qquad \square$$

Proposition 37 Consider the finite set of fractions

$$S = \left\{\frac{a_1}{b_1}, \frac{a_2}{b_2}, \ldots, \frac{a_n}{b_n}\right\}$$

with $b_k > 0$ for $k = 1, 2, \ldots, n$. Then

$$\min S \le \frac{a_1 + a_2 + \cdots + a_n}{b_1 + b_2 + \cdots + b_n} \le \max S.$$

Proof We use Proposition 36 and prove the claim by induction. By Proposition 36 we get

$$\frac{a_1}{b_1} \wedge \frac{a_2}{b_2} \le \frac{a_1 + a_2}{b_1 + b_2} \le \frac{a_1}{b_1} \vee \frac{a_2}{b_2}.$$

We assume now that

$$\frac{a_1}{b_1} \wedge \cdots \wedge \frac{a_k}{b_k} \le \frac{a_1 + \cdots + a_k}{b_1 + \cdots + b_k} \le \frac{a_1}{b_1} \vee \cdots \vee \frac{a_k}{b_k};$$

we want to show that

$$\frac{a_1}{b_1} \wedge \cdots \wedge \frac{a_k}{b_k} \wedge \frac{a_{k+1}}{b_{k+1}} \le \frac{a_1 + \cdots + a_k + a_{k+1}}{b_1 + \cdots + b_k + b_{k+1}}$$

$$\le \frac{a_1}{b_1} \vee \cdots \vee \frac{a_k}{b_k} \vee \frac{a_{k+1}}{b_{k+1}}$$

holds. Let $A = a_1 + \cdots + a_k$, $B = b_1 + \cdots + b_k$. Clearly, B is positive (because it is the sum of positive numbers. Let

$$\mathcal{A} = \frac{a_1}{b_1} \wedge \cdots \wedge \frac{a_k}{b_k} \quad \text{and} \quad \mathcal{B} = \frac{a_1}{b_1} \vee \cdots \vee \frac{a_k}{b_k}.$$

Evidently

$$\frac{a_1}{b_1} \wedge \cdots \wedge \frac{a_k}{b_k} \wedge \frac{a_{k+1}}{b_{k+1}} = \mathcal{A} \wedge \frac{a_{k+1}}{b_{k+1}}$$

and

$$\frac{a_1}{b_1} \vee \cdots \vee \frac{a_k}{b_k} \vee \frac{a_{k+1}}{b_{k+1}} = \mathcal{B} \vee \frac{a_{k+1}}{b_{k+1}}.$$

By our inductive assumption we have $\mathcal{A} \le A/B \le \mathcal{B}$. Hence

$$\mathcal{A} \wedge \frac{a_{k+1}}{b_{k+1}} \le \frac{A}{B} \wedge \frac{a_{k+1}}{b_{k+1}} \quad \text{(because } \mathcal{A} \le A/B\text{)}.$$

But

$$\frac{A}{B} \wedge \frac{a_{k+1}}{b_{k+1}} \le \frac{A + a_{k+1}}{B + b_{k+1}} \quad \text{(by Proposition 36)}.$$

Similarly,

$$\frac{A}{B} \vee \frac{a_{k+1}}{b_{k+1}} \le \mathcal{B} \vee \frac{a_{k+1}}{b_{k+1}} \quad \text{(because } A/B \le \mathcal{B}\text{)}$$

and

$$\frac{A}{B} \vee \frac{a_{k+1}}{b_{k+1}} \geq \frac{A + a_{k+1}}{B + b_{k+1}} \quad \text{(by Proposition 36).}$$

Thus

$$A \wedge \frac{a_{k+1}}{b_{k+1}} \leq \frac{A}{B} \wedge \frac{a_{k+1}}{b_{k+1}} \leq \frac{A + a_{k+1}}{B + b_{k+1}} \leq \frac{A}{B} \vee \frac{a_{k+1}}{b_{k+1}} \leq B \vee \frac{a_{k+1}}{b_{k+1}}$$

and the claim is established. □

Proposition 38 Let $n \geq 2$ and suppose that $a_1, a_2, \ldots, a_n$ are positive. Then

$$(1 + a_1)(1 + a_2) \cdots (1 + a_n) > 1 + (a_1 + a_2 + \cdots + a_n).$$

Proof The claim is easily established by induction. □

Proposition 39 Let $n \geq 2$ and let $a_1, a_2, \ldots, a_n$ be positive but less than 1. Then

$$(1 - a_1)(1 - a_2) \cdots (1 - a_n) > 1 - (a_1 + a_2 + \cdots + a_n).$$

Proof Clearly,

$$(1 - a_1)(1 - a_2) = 1 - (a_1 + a_2) + a_1a_2 > 1 - (a_1 + a_2).$$

This shows that the claim is valid for $n = 2$. We proceed by induction. Suppose that $k \geq 2$ and assume that

$$(1 - a_1) \cdots (1 - a_k) > 1 - (a_1 + \cdots + a_k).$$

Multiplying by the positive number $(1 - a_{k+1})$, we get

$$\begin{aligned}(1 - a_1) &\cdots (1 - a_k)(1 - a_{k+1}) \\ &> 1 - (a_1 + \cdots + a_k) - a_{k+1} + a_{k+1}(a_1 + \cdots + a_k). \\ &= 1 - (a_1 + \cdots + a_k + a_{k+1}).\end{aligned}$$

□

Proposition 40 If $0 < a < 1$ and n is any natural number, then

$$1 + a + a^2 + \cdots + a^n < \frac{1}{1 - a}.$$

Proof Evidently

$$(1 + a + a^2 + \cdots + a^n)(1 - a) = 1 - a^{n+1} < 1. \qquad \square$$

Proposition 41 (Bernoulli's Inequality) If $n \geq 2$ and $a > 0$ or $0 > a > -1$, then

$$(1 + a)^n > 1 + na.$$

Proof The claim follows from Propositions 38 and 39, provided that one replaces there the numbers a_j ($j = 1, \ldots, n$) by a and $-a$, respectively. If $a = 0$ or $n = 1$, one has of course equality. $\square$

Proposition 42 For $n = 2, 3, \ldots$, we have

$$\left(1 + \frac{1}{n - 1}\right)^{n-1} < \left(1 + \frac{1}{n}\right)^n < 3.$$

Proof We note that the first part of the claimed inequality is equivalent to

$$\frac{n^{n-1}}{(n - 1)^{n-1}} < \frac{(n + 1)^n}{n^n}$$

and that the latter inequality in turn is equivalent to each of the following three inequalities:

$$\frac{n - 1}{n} \frac{n^n}{(n - 1)^n} < \frac{(n + 1)^n}{n^n},$$

$$\frac{n - 1}{n} < \frac{(n^2 - 1)^n}{(n^2)^n},$$

$$\left(1 - \frac{1}{n^2}\right)^n > 1 - \frac{1}{n}.$$

But the last inequality is valid by Proposition 41; in fact, set $a = -(1/n)^2$.

To show the remaining part of the claim in the proposition, note that, for $n = 1, 2, \ldots$,

$$\left(1 + \frac{1}{n}\right)^n = 1 + \binom{n}{1}\frac{1}{n} + \cdots + \binom{n}{k}\frac{1}{n^k} + \cdots + \binom{n}{n}\frac{1}{n^n}$$

by the binomial theorem. But, for $2 \le k \le n$,

$$\binom{n}{k}\frac{1}{n^k} = \frac{1}{k!}\left(1 - \frac{1}{n}\right)\left(1 - \frac{2}{n}\right) \cdots \left(1 - \frac{k-1}{n}\right) \le \frac{1}{2\cdot 3 \cdots k} \le \frac{1}{2^{k-1}}$$

and so, for $n \ge 2$,

$$\left(1 + \frac{1}{n}\right)^n < 1 + 1 + \frac{1}{2} + \frac{1}{4} + \cdots + \frac{1}{2^{n-1}} < 1 + \frac{1}{1 - 1/2} = 3$$

by Proposition 40. Hence, for all integers $n \ge 1$, we have

$$\left(1 + \frac{1}{n}\right)^n < 3.$$

□

Proposition 43 For $n = 2, 3, \ldots$, we have

$$\left(1 + \frac{1}{n-1}\right)^n > \left(1 + \frac{1}{n}\right)^{n+1}.$$

Proof Observe that the foregoing inequality is equivalent to

$$\left(\frac{1 + 1/(n-1)}{1 + 1/n}\right)^n > 1 + \frac{1}{n},$$

which in turn is equivalent to

$$\left(1 + \frac{1}{n^2 - 1}\right)^n > 1 + \frac{1}{n};$$

but the latter inequality holds by Proposition 41 when we set $a = (n^2 - 1)^{-1}$ and note that

$$\frac{n}{n^2 - 1} > \frac{n}{n^2} = \frac{1}{n}.$$

□

Proposition 44 For any positive numbers $a_1, a_2, \ldots, a_n$, we have

$$a_1 a_2 \cdots a_n \le \left(\frac{a_1 + a_2 + \cdots + a_n}{n}\right)^n.$$

Proof (by induction) It is simple to see that the claim is valid for $n = 2$. Indeed,

$$\{(a_1 + a_2)/2\}^2 - a_1a_2 = \{(a_1 - a_2)/2\}^2 \geq 0.$$

The rest of the proof will be carries out in two steps.

Step 1: We show that the claim is valid for the case when n is a power of 2. Suppose then the claim is valid when $n = 2^k$; we want to show that it is also valid for $n = 2^{k+1}$. Using the inductive assumption that the claim holds for $n = 2^k$, we get

$$\begin{aligned} a_1 \cdots a_{2^{k+1}} &= (a_1 \cdots a_{2^k})(a_{2^k+1} \cdots a_{2^{k+1}}) \\ &\leq \left(\frac{a_1 + \cdots + a_{2^k}}{2^k}\right)^{2^k}\left(\frac{a_{2^k+1} + \cdots + a_{2^{k+1}}}{2^k}\right)^{2^k} \\ &= \left[\left(\frac{a_1 + \cdots + a_{2^k}}{2^k}\right)\left(\frac{a_{2^k+1} + \cdots + a_{2^{k+1}}}{2^k}\right)\right]^{2^k} \\ &= K. \end{aligned}$$

But $AB \leq \{(A + B)/2\}^2$ by what we know already and so, putting

$$A = \frac{a_1 + \cdots + a_{2^k}}{2^k} \quad \text{and} \quad B = \frac{a_{2^k+1} + \cdots + a_{2^{k+1}}}{2^k},$$

we see that

$$K \leq \left[\left(\frac{a_1 + \cdots + a_{2^{k+1}}}{2^{k+1}}\right)^2\right]^{2^k} = \left[\frac{a_1 + \cdots + a_{2^{k+1}}}{2^{k+1}}\right]^{2^{k+1}}.$$

Hence we have shown that the claim is valid for $n = 2^{k+1}$.

Step 2: Let $n > 2$ and n not be a power of 2, i.e., let $2^m < n < 2^{m+1}$. Put $2^{m+1} = p$. Then we adjoin to the n given numbers $a_1, \ldots, a_n$ another $p - n$ numbers $a_{n+1}, a_{n+2}, \ldots, a_p$, where we assume that each of these latter numbers equals

$$\frac{a_1 + \cdots + a_n}{n}.$$

In other words, $a_{n+1} = a_{n+2} = \cdots = a_p = a$ with

$$a = \frac{a_1 + \cdots + a_n}{n}.$$

By Step 1 (since p is a power of 2) we get

$$a_1 \cdots a_p \leq \left(\frac{a_1 + \cdots + a_p}{p}\right)^p. \tag{2}$$

We know that $a_1 a_2 \cdots a_p = a_1 a_2 \cdots a_n a^{p-n}$ and

$$\frac{a_1 + \cdots + a_p}{p} = a$$

because

$$a_1 + \cdots + a_p = (a_1 + \cdots + a_n) + (p - n)a$$

$$= na + (p - n)a = pa.$$

Thus (2) means that

$$a_1 \cdots a_n a^{p-n} \leq a^p, \quad \text{or} \quad a_1 \cdots a_n \leq a^n.$$

But this is what we wanted to show. □

Proposition 45 Let $x > 0$ and $0 < \alpha < 1$. Then

$$x^\alpha - \alpha x \leq 1 - \alpha. \tag{3}$$

Proof Differentiating the function $f(x) = x^\alpha - \alpha x$ with respect to x yields $f'(x) = \alpha(x^{\alpha-1} - 1)$. Clearly, $f'(x) > 0$ for $0 < x < 1$ and $f'(x) < 0$ for $x > 1$. Inequality (3) must therefore hold because f assumes its largest value at $x = 1$. □

Proposition 46 Let a and b be positive real numbers and α and β be such that $\alpha < 1$, $\beta < 1$, and $\alpha + \beta = 1$. Then

$$a^\alpha b^\beta \leq \alpha a + \beta b. \tag{3a}$$

Proof We take $x = a/b$ and denote $1 - \alpha$ by β; then inequality (3) evidently gives us the desired result. □

Remark To verify inequality (3a) we might also have used the fact that, for $a > 0$, $b > 0$, $\alpha < 1$, $\beta < 1$, and $\alpha + \beta = 1$,

$$\alpha \log a + \beta \log b \leq \log(\alpha a + \beta b)$$

because the logarithmic function is concave, that is, all points on an arc of its graph are situated above or on the corresponding chord.

Proposition 47 Let $a > 0$, $b > 0$, $k > 1$, $k' > 1$, and $1/k + 1/k' = 1$. Then

$$ab \leq \frac{1}{k} a^k + \frac{1}{k'} b^{k'}. \tag{3b}$$

Proof Consider inequality (3a). Setting $k = 1/\alpha$ and $k' = 1/\beta$ [hence $k' = k/(k - 1)$] and replacing in (3a) the number a by a^k and the number b by $b^{k'}$ we obtain the desired result. □

Remark To prove inequality (3b) we might also have proceeded as follows. Let $k = 1 + s$ with $s > 0$, and define $y = x^s$. But $y = x^s$ is an increasing function and $x = y^{1/s}$ is the (uniquely defined) inverse function.

A glance at Fig. 1 - 1 shows that

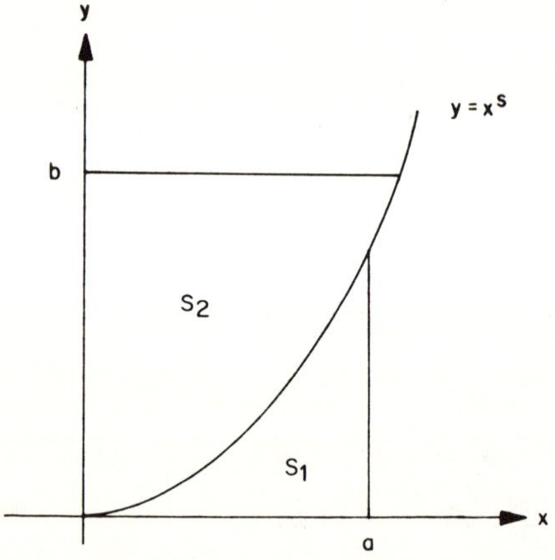

Figure 1 - 1

$$S_1 = \frac{a^k}{k} = \int_0^a x^s\,dx \qquad \text{and} \qquad S_2 = \frac{b^{k'}}{k'} = \int_0^b y^{1/s}\,dy$$

(because $1 + s^{-1} = 1 + (k - 1)^{-1} = k/(k - 1) = k'$). But the required inequality follows at once from the fact that the area of the rectangle with sides a and b is smaller than or equal to the sum of the areas S_1 and S_2. In the case when $b = a^s$ and only in this case the inequality changes into an equality.

We now show that inequality (3a) can be generalized to the case of three factors; using inequality (3a) twice we get

$$\begin{aligned} a^\alpha b^\beta c^\gamma &= a^\alpha (b^{\beta/(\beta+\gamma)} c^{\gamma/(\beta+\gamma)})^{\beta+\gamma} \\ &\le \alpha a + (\beta + \gamma) b^{\beta/(\beta+\gamma)} c^{\gamma/(\beta+\gamma)} \\ &\le \alpha a + (\beta + \gamma)\left(\frac{\beta}{\beta + \gamma} b + \frac{\gamma}{\beta + \gamma} c\right) \le \alpha a + \beta b + \gamma c; \end{aligned}$$

hence, for $a, b, c, \alpha, \beta, \gamma > 0$ and $\alpha + \beta + \gamma = 1$,

$$a^\alpha b^\beta c^\gamma \le \alpha a + \beta b + \gamma c.$$

In a completely similar manner one can make the transition from n to n + 1 factors; by induction we therefore obtain the more general inequality (note that we are changing notation):

$$a_1^{q_1} \cdots a_n^{q_n} \le q_1 a_1 + \cdots + q_n a_n$$

$$(a_1, \ldots, a_n, q_1, \ldots, q_n > 0; \quad q_1 + \cdots + q_n = 1).$$

In place of the numbers q_i we may use arbitrary numbers $p_i > 0$ by letting

$$q_i = \frac{p_i}{p_1 + \cdots + p_n}$$

so that $q_1 + \cdots + q_n = 1$. We can now write the inequality as follows:

$$\left(a_1^{p_1} \cdots a_n^{p_n}\right)^{1/\Sigma_j p_j} \le \frac{p_1 a_1 + \cdots + p_n a_n}{p_1 + \cdots + p_n} \tag{4}$$

$$(a_1, \ldots, a_n, p_1, \ldots, p_n > 0);$$

note that $\Sigma_j p_j$ stands for $\Sigma_{j=1}^n p_j = p_1 + p_2 + \cdots + p_n$. If we let $p_1 = p_2 = \cdots = p_n = 1$, we obtain the familiar inequality

$$(a_1 a_2 \cdots a_n)^{1/n} \le \frac{a_1 + a_2 + \cdots + a_n}{n}; \tag{4a}$$

it expresses the fact that the geometric mean of positive numbers is less than or equal to the arithmetic mean of the same numbers (a fact that we have already established in Proposition 44).

Proposition 48 (Hölder's Inequality) Let $a_i > 0$ and $b_i > 0$ for $i = 1, 2, \ldots, n$ and suppose that $k > 1$, $k' > 1$, and $1/k + 1/k' = 1$. Then

$$\sum_{i=1}^n a_i b_i \le \left(\sum_{i=1}^n a_i^k\right)^{1/k} \left(\sum_{i=1}^n b_i^{k'}\right)^{1/k'}. \tag{5}$$

For the special case when $k = k' = 2$, we get

$$\sum_{i=1}^n a_i b_i \le \left(\sum_{i=1}^n a_i^2\right)^{1/2} \left(\sum_{i=1}^n b_i^2\right)^{1/2};$$

this latter inequality is called the *Cauchy-Schwarz Inequality*.

Proof We first assume that

$$\sum_{i=1}^n a_i^k = \sum_{i=1}^n b_i^{k'} = 1 \tag{6}$$

and observe that the inequality to be proved will be of the form

$$\sum_{i=1}^n a_i b_i \le 1.$$

In (3b) we put successively $a = a_i$, $b = b_i$ for $i = 1, \ldots, n$ and then add up all inequalities obtained in this way. By (6) we get what we have set out to do.

The general case can be reduced to the foregoing special case if we take in place of the numbers a_i, b_i the numbers

$$a_i' = \frac{a_i}{(\Sigma_{j=1}^n a_j^k)^{1/k}}, \qquad b_i' = \frac{b_i}{(\Sigma_{j=1}^n b_j^{k'})^{1/k'}}$$

for which condition (6) is fulfilled. By what has been shown

$$\sum_{i=1}^n a_i' b_i' \le 1$$

holds and this is equivalent to (5). □

Proposition 49 (Minkowski's Inequality) Let $a_i > 0$ and $b_i > 0$ for $i = 1, \ldots, n$ and suppose that $k > 1$. Then

$$\left(\sum_{i=1}^n (a_i + b_i)^k\right)^{1/k} \le \left(\sum_{i=1}^n a_i^k\right)^{1/k} + \left(\sum_{i=1}^n b_i^k\right)^{1/k}. \tag{7}$$

Proof We observe that

$$\sum_{i=1}^n (a_i + b_i)^k = \sum_{i=1}^n a_i(a_i + b_i)^{k-1} + \sum_{i=1}^n b_i(a_i + b_i)^{k-1}.$$

Applying to the last two sums inequality (5), we get (since $1/k + 1/k' = 1$)

$$\sum_{i=1}^n (a_i + b_i)^k$$

$$\le \left(\sum_{i=1}^n a_i^k\right)^{1/k}\left(\sum_{i=1}^n (a_i + b_i)^{(k-1)k'}\right)^{1/k'}$$

$$+ \left(\sum_{i=1}^n b_i^k\right)^{1/k}\left(\sum_{i=1}^n (a_i + b_i)^{(k-1)k'}\right)^{1/k'}$$

$$= \left[\left(\sum_{i=1}^n a_i^k\right)^{1/k} + \left(\sum_{i=1}^n b_i^k\right)^{1/k}\right]\left(\sum_{i=1}^n (a_i + b_i)^k\right)^{1/k'}$$

and finally arrive at inequality (7) by division (using the last factor as divisor). □

Definition Let f be a real-valued function defined on an interval (a,b). Then f is said to be *convex* if for each x_1, x_2 in (a,b) we have

$$f(q_1x_1 + q_2x_2) \le q_1f(x_1) + q_2f(x_2) \tag{8}$$

regardless of how the positive numbers q_1 and $q_2 = 1 - q_1$ are chosen. The function f is said to be *concave* if the inequality in (8) is reversed.

Proposition 50 (Jensen's Inequality) If f is a convex function on (a,b), then

$$f(q_1x_1 + \cdots + q_nx_n) \le q_1f(x_1) + \cdots + q_nf(x_n) \tag{9}$$

$$(q_1, \ldots, q_n > 0;\ q_1 + \cdots + q_n = 1)$$

holds for any points $x_1, \ldots, x_n$ of the interval (a,b).

Proof We note that in case n = 2, we are back to the definition of convexity. We therefore assume that the inequality in question is true for some $n \ge 2$ and show that it will also be true for n + 1. In other words, we pick n + 1 points in (a,b), namely, $x_1, \ldots, x_n, x_{n+1}$, and we select n + 1 positive numbers $q_1, \ldots, q_n, q_{n+1}$ such that $q_1 + \cdots + q_n + q_{n+1} = 1$, and we seek to establish that

$$\begin{aligned} &f(q_1x_1 + \cdots + q_nx_n + q_{n+1}x_{n+1}) \\ &\le q_1f(x_1) + \cdots + q_nf(x_n) + q_{n+1}f(x_{n+1}). \end{aligned} \tag{10}$$

To this end we replace in the left-hand side of the above inequality the sum $q_nx_n + q_{n+1}x_{n+1}$ by the sum

$$(q_n + q_{n+1})\left(\frac{q_n}{q_n + q_{n+1}}\, x_n + \frac{q_{n+1}}{q_n + q_{n+1}}\, x_{n+1}\right).$$

In this way we can use inequality (9) and see that the expression on the left-hand side of (10) is smaller than or equal to

$$q_1f(x_1) + \cdots + (q_n + q_{n+1})\ f\left(\frac{q_n}{q_n + q_{n+1}}\, x_n + \frac{q_{n+1}}{q_n + q_{n+1}}\, x_{n+1}\right).$$

We now only have to apply to the values of the function in the last summand the basic inequality (8) in order to obtain (10). Hence (9) is proved completely by the Principle of Mathematical Induction. □

Remark Putting

$$q_1 = \frac{p_i}{p_1 + \cdots + p_n},$$

where $p_1, \ldots, p_n$ are arbitrary positive numbers, Jensen's Inequality assumes the form

$$f\left(\frac{\Sigma\, p_i x_i}{\Sigma\, p_i}\right) \le \frac{\Sigma\, p_i f(x_i)}{\Sigma\, p_i}. \tag{9a}$$

In case of a concave function f the inequality sign in (9a) must of course be reversed.

We conclude by giving a few applications of inequality (9a).

1. Let $f(x) = x^k$ with $x > 0$ and $k > 1$; f is clearly a convex function. Thus

$$\left(\frac{\Sigma\, p_i x_i}{\Sigma\, p_i}\right)^k \le \frac{\Sigma\, p_i x_i^k}{\Sigma\, p_i}$$

holds. This inequality is equivalent to

$$(\Sigma\, p_i x_i)^k \le (\Sigma\, p_i)^{k-1}(\Sigma\, p_i x_i^k).$$

Replacing p_i by $b_i^{k/(k-1)}$ and x_i by $a_i b_i^{-1/(k-1)}$, we obtain Hölder's Inequality (see Proposition 48)

$$\Sigma\, a_i b_i \le (\Sigma\, a_i^k)^{1/k}(\Sigma\, b_i^{k/(k-1)})^{(k-1)/k}.$$

2. Let $f(x) = \log x$ with $x > 0$; f is a concave function and so

$$\frac{\Sigma\, p_i \log x_i}{\Sigma\, p_i} \le \log\left(\frac{\Sigma\, p_i x_i}{\Sigma\, p_i}\right).$$

Taking antilogarithms we obtain

$$\left(x_1^{p_1} \cdots x_n^{p_n}\right)^{1/\Sigma p_i} \le \frac{\Sigma p_i x_i}{\Sigma p_i},$$

i.e., inequality (4).

3. Let $f(x) = x \log x$ with $x > 0$; f is a convex function. We obtain

$$\left(\frac{\Sigma p_i x_i}{\Sigma p_i}\right) \log \left(\frac{\Sigma p_i x_i}{\Sigma p_i}\right) \le \frac{\Sigma p_i x_i \log x_i}{\Sigma p_i}.$$

Multiplying by Σp_i and then passing to antilogarithms, we get

$$\frac{\Sigma p_i x_i}{\Sigma p_i} \le \left(x_1^{p_1 x_1} \cdots x_n^{p_n x_n}\right)^{1/\Sigma p_i x_i}.$$

Letting $p_i = x_i^{-1}$, we obtain

$$\frac{n}{\Sigma 1/x_i} \le (x_1 \cdots x_n)^{1/n}$$

showing that the harmonic mean of positive numbers is less than or equal to the geometric mean of the same numbers.

4. *Sequences of Real Numbers*

<u>*Definition*</u> By a *sequence of real numbers* (or *sequence*, for short) we mean a function f defined on the set N of positive integers. If $f(n) = x_n$, for $n \in N$, it is customary to denote the sequence f by the symbol $(x_n)_{n=1}^{\infty}$, or sometimes by $x_1, x_2, \ldots$. The range of f, that is, the values x_n, are called the terms of the sequence.

<u>*Definition*</u> A sequence of real numbers $(x_n)_{n=1}^{\infty}$ is said to *converge* if there is a real number x with the following property: For every $\varepsilon > 0$ there is an integer n_0, dependent on ε, such that $n \ge n_0$ implies that $|x_n - x| < \varepsilon$. This property we shall sometimes express by

saying that $|x_n - x| < \varepsilon$ for sufficiently large n. We call x the *limit of* $(x_n)_{n=1}^{\infty}$ and write $\lim_{n\to\infty} x_n = x$ or $x_n \to x$ as $n \to \infty$.

If $(x_n)_{n=1}^{\infty}$ does not converge, it is said to *diverge*. If for any real number $M > 0$ there exists a positive integer n_0 such that, for $n \geq n_0$, $x_n > M$, then we say that $(x_n)_{n=1}^{\infty}$ *properly diverges to* $+\infty$, or x_n approaches $+\infty$. In a similar way one defines *proper divergence to* $-\infty$.

Proposition 51 If a sequence has a limit, then this limit is unique.

Proof Suppose the contrary; let $x_n \to x$ and $x_n \to x'$ with $x \neq x'$. Let $\varepsilon = \frac{1}{2}|x - x'|$. Since $x_n \to x$ as $n \to \infty$, there exists $n_1 \in N$ such that $|x_n - x| < \varepsilon$ for $n \geq n_1$. Similarly, since $x_n \to x'$, there exists $n_2 \in N$ such that $|x_n - x'| < \varepsilon$ for $n \geq n_2$. Let $n_0 = \max\{n_1, n_2\}$. Then both $|x_n - x|$ and $|x_n - x'|$ are less than ε for $n \geq n_0$. Thus

$$|x - x'| = |(x_n - x') - (x_n - x)| \leq |x_n - x'| + |x_n - x|$$
$$< 2\varepsilon = |x - x'|$$

which implies $|x - x'| < |x - x'|$. This contradiction shows that $x = x'$. □

Proposition 52 If $(x_n)_{n=1}^{\infty}$ is convergent, then the range of the sequence is bounded.

Proof Suppose $x_n \to x$ as $n \to \infty$. Then there is an integer n_0 such that $n > n_0$ implies $|x_n - x| < 1$. We put

$$r = \max\left\{1, |x_1 - x|, \ldots, |x_n - x|\right\}.$$

Then $|x_n - x| \leq r$ for $n = 1, 2, \ldots$, i.e., $x - r \leq x_n \leq x + r$ for $n = 1, 2, \ldots$. □

Definition A sequence of bounded range is said to be a *bounded sequence*.

Proposition 53 Let $(x_n)_{n=1}^{\infty}$ and $(y_n)_{n=1}^{\infty}$ be two convergent sequences with limits x and y, respectively. Then, as $n \to \infty$,

(i) $x_n + y_n \to x + y$;

(ii) $cx_n \to cx$, $c + x_n \to c + x$, for any real number c;

(iii) $x_n y_n \to xy$;

(iv) $1/x_n \to 1/x$, provided $x_n \neq 0$ for $n = 1, 2, \ldots$ and $x \neq 0$.

Proof

(i) Given $\varepsilon > 0$, there exist integers n', n" such that

$$n \geq n' \text{ implies } |x_n - x| < \varepsilon/2,$$

$$n \geq n'' \text{ implies } |y_n - y| < \varepsilon/2.$$

If $n_0 = \max\{n',n''\}$, then $n \geq n_0$ implies

$$|(x_n + y_n) - (x + y)| \leq |x_n - x| + |y_n - y| < \varepsilon.$$

This proves (i). The verification of (ii) is trivial.

(iii) We use the identity

$$x_n y_n - xy = (x_n - x)(y_n - y) + x(y_n - y) + y(x_n - x). \tag{11}$$

Given $\varepsilon > 0$, there are integers n', n" such that

$$n \geq n' \text{ implies } |x_n - x| < \sqrt{\varepsilon},$$

$$n \geq n'' \text{ implies } |y_n - y| < \sqrt{\varepsilon}.$$

If we take $n_0 = \max\{n',n''\}$, $n \geq n_0$ implies $|(x_n - x)(y_n - y)| < \varepsilon$, so that $(x_n - x)(y_n - y) \to 0$ as $n \to \infty$. We now apply (i) and (ii) to (11) and conclude that $x_n y_n - xy \to 0$ as $n \to \infty$.

(iv) Choosing m such that $|x_n - x| < \frac{1}{2}|x|$ if $n \geq m$, we see that $|x_n| > \frac{1}{2}|x|$ for $n \geq m$, i.e., $|x_n x| > \frac{1}{2}|x|^2$ for $n \geq m$. Given $\varepsilon > 0$, there exists an integer $k_0 > m$ such that $n \geq k_0$ implies $|x_n - x| < \frac{1}{2}|x|^2\varepsilon$. Hence, for $n \geq k_0$,

$$\left|\frac{1}{x_n} - \frac{1}{x}\right| = \left|\frac{x_n - x}{x_n x}\right| < \frac{2}{|x|^2}|x_n - x| < \varepsilon.$$

□

Definition A sequence $(x_n)_{n=1}^{\infty}$ is said to be

(i) *monotonically increasing* if, for $n = 1, 2, \ldots, x_n \le x_{n+1}$;

(ii) *monotonically decreasing* if, for $n = 1, 2, \ldots, x_n \ge x_{n+1}$.

The class of *monotonic sequences* consists of the increasing and the decreasing sequences.

Proposition 54 Suppose $(x_n)_{n=1}^{\infty}$ is monotonic. Then $(x_n)_{n=1}^{\infty}$ converges if and only if it is bounded.

Proof From Proposition 52 we know that a convergent sequence is bounded. To show the remaining part of the claim we suppose that $x_n \le x_{n+1}$ for $n = 1, 2, \ldots$ (the verification is analogous in the other case; premultiplication by -1 turns a monotonically decreasing sequence into a monotonically increasing one). Let E be the range of the sequence $(x_n)_{n=1}^{\infty}$, i.e., the set $\{x_n : n \in N\}$. If $(x_n)_{n=1}^{\infty}$ is bounded, let x be the least upper bound of E. The sequence $(x_n)_{n=1}^{\infty}$ is obviously bounded (by x_1) below. Then, for $n = 1, 2, \ldots, x_n \le x$. For every $\varepsilon > 0$, there exists an integer n_0 such that $x - \varepsilon < x_{n_0} \le x$, for otherwise $x - \varepsilon$ would be an upper bound of E. Since the sequence $(x_n)_{n=1}^{\infty}$ increases, $n \ge n_0$ implies $x - \varepsilon < x_n \le x$, which shows that the sequence $(x_n)_{n=1}^{\infty}$ converges (to x). □

Definition An interval I is called *closed* if it contains its endpoints and *open* if it does not; I is *bounded* if it has finite length.

Proposition 55 Let $([a_n,b_n])_{n=1}^{\infty}$ be a sequence of closed, bounded intervals such that $[a_n,b_n] \supset [a_{n+1},b_{n+1}]$ for $n = 1, 2, \ldots$ and $b_n - a_n \to 0$ as $n \to \infty$. Then there is one and only one point common to all the intervals in the (nested) sequence $([a_n,b_n])_{n=1}^{\infty}$

Proof By the nestedness of the intervals,

$$a_1 \le a_n \le a_{n+1} \le b_{n+1} \le b_n \le b_1$$

for $n = 1, 2, \ldots$. Thus the sequences $(a_n)_{n=1}^{\infty}$ and $(b_n)_{n=1}^{\infty}$ are monotonic and bounded and so converge to some a and some b in R^1, respectively, by Proposition 54.

To show that $a = b$ and thus prove the required result, we note that

$$|b - a| \le |b - b_n| + |b_n - a_n| + |a_n - a|.$$

But $b_n - a_n \to 0$, $a_n \to a$, and $b_n \to b$ as $n \to \infty$. Hence, given any $\varepsilon > 0$, we can find n_0 such that for all $n > n_0$

$$|b - b_n| < \frac{\varepsilon}{3}, \quad |b_n - a_n| < \frac{\varepsilon}{3}, \quad |a_n - a| < \frac{\varepsilon}{3},$$

and thus $|b - a| < \varepsilon$. Since ε is an arbitrary positive number and $b - a$ is fixed, we must have $b - a = 0$, i.e., $a = b$. □

We shall now consider several examples illustrating the usefulness of the foregoing propositions.

Example 1 Let

$$a_n = \left(1 + \frac{1}{n}\right)^n \quad \text{and} \quad b_n = \left(1 + \frac{1}{n}\right)^{n+1}$$

and observe that

$$b_n - a_n = \frac{1}{n} a_n \to 0 \quad \text{as} \quad n \to \infty.$$

Using Propositions 42 and 43 we see that we have a nested sequence of intervals $[a_n, b_n]$; the single point common to all these closed intervals is e, the base of the natural logarithms. Note also that Propositions 42 and 54 give us the convergence of the sequence $(a_n)_{n=1}^{\infty}$ and that a similar remark applies to the sequence $(b_n)_{n=1}^{\infty}$.

Example 2 Let

$$a_n = 1 + \frac{1}{2} + \cdots + \frac{1}{n - 1} - \log n$$

and

$$b_n = 1 + \frac{1}{2} + \cdots + \frac{1}{n - 1} + \frac{1}{n} - \log n.$$

It is clear that $b_n - a_n = 1/n$. To see that $a_{n+1} > a_n$ and $b_{n+1} < b_n$ we note that

$$\frac{1}{n+1} < \log \frac{n+1}{n} < \frac{1}{n};$$

indeed, for $n < x < n + 1$, we have $1/(n + 1) < 1/x < 1/n$, and so

$$\int_n^{n+1} \frac{dx}{n+1} < \int_n^{n+1} \frac{dx}{x} < \int_n^{n+1} \frac{dx}{n}.$$

Thus

$$1 + \frac{1}{2} + \cdots + \frac{1}{n} - \log n$$

converges as $n \to \infty$; its limit is known as *Euler's constant* C, where $C = 0.5772156649\ldots$.

Example 3 Let $a_1 > b_1 > 0$ be given. We form the numbers

$$a_2 = \frac{a_1 + b_1}{2} \quad \text{and} \quad b_2 = \sqrt{a_1 b_1},$$

$$a_3 = \frac{a_2 + b_2}{2} \quad \text{and} \quad b_3 = \sqrt{a_2 b_2},$$

....

$$a_{n+1} = \frac{a_n + b_n}{2} \quad \text{and} \quad b_{n+1} = \sqrt{a_n b_n},$$

....

We observe that $a_1 > a_2 > b_2 > b_1$ and that in general

$$a_1 > a_2 > \cdots > a_n > a_{n+1} > b_{n+1} > b_n > \cdots > b_2 > b_1.$$

and hence $(a_n)_{n=1}^{\infty}$ is a monotonically decreasing bounded sequence and $(b_n)_{n=1}^{\infty}$ is a monotonically increasing bounded sequence because the a_n's and the b_n's are in fact the consecutive arithmetic and geometric means of the initially given numbers a_1 and b_1 with $a_1 > b_1 > 0$. Indeed, it is evident that $a_1 > a_2$ and $b_2 > b_1$ (because $a_1 > b_1 > 0$). To see that $a_2 > b_2$ we might proceed as in the beginning of the proof of Proposition 44 (or simply invoke Proposition 44); but there is a more interesting way of looking at it. Let a triangle ABC be inscribed in a semicircle with AB being the diameter (see Fig. 1 - 2).

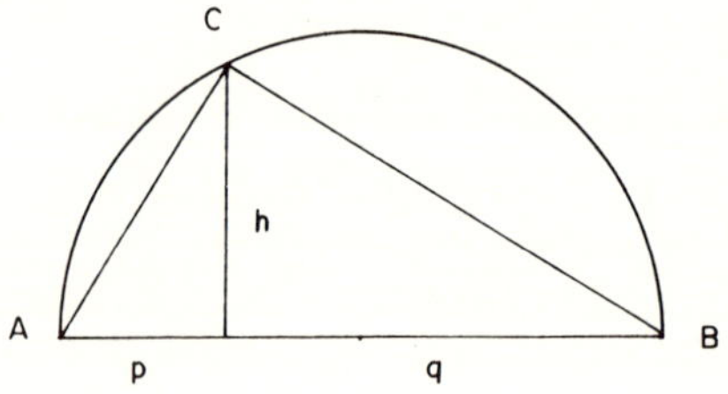

Figure 1 - 2

The altitude h divides the base AB into two segments of length p and q. By similarity of triangles we see that $h/p = q/h$ or $h^2 = pq$. But h is clearly less than or equal to $(p + q)/2$, the radius of the semicircle. In fact, equality can only occur when $p = q$.

It is easy to see that

$$a_2 - b_2 < \frac{a_1 - b_1}{2}, \quad a_3 - b_3 < \frac{a_2 - b_2}{2} < \frac{a_1 - b_1}{4}$$

and, in general,

$$a_{n+1} - b_{n+1} < \frac{a_1 - b_1}{2^n}.$$

We can therefore apply Proposition 55; the sequences $(a_n)_{n=1}^{\infty}$ and $(b_n)_{n=1}^{\infty}$ converge to one and the same limit $L = L(a_1,b_1)$.

We shall analyze this number L a bit more closely. Let

$$G = \int_0^{\pi/2} \frac{dx}{\sqrt{a_1^2 \cos^2 x + b_1^2 \sin^2 x}} \qquad (a_1 > b_1 > 0).$$

We put

$$\sin x = \frac{2a_1 \sin t}{(a_1 + b_1) + (a_1 - b_1) \sin^2 t};$$

as t changes from 0 to $\pi/2$, so grows x from 0 to $\pi/2$. Differentiation gives

$$\cos x\, dx = 2a_1 \frac{(a_1 + b_1) - (a_1 - b_1) \sin^2 t}{[(a_1 + b_1) + (a_1 - b_1) \sin^2 t]^2} \cos t\, dt.$$

But

$$\cos x = \frac{\sqrt{(a_1 + b_1)^2 - (a_1 - b_1)^2 \sin^2 t}}{(a_1 + b_1) + (a_1 - b_1) \sin^2 t} \cos t,$$

and thus

$$dx = 2a_1 \frac{(a_1 + b_1) - (a_1 - b_1) \sin^2 t}{(a_1 + b_1) + (a_1 - b_1) \sin^2 t}$$

$$\times \frac{dt}{\sqrt{(a_1 + b_1)^2 - (a_1 - b_1)^2 \sin^2 t}}.$$

On the other hand

$$\sqrt{a_1^2 \cos^2 x + b_1^2 \sin^2 x} = a_1 \frac{(a_1 + b_1) - (a_1 - b_1) \sin^2 t}{(a_1 + b_1) + (a_1 - b_1) \sin^2 t}$$

and thus

$$\frac{dx}{\sqrt{a_1^2 \cos^2 x + b_1^2 \sin^2 x}} = \frac{dt}{\sqrt{[(a_1 + b_1)/2]^2 \cos^2 t + a_1 b_1 \sin^2 t}}.$$

Putting

$$a_2 = \frac{a_1 + b_1}{2} \quad \text{and} \quad b_2 = \sqrt{a_1 b_1}$$

we get

$$G = \int_0^{\pi/2} \frac{dx}{\sqrt{a_1^2 \cos^2 x + b_1^2 \sin^2 x}}$$

$$= \int_0^{\pi/2} \frac{dt}{\sqrt{a_2^2 \cos^2 t + b_2^2 \sin^2 t}}.$$

By repeated application of this transformation, we get

$$G = \int_0^{\pi/2} \frac{dx}{\sqrt{a_n^2 \cos^2 x + b_n^2 \sin^2 x}} \qquad (n = 1, 2, 3, \ldots),$$

where a_n and b_n are defined by the recursive formula

$$a_n = \frac{a_{n-1} + b_{n-1}}{2}, \quad b_n = \sqrt{a_{n-1}b_{n-1}}.$$

As we know already, these two sequences converge to the common value $L = L(a_1,b_1)$. It is easy to see that

$$\frac{\pi}{2a_n} < G < \frac{\pi}{2b_n};$$

passage to the limit as $n \to \infty$ gives

$$G = \frac{\pi}{2L(a_1,b_1)} \quad \text{or} \quad L(a_1,b_1) = \frac{\pi}{2G}.$$

We consider an application; we calculate the integral

$$G = \int_0^{\pi/2} \frac{dx}{\sqrt{1 + \cos^2 x}} = \int_0^{\pi/2} \frac{dx}{\sqrt{2 \cos^2 x + \sin^2 x}}.$$

Here $a_1 = \sqrt{2}$ and $b_1 = 1$; the numerical sequences $(a_n)_{n=1}^{\infty}$ and $(b_n)_{n=1}^{\infty}$ converge rapidly to L in this case and a_5 and a_6 are approximately equal to 1.198154. Hence we may put L = 1.198154 and obtain the approximate value

$$G = \frac{\pi}{2L} \sim 1.3110138.$$

Example 4 Let $a_1 > b_1 > 0$ be two given numbers. We form the consecutive arithmetic and harmonic means:

$$a_2 = \frac{a_1 + b_1}{2} \quad \text{and} \quad b_2 = \frac{2a_1b_1}{a_1 + b_1},$$

$$a_3 = \frac{a_2 + b_2}{2} \quad \text{and} \quad b_3 = \frac{2a_2b_2}{a_2 + b_2},$$

....

$$a_{n+1} = \frac{a_n + b_n}{2} \quad \text{and} \quad b_{n+1} = \frac{2a_nb_n}{a_n + b_n},$$

....

Since

$$\frac{a_1 + b_1}{2} > \sqrt{a_1b_1} \quad \text{for} \quad a_1 \neq b_1,$$

we have

$$\frac{a_1 + b_1}{2} > \frac{2a_1b_1}{a_1 + b_1}.$$

It is easy to see that $a_n > a_{n+1} > b_{n+1} > b_n$ and so the sequences $(a_n)_{n=1}^{\infty}$ and $(b_n)_{n=1}^{\infty}$ converge. Passing to the limit in the equation

$$a_{n+1} = \frac{a_n + b_n}{2}$$

we see that

$$A = \frac{A + B}{2},$$

where A is the limit of $(a_n)_{n=1}^{\infty}$ and B is the limit of $(b_n)_{n=1}^{\infty}$; we can see at once that A = B. This common limit is called the *arithmetic-harmonic mean of* a_1 and a_2.

Example 5 Let c for the moment be an arbitrary real number and put

$$x_1 = \frac{c}{2}$$

and

$$x_{n+1} = \frac{c}{2} + \frac{x_n^2}{2} \quad \text{for} \quad n = 1, 2, 3, \ldots.$$

We observe that if the limit of $(x_n)_{n=1}^{\infty}$ exists, say $x_n \to a$ as $n \to \infty$, then

$$a = \frac{c}{2} + \frac{a^2}{2}, \quad \text{or} \quad a^2 - 2a + c = 0.$$

Thus

$$a = 1 \pm \sqrt{1 - c},$$

and so the sequence $(x_n)_{n=1}^{\infty}$ can not converge for $c > 1$.

(a) We assume now that $0 < c \leq 1$. Then $x_n > 0$. It is easy to verify that $x_n < 1$ and $x_n < x_{n+1}$ for all $n \in N$. Thus $\lim_{n\to\infty} x_n$ exists and cannot be larger than 1; hence the limit is equal to $1 - \sqrt{1 - c}$.

(b) Suppose that $-3 \leq c < 0$. Then $x_n \geq c/2$ and $x_n < 0$ for all $n \in N$. The validity of the claim that $x_n \geq c/2$ for all $n \in N$ is clear. We only have to look more closely at the claim that $x_n < 0$ for all $n \in N$. This is indeed true for $n = 1$. We assume that the claim is true for some n; then $|x_n| \leq |c|/2$,

$$x_n^2 \leq \frac{|c|^2}{4} < c \qquad (\text{because } |c|/4 < 1)$$

and x_{n+1} is seen to have the same sign as $c/2$, that is, x_{n+1} is negative.

At any rate the sequence $(x_n)_{n=1}^{\infty}$ is bounded. However, the sequence is not monotonic. Noting that

$$x_{2k+1} - x_{2k-1} = \frac{1}{2}(x_{2k}^2 - x_{2k-2}^2)$$

and

$$x_{2k+2} - x_{2k} = \frac{1}{2}(x_{2k+1}^2 - x_{2k-1}^2),$$

it can easily be established by induction that

$$x_{2k+1} > x_{2k-1} \quad \text{and} \quad x_{2k+2} < x_{2k} \quad \text{for } k = 1, 2, 3, \ldots.$$

In other words,

$$a' = \lim_{n\to\infty} x_{2k-1} \quad \text{and} \quad a'' = \lim_{n\to\infty} x_{2k}$$

exist. We now show that $a' = a''$. In

$$x_{n+1} = \frac{c}{2} + \frac{x_n^2}{2}$$

we let $n \to \infty$; we first let $n \to \infty$ with n even, and then with n odd, and we obtain

$$a' = \frac{c}{2} + \frac{(a'')^2}{2}, \quad a'' = \frac{c}{2} + \frac{(a')^2}{2}.$$

Thus

$$a' - a'' = \frac{1}{2}\{(a'')^2 - (a')^2\}$$

and so

$$(a' - a'')(a' + a'' + 2) = 0.$$

As we shall see at once, the second factor cannot vanish for $c > -3$, and so $a' = a''$ must hold. Indeed, if we set $a'' = -a' - 2$ in

$$a'' = \frac{c}{2} + \frac{(a')^2}{2},$$

we would get

$$(a')^2 + 2a' + (4 + c) = 0$$

which has no real root for $c > -3$.

Finally, for $c = -3$ both factors $(a' - a'')$ and $(a' + a'' + 2)$ vanish because then $a' = a'' = -1$.

Thus in all cases $a' = a''$. If we denote this common value by a we see that

$$a = 1 - \sqrt{1 - c}$$

because the limit of the negative-valued sequence $(x_n)_{n=1}^{\infty}$ cannot be (strictly) positive.

Definition Given a sequence $(x_n)_{n=1}^{\infty}$, consider a sequence $(n_k)_{k=1}^{\infty}$ of positive integers, such that $n_1 < n_2 < n_3 < \cdots$. Then the sequence $(x_{n_k})_{k=1}^{\infty}$ is called a *subsequence of* $(x_n)_{n=1}^{\infty}$. If $(x_{n_k})_{k=1}^{\infty}$ converges, its limit is called a *subsequential limit of* $(x_n)_{n=1}^{\infty}$.

Remark It is clear that $(x_n)_{n=1}^{\infty}$ converges to x if and only if every subsequence of it converges to x.

Proposition 56 (Bolzano-Weierstrass Theorem) Every bounded sequence $(x_n)_{n=1}^{\infty}$ of real numbers contains a convergent subsequence $(x_{n_k})_{k=1}^{\infty}$.

Proof If the range of the sequence $(x_n)_{n=1}^{\infty}$ is finite, the assertion is trivial; in this case there is at least one member of the range, say x, and a sequence $(n_k)_{k=1}^{\infty}$ with $x_{n_1} = x_{n_2} = \cdots = x$ and the subsequence $(x_{n_k})_{k=1}^{\infty}$ so obtained evidently converges.

If the range of the sequence $(x_n)_{n=1}^{\infty}$ is infinite, all terms x_n will be situated in an interval [a,b] of finite length on the number line. We bisect the interval [a,b] into the subintervals

$$[a,(a + b)/2] \quad \text{and} \quad [(a + b)/2,b].$$

Then at least one of these two new intervals must contain infinitely many members of the range of the sequence $(x_n)_{n=1}^{\infty}$ because otherwise the interval [a,b] could not have contained infinitely many terms of the sequence. Let us denote by $[a_1,b_1]$ that half-interval of [a,b] which contains infinitely many terms of $(x_n)_{n=1}^{\infty}$ (if both half-intervals have this property, then let $[a_1,b_1]$ be one of them).

Similarly we bisect the interval $[a_1,b_1]$; let $[a_2,b_2]$ be a half-interval of $[a_1,b_1]$ that contains infinitely many members of the range of $(x_n)_{n=1}^{\infty}$. We continue this process indefinitely; after k steps we obtain an interval $[a_k,b_k]$ that contains infinitely many members of the range of $(x_n)_{n=1}^{\infty}$.

It is clear that, for k = 1, 2, ...,

$$b_k - a_k = \frac{b - a}{2^k}$$

and thus $(b_k - a_k) \to 0$ as $k \to \infty$. Hence, by Proposition 55, there is one and only one point, say c, common to all these intervals and both $(a_k)_{k=1}^{\infty}$ and $(b_k)_{k=1}^{\infty}$ tend to c as $k \to \infty$.

We now construct inductively a subsequence $(x_{n_k})_{k=1}^{\infty}$ as follows: For x_{n_1} we take any of the terms x_n of our sequence $(x_n)_{n=1}^{\infty}$ which is situated in $[a_1,b_1]$. For x_{n_2} we take any of the terms x_n of our sequence $(x_n)_{n=1}^{\infty}$ which follows x_{n_1} and which is situated in $[a_2,b_2]$. In general we take for x_{n_k} any of the terms x_n of our sequence $(x_n)_{n=1}^{\infty}$ which follows all preceding terms $x_{n_1}, x_{n_2}, \ldots, x_{n_{k-1}}$ and which is situated in $[a_k,b_k]$. This selection is possible because each of the intervals $[a_k,b_k]$ contains infinitely many terms of our sequence $(x_n)_{n=1}^{\infty}$.

Since $a_k \le x_{n_k} \le b_k$ and both a_k and b_k tend to c as $k \to \infty$ it is clear that x_{n_k} tends to c as $k \to \infty$. Indeed, since a_k tends to c as $k \to \infty$, for any $\varepsilon > 0$, there is an integer n_0 such that $c - \varepsilon < a_k < c + \varepsilon$ holds for $k \ge n_0$. Similarly, there is an integer n_{00} such that $c - \varepsilon < b_k < c + \varepsilon$ holds for $k \ge n_{00}$. Let $n' = \max\{n_0,n_{00}\}$; then

$$c - \varepsilon < a_k \le x_{n_k} \le b_k < c + \varepsilon$$

holds for $k \ge n'$. Hence, for $k > n'$

$$c - \varepsilon < x_{n_k} < c + \varepsilon \quad \text{or} \quad |x_{n_k} - c| < \varepsilon$$

and so $x_{n_k} \to c$ as $k \to \infty$. □

Remark In picking the subsequence $(x_{n_k})_{k=1}^{\infty}$ we have used the Axiom of Choice; if Ω is any nonempty family of disjoint nonempty sets S, then a rule of selection that chooses a member from each set S of the collection Ω would exist follows only by means of an added postulate that such a choice function does indeed exist.

In the following proposition we consider a closed, bounded interval $[a,b]$ together with an arbitrary set Σ of open intervals σ which may have either a finite or an infinite number of such intervals σ. We say that the set Σ *covers* the interval $[a,b]$, if for each point x of $[a,b]$ we can find an interval σ of Σ such that σ contains the point x, that is, if $[a,b] \subset \cup\{\sigma: \sigma \in \Sigma\}$

Proposition 57 (Heine-Borel Theorem) If a closed, bounded interval $[a,b]$ is covered by an infinite set Σ of open intervals σ, then one may select from Σ a finite collection

$$\Sigma^* = \{\sigma_1, \sigma_2, \ldots, \sigma_n\}$$

such that Σ^* also covers $[a,b]$.

Proof 1 (Indirect, Using the Bolzano-Weierstrass Bisection Method) We assume that the interval $[a,b]$ cannot be covered by a finite number of open intervals σ of the collection Σ. We bisect $[a,b]$. Then at least one of the half-intervals $[a,(a + b)/2]$ or $[(a + b)/2,b]$ cannot be covered by a finite number of members of Σ (for if $\sigma_1, \sigma_2, \ldots, \sigma_m$ covered one half-interval and $\sigma_{m+1}, \sigma_{m+2}, \ldots, \sigma_n$ the other half-interval, then one could take the collection $\sigma_1, \sigma_2, \ldots, \sigma_n$ to get a cover for all of $[a,b]$). Let us denote by $[a_1,b_1]$ that half-interval of $[a,b]$ which cannot be covered by finitely many members σ of Σ. Now we repeat the described reasoning on $[a_1,b_1]$ and obtain by the process of bisection a half-interval $[a_2,b_2]$ of $[a_1,b_1]$ which cannot be covered by a finite collection of open intervals σ belonging to Σ.

Continuing this process indefinitely, we obtain a nested sequence $([a_n,b_n])_{n=1}^{\infty}$ of closed, bounded intervals; each interval is a half of its predecessor. All of these intervals are such that none of them can be covered by a finite collection of open intervals σ belonging to Σ. By Proposition 55 there is one and only one point, say c, common to all intervals $[a_n,b_n]$ with $n = 1, 2, \ldots$ and the monotonic sequences $(a_n)_{n=1}^{\infty}$ and $(b_n)_{n=1}^{\infty}$ converge to this point c.

The point c, like every other point of $[a,b]$, is situated in one of the intervals σ, say the interval $\sigma_0 = (\alpha,\beta)$ with $\alpha < c < \beta$. Since $a_n \to c$ and $b_n \to c$ as $n \to \infty$, it is clear that a_n and b_n are both in the open interval (α,β) for all sufficiently large n; thus, in fact, the single open interval $(\alpha,\beta) = \sigma_0$ covers all $[a_n,b_n]$ beginning with some value n_0 of the index n. But this is a contradiction and the proof is complete. □

Before proceeding to a second proof of Proposition 57 we consider a lemma.

Lemma Let $\bar{m}$ be an upper bound of a nonempty set M of real numbers. Then $\bar{m}$ is the supremum of M if and only if, for any $\varepsilon > 0$, there is an $m \in M$ such that $m > \bar{m} - \varepsilon$.

Proof Let $\bar{m}$ be the supremum of M. Suppose that $\varepsilon > 0$ be given. Then $\bar{m} - \varepsilon$ cannot be the supremum of M, hence there is an $m \in M$ such that $m > \bar{m} - \varepsilon$. On the other hand, suppose $\bar{m}$ is not the supremum of M. Then there is some $\bar{m}_0 < \bar{m}$ such that $\bar{m}_0$ is the supremum of M. We have to find an $\varepsilon > 0$ such that $m \leq \bar{m} - \varepsilon$ for all $m \in M$. But we only need to put $\varepsilon = \bar{m} - \bar{m}_0$. □

Proof 2 (of Proposition 57) (Direct Method) We consider the points x* of [a,b] with the property that the closed interval [a,x*] can be covered by finitely many members σ of the infinite set Σ of open intervals. Such points x* do exist; since, for example, the point a is contained in one of the open intervals σ, all points sufficiently close to a are of course members of the set X* of "star points".

Our task is to show that the point b is a "star point" as well.

Since for all x* we have $x^* \leq b$, it follows that $\sup X^* = c \leq b$; the supremum of X* exists because the set X* of all "star points" is nonempty and bounded above (by b). The point c, like any other point of the interval [a,b], belongs to some open interval $\sigma_0 = (\alpha,\beta)$ with $\alpha < c < \beta$ and $\sigma_0 \in \Sigma$. Since c is the least upper bound of the set X* of all "star points", there exists by the foregoing lemma a "star point" x_0^* such that $\alpha < x_0^* \leq c$. The closed interval $[a,x_0^*]$ can be covered by finitely many open intervals σ (because x_0^* is a "star point") of the set Σ. If we adjoin the single open interval σ_0, then the entire interval $[a,c]$ is seen to be covered by finitely many members σ of the set Σ. Thus c is a "star point" as well.

It is clear that c cannot be smaller than b, for if this were the case, there would exist a "star point" between c and b, contrary to the definition of c as the supremum of the set of all "star points". Thus $b = c$ and b is a "star point"; this means however that the entire interval $[a,b]$ can be covered by a finite subset Σ^* of the infinite set Σ. □

Definition A sequence $(x_n)_{n=1}^{\infty}$ of real numbers is said to be a *Cauchy sequence* if for every $\varepsilon > 0$ there is an integer n_0 such that $|x_n - x_{n'}| < \varepsilon$ if $n \geq n_0$ and $n' \geq n_0$.

Proposition 58 A sequence $(x_n)_{n=1}^{\infty}$ of real numbers is convergent if and only if it is a Cauchy sequence.

Proof We first show that the condition is necessary. Suppose the sequence $(x_n)_{n=1}^{\infty}$ converges to a. Then, for any $\varepsilon > 0$, we have $|x_n - a| < \varepsilon/2$.

Thus for two integers $n \geq n_0$ and $n' \geq n_0$ we have simultaneously $|x_n - a| < \varepsilon/2$ and $|a - x_{n'}| < \varepsilon/2$; from this we get that

$$|x_n - x_{n'}| = |(x_n - a) + (a - x_{n'})|$$

$$\leq |x_n - a| + |a - x_{n'}| < \frac{\varepsilon}{2} + \frac{\varepsilon}{2} = \varepsilon.$$

This shows that convergent sequences are Cauchy sequences.

The proof of the sufficiency of the condition is somewhat more demanding. Suppose that $(x_n)_{n=1}^{\infty}$ is a Cauchy sequence; we have to show that it converges. To do this, we define a set L of real numbers as follows: L consists of all those real numbers λ such that, beginning with some value of the index n, the inequality $x_n > \lambda$ holds. We also define a set U of real numbers by letting υ be in U if and only if υ does not belong to the set L. We show first that the sets L and U are nonempty. If $\varepsilon > 0$ is any preassigned real number, we can find on account of the Cauchy condition an integer n_0 depending on ε such that for $n \geq n_0$ and $n' \geq n_0$ the condition $|x_n - x_{n'}| < \varepsilon$ is satisfied, that is,

$$x_{n'} - \varepsilon < x_n < x_{n'} + \varepsilon$$

holds. Evidently, for $n' \geq n_0$, every number $x_{n'} - \varepsilon$ belongs to the set L because $x_{n'} - \varepsilon$ is smaller than x_n for sufficiently large n (namely, for $n \geq n_0$). Since, on the other hand (for the same n), the number x_n remains smaller than every number of the form $x_{n'} + \varepsilon$ (for $n' \geq n_0$), it is clear that no such number can belong to L; thus such a number belongs to U.

From the definition of the sets L and U we see immediately that every real number belongs to either L or U. Moreover, it is clear that every number λ (in L) is smaller than every number υ (in U); for if $\lambda > \upsilon$ were true, then the variable x_n would exceed υ beginning with some value of the index n, contrary to the definitiion of the numbers υ. Hence we may apply Dedekind's Theorem (see Proposition 29); according to this theorem there exists a unique real number a wedged between the members of the sets L and U: $\lambda \leq a \leq \upsilon$.

As we have observed already, for every $n' \geq n_0$ the number $x_{n'} - \varepsilon$ is one of the numbers λ and the number $x_{n'} + \varepsilon$ is one of the numbers υ. Thus, in particular,

$$x_{n'} - \varepsilon \leq a \leq x_{n'} + \varepsilon$$

or

$$|a - x_{n'}| = |x_{n'} - a| \leq \varepsilon$$

for every $n' \geq n_0$. By the definition of limit, however, this means that

$$a = \lim_{n \to \infty} x_n$$

and the proof is finished. □

As an application of Proposition 58 we consider a constructive solution of *Kepler's Equation*. Kepler's Equation serves to determine the position of a planet on its path; the equation in question reads:

$$x = q \sin x + a;$$

a and q are given numbers and q is such that $0 < q < 1$ while x is the unknown.

We pick an arbitrary number x_0 and form the sequence of numbers

$$x_1 = q \sin x_0 + a,$$

$$x_2 = q \sin x_1 + a,$$

....

$$x_n = q \sin x_{n-1} + a,$$

$$x_{n+1} = q \sin x_n + a,$$

....

Subtraction of the first equation from the second equation gives

$$\begin{aligned} x_2 - x_1 &= q(\sin x_1 - \sin x_0) \\ &= 2q \sin \frac{x_1 - x_0}{2} \cos \frac{x_1 + x_0}{2}. \end{aligned}$$

Noting that $|\sin t| \le |t|$ and $|\cos t| \le 1$, we obtain

$$|x_2 - x_1| \le 2q \frac{|x_1 - x_0|}{2} = q|x_1 - x_0|.$$

In the same way we get

$$|x_3 - x_2| \le q|x_2 - x_1|;$$

this in turn yields

$$|x_3 - x_2| \le q^2|x_1 - x_0|,$$

since

$$|x_2 - x_1| \le q|x_1 - x_0|.$$

Continuing this process, we obtain, for n = 1, 2, 3, ...,

$$|x_{n+1} - x_n| \le q^n |x_1 - x_0|.$$

We now consider the difference $x_m - x_n$ and assume without loss of generality that $m > n$; we get

$$x_m - x_n = (x_m - x_{m-1}) + (x_{m-1} - x_{m-2}) + (x_{m-2} - x_{m-3}) + \cdots + (x_{n+1} - x_n).$$

Using the triangle inequality and the summation formula for geometric progressions, we obtain

$$\begin{aligned} |x_m - x_n| &\le |x_m - x_{m-1}| + |x_{m-1} - x_{m-2}| + |x_{m-2} - x_{m-3}| \\ &\quad + \cdots + |x_{n+1} - x_n| \\ &\le (q^{m-1} + q^{m-2} + q^{m-3} + \cdots + q^n)|x_1 - x_0| \\ &= q^n \frac{1 - q^{m-n}}{1 - q} |x_1 - x_0|. \end{aligned}$$

As $n \to \infty$, $q^n \to 0$ (because $0 < q < 1$); the factor $|x_1 - x_0|$ remains constant; the fraction $(1 - q^{m-n})/(1 - q)$ is situated between 0 and $1/(1 - q)$, hence is bounded. Thus, as n grows unboundedly, for arbitrary $m > n$ the difference $x_m - x_n$ tends to zero. This shows that $(x_n)_{n=1}^{\infty}$ is a Cauchy sequence and so

$$t = \lim_{n\to\infty} x_n$$

exists.

In the equation

$$x_{n+1} = q \sin x_n + a$$

we now pass to the limit as $n \to \infty$. But $\sin x_n \to \sin t$ as $n \to \infty$ for

$$|\sin x_n - \sin t| = 2\left|\sin \frac{x_n - t}{2}\right| \left|\sin \frac{x_n + t}{2}\right| \le |x_n - t|.$$

This shows that

$$t = q \sin t + a$$

and therefore t is a root of Kepler's equation.

This root is independent of the initial term x_0. Indeed, if there were another root, say t', of the Kepler equation, that is, if

$$t' = q \sin t' + a$$

holds, then

$$t' - t = q(\sin t' - \sin t) = 2q \sin \frac{t' - t}{2} \cos \frac{t' + t}{2}$$

and $|t' - t| \le q|t' - t| < |t' - t|$ would follow; but this is absurd. Thus $t' = t$.

To compute limits of indeterminate forms of the type ∞/∞ we can sometimes use the following result.

Proposition 59 (Theorem of Stolz) Let $(x_n)_{n=1}^{\infty}$ and $(y_n)_{n=1}^{\infty}$ be two sequences of real numbers. Suppose that the second sequence properly diverges to $+\infty$ and that it is strictly increasing for all sufficiently large n. Then

$$\lim_{n\to\infty} \frac{x_n}{y_n} = \lim_{n\to\infty} \frac{x_n - x_{n-1}}{y_n - y_{n-1}}$$

provided that the limit on the right-hand side exists (be it finite or infinite).

Proof We assume first that the limit is finite, that is,

$$\lim_{n\to\infty} \frac{x_n - x_{n-1}}{y_n - y_{n-1}} = r$$

where r is a finite real number. Then, for any $\varepsilon > 0$, there exists a natural number k such that, for $n \ge k$, we have

$$\left|\frac{x_n - x_{n-1}}{y_n - y_{n-1}} - r\right| < \frac{\varepsilon}{2} \quad \text{and} \quad y_n - y_{n-1} > 0.$$

We see therefore that all fractions

$$\frac{x_{k+1} - x_k}{y_{k+1} - y_k}, \quad \frac{x_{k+2} - x_{k+1}}{y_{k+2} - y_{k+1}}, \quad \ldots, \quad \frac{x_{n-1} - x_{n-2}}{y_{n-1} - y_{n-2}}, \quad \frac{x_n - x_{n-1}}{y_n - y_{n-1}}$$

are situated between $r - \varepsilon/2$ and $r + \varepsilon/2$. By Proposition 37, the fraction

$$\frac{x_n - x_k}{y_n - y_k}$$

must also be between $r - \varepsilon/2$ and $r + \varepsilon/2$ and hence

$$\left|\frac{x_n - x_k}{y_n - y_k} - r\right| < \frac{\varepsilon}{2}.$$

We now use the identity

$$\frac{x_n}{y_n} - r = \frac{x_k - ry_k}{y_n} + \left(1 - \frac{y_k}{y_n}\right)\left(\frac{x_n - x_k}{y_n - y_k} - r\right)$$

and we get that

$$\left|\frac{x_n}{y_n} - r\right| \le \left|\frac{x_k - ry_k}{y_n}\right| + \left|\frac{x_n - x_k}{y_n - y_k} - r\right|$$

(noting that $y_k/y_n \to 0$ as $n \to \infty$). We know already that the second summand on the right-hand side of the foregoing inequality is smaller than $\varepsilon/2$ for $n \ge k$; the first summand (whose numerator is a fixed quantity) also becomes smaller than $\varepsilon/2$ for $n \ge n'$ because $y_n \to \infty$. Choosing $n' \ge k$, we see that, for $n \ge n'$,

$$\left|\frac{x_n}{y_n} - r\right| < \varepsilon.$$

This then proves the proposition for the considered special case.

To finish the proof, we observe that the case of an infinite limit easily reduces to the case of a finite limit. Take for example the case where $(x_n - x_{n-1})/(y_n - y_{n-1}) \to \infty$ as $n \to \infty$, that is, the sequence properly diverges to $+\infty$. Then for all sufficiently large n, $x_n - x_{n-1} > y_n - y_{n-1}$; hence $x_n \to \infty$ with $y_n \to \infty$ and the sequence

$(x_n)_{n=1}^{\infty}$ must be strictly increasing for all sufficiently large n. Therefore we can use what we have proved in the beginning of this proof and apply that result to the reciprocal expression, namely y_n/x_n:

$$\lim_{n \to \infty} \frac{y_n}{x_n} = \lim_{n \to \infty} \frac{y_n - y_{n-1}}{x_n - x_{n-1}} = 0$$

and we conclude that

$$\lim_{n \to \infty} \frac{x_n}{y_n} = \infty.$$

□

Corollary Let $(a_n)_{n=1}^{\infty}$ converge to a as $n \to \infty$, i.e., let a be the limit of the sequence $(a_n)_{n=1}^{\infty}$ in case the sequence is convergent or properly divergent to infinity. Then the sequence of arithmetic means

$$b_n = \frac{a_1 + a_2 + \cdots + a_n}{n}$$

also tends to a as $n \to \infty$.

Proof We set

$$x_n = a_1 + a_2 + \cdots + a_n \quad \text{and} \quad y_n = n$$

and apply Proposition 59. Noting that

$$x_n - x_{n-1} = a_n \quad \text{and} \quad y_n - y_{n-1} = 1,$$

everything is clear. □

Proposition 60 If $a_n \to 0$ and $b_n \to 0$, and if, in addition, $(b_n)_{n=1}^{\infty}$ is strictly decreasing for all sufficiently large n, then

$$\lim_{n \to \infty} \frac{a_n}{b_n} = \lim_{n \to \infty} \frac{a_n - a_{n+1}}{b_n - b_{n+1}}$$

provided that the second quotient is convergent or properly divergent.

Proof Suppose first that the limit is finite and equal to s; then if $\varepsilon > 0$ is given, m can be found so that

$$s - \varepsilon < \frac{a_n - a_{n+1}}{b_n - b_{n+1}} < s + \varepsilon \quad \text{and} \quad b_n - b_{n+1} > 0$$

if $n \geq m$; thus, for $n \geq m$,

$$(s - \varepsilon)(b_n - b_{n+1}) < a_n - a_{n+1} < (s + \varepsilon)(b_n - b_{n+1}).$$

Changing n to $n + 1, n + 2, \ldots, n + p - 1$ and adding the results, we find

$$(s - \varepsilon)(b_n - b_{n+p}) < a_n - a_{n+p} < (s + \varepsilon)(b_n - b_{n+p}).$$

Taking the limit as $p \to \infty$, we obtain

$$(s - \varepsilon)b_n \leq a_n \leq (s + \varepsilon)b_n$$

because by assumption $a_{n+p} \to 0$ and $b_{n+p} \to 0$. Since b_n is positive, we have, for $n \geq m$

$$\left|\frac{a_n}{b_n} - s\right| \leq \varepsilon \quad \text{or} \quad \lim_{n\to\infty} \frac{a_n}{b_n} = s.$$

On the other hand, if $(a_n - a_{n+1})/(b_n - b_{n+1})$ properly diverges to ∞, we can find m so that $(a_n - a_{n+1})/(b_n - b_{n+1}) > k$, for $n \geq m$, however large k may be. By the same argument as before, we get

$$a_n - a_{n+p} > k(b_n - b_{n+p}),$$

which leads to $a_n \geq kb_n$, or $a_n/b_n \geq k$ for $n \geq m$. Thus

$$\lim_{n\to\infty} \frac{a_n}{b_n} = \infty. \qquad \square$$

Definition Let $(x_n)_{n=1}^{\infty}$ be a sequence of real numbers. A real number a is said to be an *accumulation point of* $(x_n)_{n=1}^{\infty}$ if for any $\varepsilon > 0$ we have:

$$|x_n - a| < \varepsilon \quad \text{for infinitely many } n$$

(that is, if for every index k we can find an x_n with $n \geq k$ such that $|x_n - a| < \varepsilon$ holds).

Clearly, every accumulation point of a subsequence of $(x_n)_{n=1}^{\infty}$ is an accumulation point of $(x_n)_{n=1}^{\infty}$. The limit of a convergent sequence is an accumulation point of this sequence, in fact, the only one.

Proposition 61 If a is an accumulation point of the sequence $(x_n)_{n=1}^{\infty}$, then there is a subsequence $(x_{n_j})_{j=1}^{\infty}$ that converges to a.

Proof We construct such a subsequence by induction; in fact, the subsequence will be such that, for all k = 1, 2, ...,

$$|x_{n_k} - a| < \frac{1}{k}.$$

Evidently $x_{n_k} \to a$ as $k \to \infty$ will then hold.

We choose n_1 such that $|x_{n_1} - a| < 1$. Suppose that n_k is already determined such that $|x_{n_k} - a| < 1/k$. Since a is an accumulation point of $(x_n)_{n=1}^{\infty}$, we have $|x_n - a| < \varepsilon$ for infinitely many n. Taking $\varepsilon = 1/(k + 1)$, we see that there is an integer $n = n_{k+1} > n_k$ such that $|x_{n_{k+1}} - a| < 1/(k + 1)$. □

Remark From Proposition 56 (Bolzano-Weierstrass Theorem) we know that every bounded sequence has at least one accumulation point.

Examples:

1. The sequence $(\{-1\}^n)_{n=1}^{\infty}$ has exactly two accumulation points, namely -1 and +1.

2. The sequence $(x_n)_{n=1}^{\infty}$ defined by

$$x_n = \frac{n - \frac{(k-1)k}{2}}{k+1} \quad \text{for} \quad \frac{(k-1)k}{2} < n \leq \frac{k(k+1)}{2}$$

with k = 1, 2, ... has all points of the closed interval [0,1] as accumulation points because the range of this sequence consists of all rational points between 0 and 1.

Proposition 62 Suppose that the set A of all accumulation points of a bounded sequence $(x_n)_{n=1}^{\infty}$ is nonempty. Then A contains both sup A and inf A.

Proof It will be enough to take the case where $(x_n)_{n=1}^{\infty}$ is bounded above and A is nonempty. Let $\alpha = \sup A$; this means that for every $\varepsilon > 0$: (i) at least one element a of A satisfies the inequality $a > \alpha - \varepsilon$, and (ii) every element a of A satisfies the inequality $a < \alpha + \varepsilon$.

Thus, by (i) there is some $a \in A$ with $a > \alpha - \varepsilon$. Since $a \in A$ means that a is an accumulation point of $(x_n)_{n=1}^{\infty}$ and because $\alpha - \varepsilon$ is a point to the left of a on the number line, we have that

$$x_n > \alpha - \varepsilon \quad \text{for infinitely many } n.$$

Since $\alpha + \varepsilon$ is to the right of every point a of A by (ii), the inequality $x_n \geq \alpha + \varepsilon$ can hold at most for a finite number of indices n and so

$$x_n < \alpha + \varepsilon \quad \text{for all sufficiently large } n.$$

Thus $\alpha \in A$. □

Definition The real numbers sup A and inf A occuring in Proposition 62 are called the *limit superior* and the *limit inferior of the sequence* $(x_n)_{n=1}^{\infty}$, respectively; they are denoted by

$$\overline{\lim_{n\to\infty}}\, x_n \quad \text{and} \quad \underline{\lim_{n\to\infty}}\, x_n.$$

We observe that the numbers $\bar{a} = \overline{\lim}\, x_n$ and $\underline{a} = \underline{\lim}\, x_n$ are characterized by the following properties:

For any $\varepsilon > 0$

$$x_n > \bar{a} - \varepsilon \quad \text{for infinitely many } n,$$

$x_n < \bar{a} + \varepsilon$ for all sufficiently large n,

$x_n < \underline{a} + \varepsilon$ for infinitely many n,

$x_n > \underline{a} - \varepsilon$ for all sufficiently large n.

For a bounded sequence $(x_n)_{n=1}^{\infty}$ the $\overline{\lim}\, x_n$ and $\underline{\lim}\, x_n$ exist and $\underline{\lim}\, x_n \le \overline{\lim}\, x_n$ holds; equality occurs precisely when the sequence converges (and then $\lim x_n = \underline{\lim}\, x_n = \overline{\lim}\, x_n$). It is also easy to see that $\underline{\lim}\, x_n = -\overline{\lim}\,(-x_n)$.

Evidently, $\bar{a} = \overline{\lim}\, x_n$ may also be characterized by the properties:

(i) if $a < \bar{a}$, then $a < x_n$ for infinitely many n, and

(ii) if $\bar{a} < b$, then $x_n < b$ for all sufficiently large n.

Similar conditions characterize $\underline{a} = \underline{\lim}\, x_n$.

We now show that we could also have defined $\bar{a} = \overline{\lim}\, x_n$ and $\underline{a} = \underline{\lim}\, x_n$ by putting

$$\bar{a} = \inf_{k\ge 1} \sup\{x_k, x_{k+1}, \dots\} \quad \text{and} \quad \underline{a} = \sup_{k\ge 1} \inf\{x_k, x_{k+1}, \dots\}.$$

We need only consider the first of these, i.e.,

$$\bar{a} = \inf_{k\ge 1} \sup_{n\ge k} x_n.$$

From condition (i) we see that if $a < \bar{a}$, then $a < \sup\{x_n : n \ge k\}$ for every k, and therefore

$$a \le \inf_{k\ge 1} \sup_{n\ge k} x_n.$$

This implies that

$$\bar{a} \le \inf_{k\ge 1} \sup_{n\ge k} x_n$$

because of the arbitrariness of the a. From condition (ii) we see see that $\bar{a} < b$ implies $\sup\{x_n : n \ge k\} \le b$ if k is sufficiently large.

Thus

$$\inf_{k\geq 1} \sup_{n\geq k} x_n \leq b;$$

because of the arbitrariness of b this implies

$$\inf_{k\geq 1} \sup_{n\geq k} x_n \leq \bar{a}.$$

Thus

$$\bar{a} = \inf_{k\geq 1} \sup_{n\geq k} x_n.$$

Conversely, let

$$\bar{a} = \inf_{k\geq 1} \sup_{n\geq k} x_n.$$

If $a < \bar{a}$, then $a < \sup\{x_n : k \geq 1\}$ for every k, so that to each k there corresponds some n with $k \leq n$ and $a < x_n$. This, however, is condition (i). If $\bar{a} < b$, this means that $\sup\{x_n : n \geq k\} < b$ for some k. But then $x_n < b$ if $n \geq k$. This, however, is condition (ii). □

Remark In the extended real number system R_e^1 the limit superior and the limit inferior exist for an arbitrary sequence of real numbers (be the sequence bounded or not). The extended real number $+\infty$ is the limit superior of $(x_n)_{n=1}^{\infty}$ if and only if, given M and k, there is an $n \geq k$ such that $x_n > M$; this occurs if and only if there is a subsequence of $(x_n)_{n=1}^{\infty}$ that properly diverges to $+\infty$. The extended real number $-\infty$ is the limit inferior of $(x_n)_{n=1}^{\infty}$ if and only if the sequence properly diverges to $-\infty$.

It is clear that the limit superior of a sequence $(x_n)_{n=1}^{\infty}$ can also be described thus: $\bar{a} = \overline{\lim}\, x_n$ if and only if, given any $\varepsilon > 0$, the following two conditions are satisfied:

(i) there exists a positive integer k such that $x_n < \bar{a} + \varepsilon$ for all $n \geq k$, and

(ii) given any positive integer k, there exists an integer $n \geq k$ such that $x_n > \bar{a} - \varepsilon$.

If $(x_n)_{n=1}^{\infty}$ and $(y_n)_{n=1}^{\infty}$ are two sequences, then

$$\underline{\lim}\, x_n + \underline{\lim}\, y_n \le \underline{\lim}\,(x_n + y_n) \le \overline{\lim}\, x_n + \underline{\lim}\, y_n$$
$$\le \overline{\lim}\,(x_n + y_n) \le \overline{\lim}\, x_n + \overline{\lim}\, y_n;$$

if $x_n \ge 0$ and $y_n \ge 0$, $\overline{\lim}\,(x_n y_n) \le (\overline{\lim}\, x_n)(\overline{\lim}\, y_n)$ provided that the product $(\overline{\lim}\, x_n)(\overline{\lim}\, y_n)$ is not of the form $0 \cdot \infty$. We leave the verification of these facts to the reader.

5. *Series of Real Numbers*

Definition Let

$$\sum_{n=1}^{\infty} a_n = a_1 + a_2 + a_3 + \cdots$$

be a series of real numbers with partial sums

$$s_n = a_1 + \cdots + a_n \quad (n \in N).$$

If the sequence $(s_n)_{n=1}^{\infty}$ of partial sums converges to a (finite) real number A, we say that the series $\Sigma_{n=1}^{\infty} a_n$ *converges to* A. If $(s_n)_{n=1}^{\infty}$ diverges properly to infinity, we say that $\Sigma_{n=1}^{\infty} a_n$ *diverges*.

Proposition 63 If $\Sigma_{n=1}^{\infty} a_n$ converges to A and $\Sigma_{n=1}^{\infty} b_n$ converges to B, then $\Sigma_{n+1}^{\infty}(a_n + b_n)$ converges to A + B. Also, if c is any (finite) real number, then the series $\Sigma_{n=1}^{\infty} ca_n$ converges to cA.

Proof The claims of the proposition are immediate consequences of Proposition 53. □

Proposition 64 $\Sigma_{n=1}^{\infty} a_n$ converges if and only if for any $\varepsilon > 0$ there exists an integer n_0 such that

$$\left|\sum_{k=n}^{m} a_k\right| < \varepsilon$$

if $m \ge n \ge n_0$.

Proof The proposition is a direct consequence of Proposition 58. □

Taking m = n in Proposition 64 we get $|a_n| < \varepsilon$ for $n \geq n_0$; thus the following proposition gives a necessary (but by no means sufficient) condition that a series be convergent.

Proposition 65 If $\Sigma_{n=1}^{\infty} a_n$ is a convergent series, then $a_n \to 0$ as $n \to \infty$.

Proposition 66 If $\Sigma_{n=1}^{\infty} a_n$ is a series of nonnegative numbers with $s_n = a_1 + \cdots + a_n$, where $n \in N$, then $\Sigma_{n=1}^{\infty} a_n$ converges if and only if the sequence $(s_n)_{n=1}^{\infty}$ is bounded.

Proof The claim follows at once from Proposition 54. □

Remark Observe that the *harmonic series* $\Sigma_{n=1}^{\infty} (1/n)$ diverges; for details see Example 2 after Propositition 55. In the case of the harmonic series the terms of the series tend to 0 but the sequence of partial sums is unbounded. For further comment on this series see Exercise 18 at the end of this chapter.

Proposition 67 We have the following comparison tests:

(i) If $|a_n| \leq c_n$ for $n \geq n_0$, where n_0 is some fixed integer, and if $\Sigma_{n=1}^{\infty} c_n$ converges, then $\Sigma_{n=1}^{\infty} a_n$ converges.

(ii) If $a_n \geq d_n \geq 0$ for $n \geq 0$, and if $\Sigma_{n=1}^{\omega} d_n$ diverges then the series $\Sigma_{n=1}^{\infty} a_n$ diverges.

Proof By Proposition 65, given any $\varepsilon > 0$, there exists some integer $n_{00} \geq n_0$ such that $m \geq n \geq n_{00}$ implies $\Sigma_{k=n}^{m} c_k < \varepsilon$. Hence

$$\left|\sum_{k=n}^{m} a_k\right| \leq \sum_{k=n}^{m} |a_k| \leq \sum_{k=n}^{m} c_k < \varepsilon$$

and (i) follows. Part (ii) follows from Proposition 66. □

Proposition 68 Let $a_1 \geq a_2 \geq a_3 \geq \cdots \geq 0$. Then the series $\Sigma_{n=1}^{\infty} a_n$ converges if and only if the following series converges

$$\sum_{k=0}^{\infty} 2^k \cdot a_{2^k} = a_1 + 2a_2 + 4a_4 + 8a_8 + \cdots.$$

Proof By Proposition 66 it will be sufficient to consider the boundedness of the sequence of partial sums. Let

$$s_n = a_1 + \cdots + a_n \quad \text{and} \quad t_k = a_1 + 2a_2 + \cdots + 2^k \cdot a_{2^k}.$$

For $n < 2^k$,

$$\begin{aligned} s_n &\leq a_1 + (a_2 + a_3) + \cdots + \left(a_{2^k} + \cdots + a_{2^{k+1}-1}\right) \\ &\leq a_1 + 2a_2 + \cdots + 2^k \cdot a_{2^k} \\ &= t_k; \end{aligned}$$

For $n > 2^k$,

$$\begin{aligned} s_n &\geq a_1 + a_2 + (a_3 + a_4) + \cdots + \left(a_{2^{k-1}+1} + \cdots + a_{2^k}\right) \\ &\geq \frac{1}{2} a_1 + a_2 + 2a_4 + \cdots + 2^{k-1} \cdot a_{2^k} \\ &= \frac{1}{2} t_k . \end{aligned}$$

Thus the sequences $(s_n)_{n=1}^{\infty}$ and $(t_k)_{k=1}^{\infty}$ are either both bounded or both unbounded. □

Examples

1. If $0 \leq x < 1$, then the *geometric series*

$$\sum_{n=0}^{\infty} x^n = \frac{1}{1 - x};$$

if $x \geq 1$, the series diverges. Indeed, if $x \neq 0$, then

$$s_n = \sum_{k=0}^{n} x^k = \frac{1 - x^{n+1}}{1 - x},$$

and the result follows if we let $n \to \infty$.

2. The series

$$\sum_{n=1}^{\infty} \frac{1}{n^p}$$

converges if $p > 1$ and diverges if $p \leq 1$. Indeed, if $p \leq 0$, divergence follows from Proposition 65. If $p > 0$, Proposition 68 is applicable, and we are led to the series

$$\sum_{k=0}^{\infty} 2^k \frac{1}{2^{kp}} = \sum_{k=0}^{\infty} 2^{(1-p)k}.$$

But $2^{1-p} < 1$ if and only if $1 - p < 0$, and the claim follows by comparison with the geometric series taking $x = 2^{1-p}$. The case when p equals 1 we have already considered in the remark following Proposition 66.

Proposition 69 Let $a_1 \geq a_2 \geq a_3 \geq \cdots \geq 0$. If $\Sigma_{n=1}^{\infty} a_n$ converges, then $na_n \to 0$ as $n \to \infty$.

Proof Let $s_n = a_1 + \cdots + a_n$. If $\Sigma_{n=1}^{\infty} a_n = A$, then

$$\lim_{n \to \infty} s_n = A = \lim_{n \to \infty} s_{2n}.$$

Thus $s_{2n} - s_n \to 0$ as $n \to \infty$. But

$$s_{2n} - s_n = a_{n+1} + a_{n+2} + \cdots + a_{2n} \geq n \cdot a_{2n},$$

and so $0 \leq na_{2n} \leq s_{2n} - s_n$. Thus $na_{2n} \to 0$ as $n \to \infty$ and so $2na_{2n} \to 0$ as $n \to \infty$. But $a_{2n+1} \leq a_{2n}$ and so

$$(2n + 1)a_{2n+1} \leq \left(\frac{2n + 1}{2n}\right)\left(2na_{2n}\right).$$

Noting that the factors $(2n + 1)/2n$ and $2na_{2n}$ tend to 1 and 0, respectively, we see that $(2n + 1)a_{2n+1}$ tends to 0. We therefore have that na_n tends to zero as $n \to \infty$. □

Proposition 70 If $a_1 \geq a_2 \geq a_3 \geq \cdots \geq 0$ and $a_n \to 0$ as $n \to \infty$, then the *alternating series*

$$\sum_{n=1}^{\infty} (-1)^{n+1} a_n$$

converges.

Proof Consider the sequence $(s_n)_{n=1}^{\infty}$ of partial sums of the series $\Sigma_{n=1}^{\infty} (-1)^{n+1} a_n$. The subsequences

$$(s_{2n})_{n=1}^{\infty} \quad \text{and} \quad (s_{2n-1})_{n=1}^{\infty}$$

are bounded by 0 and a_1 and form a nested sequence of closed intervals

$$([s_{2n}, s_{2n-1}])_{n=1}^{\infty}.$$

But $s_{2n-1} - s_{2n} = a_{2n} \to 0$ as $n \to \infty$ and so Proposition 55 applies. □

Proposition 71 If $\Sigma_{n=1}^{\infty} a_n$ is a divergent series of positive numbers, then there is a sequence $(c_n)_{n=1}^{\infty}$ of positive numbers which converges to 0 but for which $\Sigma_{n=1}^{\infty} c_n a_n$ still diverges.

Proof Let $s_n = a_1 + \cdots + a_n$. We first establish that the series

$$\sum_{k=1}^{\infty} \frac{s_{k+1} - s_k}{s_{k+1}}$$

diverges. Since $(s_k)_{k=1}^{\infty}$ properly diverges to infinity, we can, for any $m \in N$, choose $n \in N$ such that $s_{n+1} > 2s_m$. Moreover, $(s_k)_{k=1}^{\infty}$ is monotonically increasing. Thus

$$\sum_{k=m}^{n} \frac{s_{k+1} - s_k}{s_{k+1}} \geq \sum_{k=m}^{n} \frac{s_{k+1} - s_k}{s_{n+1}} = \frac{s_{n+1} - s_m}{s_{n+1}} > \frac{s_{n+1} - \frac{1}{2}s_{n+1}}{s_{n+1}} = \frac{1}{2}.$$

Hence, for any $m \in N$ there exists $n \in N$ such that

$$\sum_{k=m}^{n} \frac{s_{k+1} - s_k}{s_{k+1}} \geq \frac{1}{2}.$$

This means that the partial sums of the series

$$\sum_{k=1}^{\infty} \frac{s_{k+1} - s_k}{s_{k+1}}$$

do not form a Cauchy sequence (see Proposition 64) and hence

$$\sum_{k=1}^{\infty} \frac{s_{k+1} - s_k}{s_{k+1}} = \infty.$$

But $s_{k+1} - s_k = a_{k+1}$ and so

$$\sum_{k=1}^{\infty} \frac{a_{k+1}}{s_{k+1}} = \sum_{k=2}^{\infty} \frac{a_k}{s_k} = \infty.$$

Take $c_k = 1/s_k$; then $c_k \to 0$ as $k \to \infty$ and $\Sigma_{k=2}^{\infty} c_k a_k = \infty$. □

Proposition 72 (Root Test) Given $\Sigma_{n=1}^{\infty} a_n$, let

$$c = \overline{\lim_{n \to \infty}} \sqrt[n]{|a_n|}.$$

Then

(i) if $c < 1$, $\Sigma_{n=1}^{\infty} a_n$ converges;

(ii) if $c > 1$, $\Sigma_{n=1}^{\infty} a_n$ diverges;

(iii) if $c = 1$, the test gives no information.

Proof If $c < 1$, we pick b so that $c < b < 1$, and an integer n_0 such that, for $n \geq n_0$,

$$\sqrt[n]{|a_n|} < b;$$

that is, $n \geq n_0$ implies $|a_n| < b^n$. Since $0 < b < 1$, $\Sigma_{n=1}^{\infty} b^n$ converges and the convergence of the series $\Sigma_{n=1}^{\infty} a_n$ follows by Proposition 67.

If $c > 1$, then there is a sequence $(n_k)_{k=1}^{\infty}$ such that

$$\sqrt[n_k]{|a_{n_k}|} \to c.$$

Hence $|a_n| > 1$ for infinitely many n, so that the condition $a_n \to 0$, necessary for the convergence of $\Sigma_{n=1}^{\infty} a_n$ by Proposition 65, fails to hold.

To verify part (iii), we only need to consider the series

$$\sum_{n=1}^{\infty} \frac{1}{n} \quad \text{and} \quad \sum_{n=1}^{\infty} \frac{1}{n^2};$$

for both of these series $c = 1$, but the first diverges and the second converges. □

N.B.: In Exercise 40 at the end of this chapter we shall discuss the evaluation of the series $\Sigma_{n=1}^{\infty} n^{-2}$ by elementary means. For the usual method of evaluating this series by Fourier series techniques see the example following Proposition 38 in Chapter 4.

Corollary Given a sequence $(c_n)_{n=0}^{\infty}$ of real numbers. Consider the series $\Sigma_{n=0}^{\infty} c_n x^n$ and let

$$\gamma = \overline{\lim_{n\to\infty}} \sqrt[n]{|c_n|}, \qquad R = \frac{1}{\gamma}.$$

(If $\gamma = 0$, $R = +\infty$; if $\gamma = +\infty$, $R = 0$.) Then $\Sigma_{n=0}^{\infty} c_n x^n$ converges if $|x| < R$, and diverges if $|x| > R$.

Proof We let $a_n = c_n x^n$, and apply the root test:

$$\overline{\lim_{n\to\infty}} \sqrt[n]{|a_n|} = |x| \overline{\lim_{n\to\infty}} \sqrt[n]{|c_n|} = \frac{|x|}{R}.$$ □

Remark The number R in the foregoing corollary is called the *radius of convergence* of $\Sigma_{n=0}^{\infty} c_n x^n$. This type of series, called *power series*, we shall study in detail in Section 6 of Chapter 4.

Proposition 73 (Ratio Test) The series $\Sigma_{n=1}^{\infty} a_n$

(i) converges if $\overline{\lim\limits_{n\to\infty}} \left|\frac{a_{n+1}}{a_n}\right| < 1$,

(ii) diverges if $\left|\frac{a_{n+1}}{a_n}\right| \geq 1$ for $n \geq n_0$, where n_0 is some fixed integer.

(iii) If $\underline{\lim\limits_{n\to\infty}} \left|\frac{a_{n+1}}{a_n}\right| \leq 1 \leq \overline{\lim\limits_{n\to\infty}} \left|\frac{a_{n+1}}{a_n}\right|$ the test fails.

Proof If condition (i) holds, we can find $b < 1$, and an integer k, such that, for $n \geq k$,

$$\left|\frac{a_{n+1}}{a_n}\right| < b.$$

In particular,

$$|a_{k+1}| < b|a_k|,$$

$$|a_{k+2}| < b|a_{k+1}| < b^2|a_k|,$$

$$\cdots\cdots\cdots\cdots\cdots\cdots\cdots\cdots$$

$$|a_{k+p}| < b^p|a_k|.$$

Thus, for $n \geq k$,

$$|a_n| < |a_k|b^{-k}b^n,$$

and (i) follows by Proposition 67 since $\Sigma_{n=1}^{\infty} b^n$ converges.

If $|a_{n+1}| \geq |a_n|$ for $n \geq n_0$, then a_n does not tend to 0 as $n \to \infty$, and (ii) follows by Proposition 65.

To verify part (iii), we again consider the series

$$\sum_{n=1}^{\infty} \frac{1}{n} \quad \text{and} \quad \sum_{n=1}^{\infty} \frac{1}{n^2}.$$

For each of these series we have in fact that $(a_{n+1})/(a_n) \to 1$ as $n \to \infty$, but the first series diverges and the second series converges. □

Remark The ratio test is often easier to apply than the root test, but the root test has wider scope. Whenever the ratio test indicates convergence, the root test does too; whenever the root test fails, the ratio test fails too. This is due to the general fact that for any sequence $(c_n)_{n=1}^{\infty}$ of positive numbers, we have

$$\underline{\lim}_{n\to\infty} \frac{c_{n+1}}{c_n} \le \underline{\lim}_{n\to\infty} \sqrt[n]{c_n},$$

and

$$\overline{\lim}_{n\to\infty} \sqrt[n]{c_n} \le \overline{\lim}_{n\to\infty} \frac{c_{n+1}}{c_n}.$$

For the series

$$\frac{1}{2} + \frac{1}{3} + \frac{1}{2^2} + \frac{1}{3^2} + \frac{1}{2^3} + \frac{1}{3^3} + \frac{1}{2^4} + \frac{1}{3^4} + \cdots$$

the foregoing two inequalities assume the following form:

$$0 < \frac{1}{\sqrt{3}} \quad \text{and} \quad \frac{1}{\sqrt{2}} < +\infty.$$

For this series the root test indicates convergence, while the ratio test is inconclusive.

Definition Let $\Sigma_{n=1}^{\infty} a_n$ be a series of real numbers.

(i) If $\Sigma_{n=1}^{\infty} |a_n|$ converges, we say that $\Sigma_{n=1}^{\infty} a_n$ *converges absolutely*.

(ii) If $\Sigma_{n=1}^{\infty} a_n$ converges but $\Sigma_{n=1}^{\infty} |a_n|$ diverges, we say that $\Sigma_{n=1}^{\infty} a_n$ *converges conditionally*.

Proposition 74 If $\Sigma_{n=1}^{\infty} a_n$ converges absolutely, then it converges.

Proof The claim follows from Proposition 67. □

Definition Let $(n_i)_{i=1}^{\infty}$ be a sequence of positive integers such that each positive integer occurs exactly once among the n_i (that is, $(n_i)_{i=1}^{\infty}$ is a one-to-one function from N onto N). If $\Sigma_{n=1}^{\infty} a_n$ is a series of real numbers and if, for $i \in N$,

$$b_i = a_{n_i},$$

then $\Sigma_{i=1}^{\infty} b_i$ is called a *rearrangement of* $\Sigma_{n=1}^{\infty} a_n$.

Example Consider the series

$$\sum_{n=1}^{\infty} \frac{(-1)^{n+1}}{n} = 1 - \frac{1}{2} + \frac{1}{3} - \frac{1}{4} + \frac{1}{5} - \frac{1}{6} + \cdots;$$

it converges by Proposition 70, but is not absolutely convergent. We examine the following rearrangement of this series: Let the first p positive terms be followed by the first q negative terms, then the next p positive terms be followed by the next q negative terms, etc. The resulting series converges to

$$\log\left(2\sqrt{\frac{p}{q}}\right).$$

To see this, we let

$$H_n = 1 + \frac{1}{2} + \cdots + \frac{1}{n} = \log n + C + \gamma_n,$$

where C is Euler's constant and $\gamma_n \to 0$ as $n \to \infty$ (see Example 2 following Proposition 55). Then

$$\frac{1}{2} + \frac{1}{4} + \cdots + \frac{1}{2m} = \frac{1}{2} H_m = \frac{1}{2} \log m + \frac{1}{2} C + \frac{1}{2} \gamma_m$$

and

$$1 + \frac{1}{3} + \cdots + \frac{1}{2k - 1} = H_{2k} - \frac{1}{2} H_k$$

$$= \log 2 + \frac{1}{2} \log k + \frac{1}{2} C + \gamma_{2k} - \frac{1}{2} \gamma_k.$$

The foregoing shows that the partial sums of the series in question are of the form $\log(2\sqrt{p/q}) + \delta_n$, where $(\delta_n)_{n=1}^{\infty}$ is a *null sequence*, that is, $\delta_n \to 0$ as $n \to \infty$.

We note that, for $p = q = 1$, we get

$$1 - \frac{1}{2} + \frac{1}{3} - \frac{1}{4} + \frac{1}{5} - \frac{1}{6} + \cdots = \log 2;$$

for $p = 2$, $q = 1$, we obtain

$$1 + \frac{1}{3} - \frac{1}{2} + \frac{1}{5} + \frac{1}{7} - \frac{1}{4} + \cdots = \frac{3}{2} \log 2;$$

for $p = 1$, $q = 4$, we have

$$1 - \frac{1}{2} - \frac{1}{4} - \frac{1}{6} - \frac{1}{8} + \frac{1}{3} - \frac{1}{10} - \frac{1}{12} - \frac{1}{14} - \frac{1}{16} + \frac{1}{5} - \cdots = 0,$$

and so forth.

If we separate a series $\Sigma_{n=1}^{\infty} a_n$ into the series of positive a_n and the series of negative a_n, we can exhibit an important distinction between absolutely convergent and conditionally convergent series.

If $\Sigma_{n=1}^{\infty} a_n$ is a series of real numbers, let

$$p_n = \max\{a_n, 0\} = \frac{a_n + |a_n|}{2}$$

and

$$q_n = \min\{a_n, 0\} = \frac{a_n - |a_n|}{2}.$$

Then

$$a_n = p_n + q_n.$$

Proposition 75 We have the following result:

(i) If $\Sigma_{n=1}^{\infty} a_n$ converges absolutely, then both $\Sigma_{n=1}^{\infty} p_n$ and $\Sigma_{n=1}^{\infty} q_n$ converge.

(ii) If $\Sigma_{n=1}^{\infty} a_n$ converges conditionally, then both $\Sigma_{n=1}^{\infty} p_n$ and $\Sigma_{n=1}^{\infty} q_n$ diverge.

Proof

(i) If both $\Sigma_{n=1}^{\infty} a_n$ and $\Sigma_{n=1}^{\infty} |a_n|$ converge, then so does the series $\Sigma_{n=1}^{\infty} (a_n + |a_n|)$. Thus $\Sigma_{n=1}^{\infty} 2p_n$ converges, implying that $\Sigma_{n=1}^{\infty} p_n$ converges. The convergence of $\Sigma_{n=1}^{\infty} q_n$ is established in an entirely similar manner.

(ii) Suppose $\Sigma_{n=1}^{\infty} a_n$ converges but $\Sigma_{n=1}^{\infty} |a_n|$ diverges. Then the series $\Sigma_{n=1}^{\infty} p_n$ and $\Sigma_{n=1}^{\infty} q_n$ both diverge. For if both were convergent, then $\Sigma_{n=1}^{\infty} (p_n + q_n) = \Sigma_{n=1}^{\infty} a_n$ would converge, contrary to hypothesis. Since

$$\sum_{n=1}^{k} |a_n| = \sum_{n=1}^{k} (p_n - q_n) = \sum_{n=1}^{k} p_n - \sum_{n=1}^{k} q_n.$$

divergence of $\Sigma_{n=1}^{\infty} p_n$ and convergence of $\Sigma_{n=1}^{\infty} q_n$ (or vice versa) implies divergence of $\Sigma_{n=1}^{\infty} a_n$, again contrary to hypothesis. □

Proposition 76 (Riemann's Rearrangement Theorem) The terms of any conditionally convergent series can be rearranged to give either a divergent series or a conditionally convergent series whose sum is an arbitrary preassigned number.

Proof Let $\Sigma_{n=1}^{\infty} a_n$ be a conditionally convergent series, with divergent nonnegative and nonpositive parts $\Sigma_{n=1}^{\infty} p_n$ and $\Sigma_{n=1}^{\infty} q_n$, respectively, and let c be an arbitrary (finite) real number. Let the rearrangement be determined as follows: First put down terms

$$p_1 + \cdots + p_{m_1}$$

until the total partial sum first exceeds c, then attach terms

$$q_1 + \cdots + q_{n_1}$$

until the total partial sum first falls short of c. Then attach terms

$$p_{m_1+1} + \cdots + p_{m_2}$$

until the total partial sum first exceeds c, then terms

$$q_{n_1+1} + \cdots + q_{n_2}$$

until the total partial sum first falls short of c, and so forth. Each of these steps is possible because of the divergence of $\Sigma_{n=1}^{\infty} p_n$ and $\Sigma_{n=1}^{\infty} q_n$. The resulting rearrangement of $\Sigma_{n=1}^{\infty} a_n$ converges to c since $p_n \to 0$ and $q_n \to 0$ as $n \to \infty$ (because $a_n \to 0$ as $n \to \infty$).

To see that there are rearrangements of $\Sigma_{n=1}^{\infty} a_n$ the partial sums of which properly diverge to $+\infty$, we consider a rearrangement of $\Sigma_{n=1}^{\infty} a_n$ in which we have alternately a group of positive terms followed by a single negative term. Since the series $\Sigma_{n=1}^{\infty} p_n$ diverges, its partial sums are unbounded and we can choose m_1 so large that

$$p_1 + \cdots + p_{m_1} > 1 - q_1,$$

then $m_2 > m_1$ so large that

$$p_1 + p_2 + \cdots + p_{m_1} + \cdots + p_{m_2} > 2 - q_1 - q_2$$

and generally, $m_k > m_{k-1}$ so large that

$$p_1 + p_2 + \cdots + p_{m_k} > k - q_1 - q_2 - \cdots - q_k \quad (k = 3, 4, \ldots).$$

Hence the series

$$p_1 + \cdots + p_{m_1} + q_1 + p_{m_1+1} + \cdots + p_{m_2} + q_2 + p_{m_2+1} + \cdots,$$

in which we have alternately a group of positive terms followed by a single negative term, is clearly divergent; its k-th partial sum

$$p_1 + \cdots + p_{m_1} + q_1 + \cdots + p_{m_k} + q_k$$

exceeds k.

Analogously one may obtain rearrangements of $\Sigma_{n=1}^{\infty} a_n$ that diverge to $-\infty$. □

Proposition 77 If $\Sigma_{n=1}^{\infty} a_n$ converges absolutely to A, then any rearrangement $\Sigma_{n=1}^{\infty} b_n$ of $\Sigma_{n=1}^{\infty} a_n$ also converges to A.

Proof We first prove that the proposition is true if $\Sigma_{n=1}^{\infty} a_n$ is a series of nonnegative numbers.

For each $k \in N$, let $s_k = b_1 + \cdots + b_k$. Since $b_i = a_{n_i}$ for some sequence $(n_i)_{i=1}^{\infty}$, we have

$$b_1 = a_{n_1}, \quad \ldots, \quad b_k = a_{n_k}.$$

Let $m = \max\{n_1,\ldots,n_k\}$. Then clearly

$$s_k \le a_1 + \cdots + a_m \le A.$$

Thus $\Sigma_{n=1}^{\infty} b_n$ converges to some real number B (by Proposition 66). But

$$B = \lim_{n\to\infty} s_k$$

and so $B \le A$ (that is, $\Sigma_{n=1}^{\infty} b_n \le \Sigma_{n=1}^{\infty} a_n$). However, since $\Sigma_{n=1}^{\infty} a_n$ is also a rearrangement of $\Sigma_{n=1}^{\infty} b_n$, the same reasoning with the roles of $\Sigma_{n=1}^{\infty} a_n$ and $\Sigma_{n=1}^{\infty} b_n$ reversed shows that $A \le B$. Hence $A = B$.

We now consider the general case. If $\Sigma_{n=1}^{\infty} a_n$ converges absolutely, then both $\Sigma_{n=1}^{\infty} p_n$ and $\Sigma_{n=1}^{\infty} q_n$ converge by Proposition 75; say $\Sigma_{n=1}^{\infty} p_n = P$ and $\Sigma_{n=1}^{\infty} q_n = Q$ (so that $Q \le 0$). Then $A = P + Q$. For some $(n_i)_{i=1}^{\infty}$, we have

$$b_i = a_{n_i} = p_{n_i} + q_{n_i}. \tag{12}$$

Moreover, $\Sigma_{i=1}^{\infty} p_{n_i}$ is a rearrangement of the series $\Sigma_{n=1}^{\infty} p_n$ of nonnegative terms. Hence, by the first part of this proof, $\Sigma_{i=1}^{\infty} p_{n_i}$ converges and $\Sigma_{i=1}^{\infty} p_{n_i} = P$. Similarly, $\Sigma_{i=1}^{\infty} q_{n_i} = Q$. From (12) we see that $\Sigma_{i=1}^{\infty} b_i$ converges and

$$\sum_{i=1}^{\infty} b_i = \sum_{i=1}^{\infty} p_{n_i} + \sum_{i=1}^{\infty} q_{n_i} = P + Q = A.$$

It only remains to show that $\Sigma_{i=1}^{\infty} b_i$ converges absolutely. By (12), we have, however, that

$$|b_i| \le |p_{n_i}| + |q_{n_i}| = p_{n_i} - q_{n_i}.$$

Thus, for any $k \in N$,

$$\begin{aligned} |b_1| + \cdots + |b_k| &\le \sum_{i=1}^{k} p_{n_i} - \sum_{i=1}^{k} q_{n_i} \\ &\le \sum_{i=1}^{\infty} p_{n_i} - \sum_{i=1}^{\infty} q_{n_i} \\ &= P - Q. \end{aligned}$$

The partial sums of $\Sigma_{i=1}^{\infty} |b_i|$ are thus all bounded above by $P - Q$ and hence $\Sigma_{i=1}^{\infty} |b_i| < \infty$. □

Proposition 78 If all rearrangements of $\Sigma_{n=1}^{\infty} a_n$ converge, then they all converge to the same sum.

Proof Either $\Sigma_{n=1}^{\infty} a_n$ converges absolutely, in which case Proposition 77 applies, or the series does not converge absolutely, in which case there exists a divergent rearrangement by Proposition 76. □

Definition A series is said to *converge unconditionally* if every rearrangement of it converges.

Remark To sum up, we have shown that a series converges unconditionally if and only if it converges absolutely. Also, if a series converges unconditionally, then every rearrangement of it converges to the same sum.

We conclude this section by considering the numerical calculation of e, the base of the natural logarithms. In Example 1 following Proposition 55 we defined the number e as

$$e = \lim_{n \to \infty} \left(1 + \frac{1}{n}\right)^n.$$

We now show that

$$e = \sum_{n=0}^{\infty} \frac{1}{n!},$$

where $n! = 1 \cdot 2 \cdot 3 \cdots n$ if $n \geq 1$, and $0! = 1$.

Let

$$s_n = \sum_{k=0}^{n} \frac{1}{k!} \quad \text{and} \quad t_n = \left(1 + \frac{1}{n}\right)^n.$$

By the Binomial Theorem $t_n \leq s_n$ because

$$t_n = 1 + 1 + \frac{1}{2!}\left(1 - \frac{1}{n}\right) + \frac{1}{3!}\left(1 - \frac{1}{n}\right)\left(1 - \frac{2}{n}\right) + \cdots$$

$$+ \frac{1}{n!}\left(1 - \frac{1}{n}\right)\left(1 - \frac{2}{n}\right) \cdots \left(1 - \frac{n-1}{n}\right).$$

For $n > k$ we therefore have

$$t_n > 2 + \frac{1}{2!}\left(1 - \frac{1}{n}\right) + \frac{1}{3!}\left(1 - \frac{1}{n}\right)\left(1 - \frac{2}{n}\right) + \cdots$$

$$+ \frac{1}{k!}\left(1 - \frac{1}{n}\right)\left(1 - \frac{2}{n}\right) \cdots \left(1 - \frac{k-1}{n}\right).$$

Keeping k fixed and letting $n \to \infty$, we obtain

$$e > 2 + \frac{1}{2!} + \frac{1}{3!} + \cdots + \frac{1}{k!} = s_k .$$

This inequality holds for every natural number k. Thus

$$t_n \leq s_n < e.$$

and so

$$\lim_{n \to \infty} s_n = e.$$

The rapidity with which the series

$$\sum_{n=0}^{\infty} \frac{1}{n!}$$

converges to e can be estimated as follows:

$$e - s_n = \frac{1}{(n+1)!} + \frac{1}{(n+2)!} + \frac{1}{(n+3)!} + \cdots$$

$$< \frac{1}{(n+1)!}\left(1 + \frac{1}{n+1} + \frac{1}{(n+1)^2} + \cdots\right)$$

$$= \frac{1}{n!n}.$$

For s_{10} we get approximately 2.7182818 with an error of less than 10^{-7} by which s_{10} differs from e.

6. *Systematic Fractions*

Proposition 79 Let M be a nonempty set of integers. If M is bounded above, then M contains a largest element.

Proof Since M is bounded above, it has a supremum $\bar{m}$. By the Archimedean property, there is an integer $n \in N$ such that $n > \bar{m}$. Thus, for all $m \in M$, $n - m > 0$. As m varies in M, the numbers $n - m$ form a nonempty subset of the set N of all natural numbers and therefore contains a smallest element, say $n - m_0$, by the Well-Ordering Principle. Then the number m_0 is the largest element of M. □

Definition Let y be any real number. The largest integer less than or equal to y (the existence and uniqueness of which is assured by Proposition 79), that is, the integer p such that

$$p \le y < p + 1,$$

is called the *largest integer contained in* y or the *integral part of* y and is usually denoted by [y]; the number y - [y] is called the *fractional part of* y and is usually denoted by $\{y\}$.

Let p be a fixed natural number larger than 1. Suppose that γ_0 is a given real number and let $c_0 = [\gamma_0]$. We set

$$\gamma_0 = c_0 + \frac{\gamma_1}{p}.$$

The number γ_1 so defined clearly satisfies the inequalities

$$0 \le \gamma_1 < p.$$

Next we decompose the number γ_1 into its integral and fractional parts:

$$\gamma_1 = c_1 + \frac{\gamma_2}{p},$$

where $c_1 = [\gamma_1]$ and $0 \le \gamma_2 < p$. Continuing in this manner we obtain

$$\gamma_k = c_k + \frac{\gamma_{k+1}}{p}, \qquad 0 \le \gamma_{k+1} < p \qquad (k = 0, 1, 2, \ldots).$$

This process is called the *p-adic algorithm*.

From the foregoing it is clear that c_k can only be one of the numbers in the set $\{0, 1, 2, \ldots, p - 1\}$ when $k = 1, 2, \ldots$. Of course, the integer c_0 is not subject to such restriction.

It may naturally happen that the integers c_k are all zero beginning with some index k_0. If for example γ_0 is an integer, then $\gamma_1 = 0$ and so all numbers of the succession $c_1, c_2, \ldots$ are zero. It is impossible, however, that all numbers c_k, beginning with some index k_0, can be equal to $p - 1$. Indeed, if we had $c_k = p - 1$ for all $k \ge k_0$, then we could take $k_0 > 0$ and it would follow that

$$\gamma_k = (p - 1) + \frac{\gamma_{k+1}}{p} \qquad (k = k_0, k_0 + 1, \ldots)$$

or

$$p - \gamma_k = \frac{p - \gamma_{k+1}}{p} \qquad (k = k_0, k_0 + 1, \ldots).$$

Hence

$$p - \gamma_{k_0} = \frac{p - \gamma_{k_0+1}}{p} = \frac{p - \gamma_{k_0+2}}{p^2} = \cdots$$

or

$$p - \gamma_{k_0} = \frac{p - \gamma_{k_0+j+1}}{p^{j+1}} \le \frac{p}{p^{j+1}} = \frac{1}{p^j}$$

for an arbitrarily large natural number j. Thus we would have $p - \gamma_{k_0} \le 0$, contradicting $\gamma_{k_0} < p$.

The next proposition will show that the numbers c_k are not subject to any further restrictions.

Proposition 80 If $c_0, c_1, c_2, \ldots$ is a sequence of integers such that

$$0 \le c_k \le p - 1 \quad \text{for } k = 1, 2, \ldots \text{ and}$$

$$c_k \le p - 2 \quad \text{for infinitely many } k,$$

then there is one and only one number γ_0 for which the p-adic algorithm produces the given sequence $c_0, c_1, c_2, \ldots$.

Proof Suppose there were two such numbers γ_0 and δ_0. Then the p-adic algorithm would yield two chains of equations

$$\gamma_0 = c_0 + \frac{\gamma_1}{p}, \quad \gamma_1 = c_1 + \frac{\gamma_2}{p}, \quad \gamma_2 = c_2 + \frac{\gamma_3}{p}, \quad \ldots$$

and

$$\delta_0 = c_0 + \frac{\delta_1}{p}, \quad \delta_1 = c_1 + \frac{\delta_2}{p}, \quad \delta_2 = c_2 + \frac{\delta_3}{p}, \quad \ldots$$

with

$$0 \le \gamma_k < p \quad \text{and} \quad 0 \le \delta_k < p \qquad (k = 1, 2, \ldots).$$

This would imply:

$$\gamma_0 - \delta_0 = \frac{\gamma_1 - \delta_1}{p}, \quad \gamma_1 - \delta_1 = \frac{\gamma_2 - \delta_2}{p}, \quad \gamma_2 - \delta_2 = \frac{\gamma_3 - \delta_3}{p}, \quad \ldots.$$

Hence

$$|\gamma_0 - \delta_0| = \frac{|\gamma_1 - \delta_1|}{p} = \frac{|\gamma_2 - \delta_2|}{p^2} = \cdots$$

$$= \frac{|\gamma_{n+1} - \delta_{n+1}|}{p^{n+1}} < \frac{p}{p^{n+1}} = \frac{1}{p^n}.$$

Since n can be made arbitrarily large, we conclude that $\gamma_0 = \delta_0$. This then shows that there is at most one number γ_0 of the required kind and the uniqueness is established.

To prove the existence, we must show that there is indeed a number γ_0 of the required kind. For this purpose we consider the series

$$\sum_{k=0}^{\infty} \frac{c_{i+k}}{p^k},$$

where i is any term of the sequence 0, 1, 2, ... The partial sums

$$s_i^{(n)} = \sum_{k=0}^{n} \frac{c_{i+k}}{p^k} \quad (n = 1, 2, \ldots)$$

form a monotonically increasing sequence which, on account of the estimate

$$s_i^{(n)} = c_i + \sum_{k=1}^{n} \frac{c_{i+k}}{p^k} \le c_i + \sum_{k=1}^{n} \frac{p-1}{p^k} = c_i + 1 - \frac{1}{p^n} < c_i + 1,$$

is bounded. Hence the series converges and we can set

$$\lim_{n\to\infty} s_i^{(n)} = \sum_{k=0}^{\infty} \frac{c_{i+k}}{p^k} = \gamma_i$$

for i = 0, 1, 2, ... Since

$$c_i \le s_i^{(n)} < c_i + 1,$$

we must have that

$$c_i \le \gamma_i \le c_i + 1.$$

But the second equality sign cannot occur. Indeed, since $c_k \le p - 2$ must hold for infinitely many k, it is possible to find for every index i an integer $j \ge 1$ such that $c_{i+j} \le p - 2$. Thus, for $n \ge j$,

$$s_i^{(n)} = c_i + \sum_{k=1}^{n} \frac{c_{i+k}}{p^k} - \frac{1}{p^j} = c_i + 1 - \frac{1}{p^n} - \frac{1}{p^j} < c_i + 1 - \frac{1}{p^j}.$$

This means that

$$\gamma_i = \lim_{n\to\infty} s_i^{(n)} \le c_i + 1 - \frac{1}{p^j} < c_i + 1$$

or, to sum up,

$$c_i \le \gamma_i < c_i + 1 \quad (i = 0, 1, 2, \ldots).$$

Thus c_i is the largest integer contained in γ_i. But, by the definition of $s_i^{(n)}$, we have that

$$s_i^{(n)} = c_i + \frac{s_{i+1}^{(n-1)}}{p} \quad (i = 0, 1, 2, \ldots).$$

Letting $n \to \infty$, we obtain

$$\gamma_i = c_i + \frac{\gamma_{i+1}}{p} \quad (i = 0, 1, 2, \ldots).$$

This shows that application of the p-adic algorithm to the number

$$\gamma_0 = \sum_{k=0}^{\infty} \frac{c_k}{p^k}$$

generates precisely the sequence of integers $c_0, c_1, c_2, \ldots$ and the proposition is established. □

We summarize: Every real number γ_0 can be represented uniquely in the form of a *p-adic fraction* (that is, has a unique *p-adic expansion*):

$$\gamma_0 = \sum_{k=0}^{\infty} \frac{c_k}{p^k}.$$

In analogy to the decimal fractions we can also write

$$\gamma_0 = c_0 + 0.c_1c_2c_3 \ldots.$$

The integers $c_0, c_1, c_2, c_3, \ldots$ result by use of the p-adic algorithm; they satisfy the inequalities

$$0 \le c_k \le p - 1 \quad \text{for } k \ge 1, \text{ and}$$

$$c_k \le p - 2 \quad \text{for infinitely many } k.$$

Conversely, every such p-adic fraction (or p-adic expansion) represents a real number.

Applications:

1. It is now easy to show that the set of all real numbers is an uncountably infinite set, i.e., it cannot be put in a one-to-one correspondence with the set N of natural numbers.

Suppose that R^1, the set of all real numbers, were countable; then certainly the (infinite) set of numbers x such that $0 \le x < 1$ would have to be countable and we could enumerate it in the form of a sequence $x_1, x_2, x_3, \ldots$. Taking $p \ge 2$, we could write $x_1, x_2, x_3, \ldots$ in the form of p-adic fractions:

$$x_1 = 0.c_{11}c_{12}c_{13} \ldots$$

$$x_2 = 0.c_{21}c_{22}c_{23} \ldots$$

$$x_3 = 0.c_{31}c_{32}c_{33} \ldots$$

$$\vdots$$

But the number

$$x = 0.b_1b_2b_3 \ldots$$

cannot occur in the foregoing enumeration if we set

$$\begin{aligned} b_k &= 1 \quad \text{for} \quad c_{kk} \neq 1 \\ &= 0 \quad \text{for} \quad c_{kk} = 1 \end{aligned} \qquad (k \geq 1).$$

Indeed, if x were equal to x_m, then we would have in particular that b_k would coincide with c_{kk}; but this contradicts the definition of b_k.

2. Consider the (open) square $0 < x < 1$ and $0 < y < 1$ in the x, y plane; we want to show that there is a one-to-one correspondence between the set of points situated in this square and the set of points making up one of the sides of the square.

Let (a,b) be a point of this square. Then a and b can be represented as a p-adic fraction, say with $p = 2$:

$$a = 0.a_1a_2a_3 \ldots$$

$$b = 0.b_1b_2b_3 \ldots$$

We now associate with the point (a,b) the point c on the x axis which is defined by

$$c = 0.a_1b_1a_2b_2a_3b_3 \ldots$$

The claimed one-to-one correspondence is now obvious.

3. Divide the closed interval [0,1] into three equal parts and remove the middle-third open interval (1/3,2/3); this removes all numbers in [0,1] that require $c_1 = 1$ in their triadic (i.e., $p = 3$) expansion. At the second stage, remove the middle-third open interval of each of the remaining two intervals, [0,1/3] and [2/3,1], thus eliminating all real numbers in [0,1] requiring $c_2 = 1$ in their triadic expansion. Proceeding analogously, remove at the n-th stage the union M_n of the (open) middle-thirds of the 2^{n-1} intervals present. The set

$$C = [0,1] - \bigcup_{n=1}^{\infty} M_n$$

is called *Cantor's (ternary) set*. It consists of all real numbers in [0,1] that do not require the use of "1" in their triadic expansion, i.e., c belongs to Cantor's (ternary) set C if and only if

$$c = \sum_{k=1}^{\infty} \frac{c_k}{3^k}$$

with

$$c_k = 2 \text{ or } = 0.$$

Adding up the lengths of all the open intervals we removed in the construction of the set C, it is easy to see that the resulting series converges to 1, the length of the initial interval [0,1]. Yet the set C is uncountably infinite; a proof of this claim can easily be rendered by a minor modification of the reasoning described in application 1 above.

We now turn to a generalization of p-adic expansions; this generalization is due to G. Cantor.

Let $p_1, p_2, p_3, \ldots$ be a given sequence of natural numbers such that

$$p_k \geq 2 \quad (k = 1, 2, 3, \ldots).$$

Let γ_0 be a given real number and put $c_0 = [\gamma_0]$. We set

$$\gamma_0 = c_0 + \frac{\gamma_1}{p_1}$$

and note that $0 \leq \gamma_1 < p$. Next we decompose γ_1 and put

$$\gamma_1 = c_1 + \frac{\gamma_2}{p_2},$$

where $c_1 = [\gamma_1]$ and $0 \leq \gamma_2 < p_2$. Continuing in this manner, we obtain

Cantor's algorithm:

$$\gamma_0 = c_0 + \frac{\gamma_1}{p_1} \quad \text{with} \quad 0 < \gamma_1 < p_1,$$

$$\gamma_1 = c_1 + \frac{\gamma_2}{p_2} \quad \text{with} \quad 0 < \gamma_2 < p_2,$$

$$\gamma_2 = c_2 + \frac{\gamma_3}{p_3} \quad \text{with} \quad 0 < \gamma_3 < p_3,$$

....

$$\gamma_n = c_n + \frac{\gamma_{n+1}}{p_{n+1}} \quad \text{with} \quad 0 < \gamma_{n+1} < p_{n+1},$$

and so forth. An easy calculation shows that

$$\gamma_0 = c_0 + \sum_{k=1}^{n} \frac{c_k}{p_1 p_2 \cdots p_k} + \frac{\gamma_{n+1}}{p_1 p_2 \cdots p_{n+1}}.$$

Since

$$0 \le \frac{\gamma_{n+1}}{p_1 p_2 \cdots p_{n+1}} < \frac{1}{p_1 p_2 \cdots p_n} \le \frac{1}{2^n},$$

we see that

$$\gamma_0 = c_0 + \sum_{k=1}^{\infty} \frac{c_k}{p_1 p_2 \cdots p_k}.$$

It is clear from the construction that, for $k \ge 1$, c_k can only be one of the numbers in the set $\{0, 1, 2, \ldots, p_k - 1\}$. We also note that for all sufficiently large k we have $c_k \le p_k - 2$. Indeed, if beginning with some k_0 we had $c_k = p_k - 1$ for all $k \ge k_0$, then

$$\gamma_k = (p_k - 1) + \frac{\gamma_{k+1}}{p_{k+1}} \quad (k = k_0, k_0 + 1, k_0 + 2, \ldots)$$

would follow. This in turn would imply

$$p_k - \gamma_k = \frac{p_{k+1} - \gamma_{k+1}}{p_{k+1}} \quad (k = k_0, k_0 + 1, k_0 + 2, \ldots)$$

or

$$p_{k_0} - \gamma_{k_0} = \frac{p_{k_0+1} - \gamma_{k_0+1}}{p_{k_0+1}} = \frac{p_{k_0+2} - \gamma_{k_0+2}}{p_{k_0+1}\, p_{k_0+2}} = \cdots.$$

Thus

$$p_{k_0} - \gamma_{k_0} = \frac{p_{k_0+j+1} - \gamma_{k_0+j+1}}{p_{k_0+1} \cdots p_{k_0+j+1}} \le \frac{1}{p_{k_0+1} \cdots p_{k_0+j}} \le \frac{1}{2^j}$$

and so $p_{k_0} - \gamma_{k_0} \le 0$, contradicting $\gamma_{k_0} < p_{k_0}$.

Proposition 81 If $p_1, p_2, p_3, \ldots$ is a given sequence of natural numbers all of which are larger than 1, then every real number δ_0 can be represented in a unique manner as a *Cantor series* (that is, a series of the form)

$$\gamma_0 = c_0 + \sum_{k=1}^{\infty} \frac{c_k}{p_1 p_2 \cdots p_k}$$

provided one requires that the numbers c_k are integers and satisfy the inequalities

$$0 \le c_k \le p_k - 1 \quad \text{for } k \ge 1, \text{ and}$$

$$c_k \le p_k - 2 \quad \text{for infinitely many } k.$$

One finds the numbers c_k by Cantor's algorithm. Conversely, every such series is convergent and thus represents a certain real number.

Proof That there is such a representation, namely the one produced by Cantor's algorithm, has already been shown. To prove the remaining claims of the proposition, we must show that every Cantor series converges to γ_0 and that the application of Cantor's algorithm to γ_0 produces exactly the sequence $c_0, c_1, c_2, \ldots$.

To this end we put

$$s_i^{(n)} = c_i + \sum_{k=1}^{n} \frac{c_{i+k}}{p_{i+1} \cdots p_{i+k}} \quad (i = 0, 1, \ldots; n = 1, 2, \ldots).$$

We observe that $s_i^{(n)} \geq c_i$ and that the sequence $(s_i^{(n)})_{n=1}^{\infty}$ is monotonically increasing. To show that $(s_i^{(n)})_{n=1}^{\infty}$ is bounded, we carry out a suitable estimate. Since $c_{i+k} \leq p_{i+k} - 1$ for all indices and for a certain index j we even have $c_{i+j} \leq p_{i+j} - 2$, then for $n \geq j$:

$$s_i^{(n)} \leq \sum_{k=1}^{n} \frac{p_{i+k} - 1}{p_{i+1} \cdots p_{i+k}} - \frac{1}{p_{i+1} \cdots p_{i+j}}$$

$$= c_i + \left(1 - \frac{1}{p_{i+1}}\right) + \left(\frac{1}{p_{i+1}} - \frac{1}{p_{i+1}\, p_{i+2}}\right) + \cdots$$

$$+ \left(\frac{1}{p_{i+1} \cdots p_{i+n-1}} - \frac{1}{p_{i+1} \cdots p_{i+n}}\right) - \frac{1}{p_{i+1} \cdots p_{i+j}}$$

$$= c_i + 1 - \frac{1}{p_{i+1} \cdots p_{i+n}} - \frac{1}{p_{i+1} \cdots p_{i+j}}$$

$$< c_i + 1 - \frac{1}{p_{i+1} \cdots p_{i+j}}.$$

Hence the sequence $(s_i^{(n)})_{n=1}^{\infty}$ converges and we set

$$\gamma_i = \lim_{n \to \infty} s_i^{(n)} = c_i + \sum_{k=1}^{\infty} \frac{c_{i+k}}{p_{i+1} \cdots p_{i+k}} \quad (i = 0, 1, 2, \ldots).$$

In view of the above estimate for $s_i^{(n)}$, we have

$$c_i \leq \gamma_i \leq c_i + 1 - \frac{1}{p_{i+1} \cdots p_{i+j}}$$

and consequently $c_i \leq \gamma_i < c_i + 1$; thus $c_i = [\gamma_i]$.

From the definition of $s_i^{(n)}$ we immediately get that

$$s_i^{(n)} = c_i + \frac{s_{i+1}^{(n-1)}}{p_{i+1}} \quad (i = 0, 1, 2, \ldots)$$

and therefore by passage to the limit as $n \to \infty$ we obtain

$$\gamma_i = c_i + \frac{\gamma_{i+1}}{p_{i+1}} \quad (i = 0, 1, 2, \ldots).$$

This, however, means that Cantor's algorithm applied to the number

$$\gamma_0 = c_0 + \sum_{k=1}^{\infty} \frac{c_k}{p_1 p_2 \cdots p_k}$$

generates exactly the sequence $c_0, c_1, c_2, \ldots$. □

Proposition 82 If the sequence $p_1, p_2, p_3, \ldots$ is such that each prime number divides infinitely many terms of it, then the Cantor series converges to a rational number if and only if all the numbers c_k in Proposition 81, beginning with some value of the index k, vanish.

Proof Let the sequence $p_1, p_2, p_3, \ldots$ have the property that each prime number divides infinitely many of the p_k's; this is certainly the case when $p_k = k + 1$, for example. A consequence of this assumption on the p_k's is that for each positive integer b we can find an index n so that the product $p_1 p_2 \cdots p_n$ will be divisible by b.

Now let a/b be a rational number; if one applies to it Cantor's algorithm, one obtains

$$\frac{a}{b} = c_0 + \sum_{k=1}^{n} \frac{c_k}{p_1 p_2 \cdots p_k} + \frac{\gamma_{n+1}}{p_1 p_2 \cdots p_{n+1}}.$$

Multiplication by $p_1 p_2 \cdots p_n$ gives

$$\frac{a}{b}\, p_1 p_2 \cdots p_n = c_0 p_1 p_2 \cdots p_n + \sum_{k=1}^{n} c_k \frac{p_1 p_2 \cdots p_n}{p_1 p_2 \cdots p_k} + \frac{\gamma_{n+1}}{p_{n+1}}.$$

But with suitable choice of n the product $p_1 p_2 \cdots p_n$ is divisible by b. In the sum on the right-hand side each term is an integer since all factors in the denominators cancel against factors in the numerators. Hence γ_{n+1}/p_{n+1} must be an integer; since $0 \le \gamma_{n+1} < p_{n+1}$ we have that $\gamma_{n+1} = 0$. Therefore Cantor's algorithm gives successively

$$c_{n+1} = 0, \quad \gamma_{n+2} = 0,$$

$$c_{n+2} = 0, \quad \gamma_{n+3} = 0,$$

....

The c_k's therefore vanish for all $k > n$.

Conversely, if for $k > n$ all c_k vanish, then it is evident that Cantor's series represents a rational number, for in this case we have the finite sum

$$c_0 + \sum_{k=1}^{n} \frac{c_k}{p_1 p_2 \cdots p_k}. \qquad \square$$

Example The series

$$2 + \sum_{k=1}^{\infty} \frac{1}{(k+1)!} = 2 + \sum_{k=1}^{\infty} \frac{1}{2 \cdot 3 \cdots (k+1)}$$

is clearly a Cantor series with $p_k = k + 1$. Since all numerators c_k are equal to 1, Proposition 82 shows that e, the base of the natural logarithms (see the discussion at the end of the preceding section), is an irrational number.

Remark At the end of the preceding section we derived the estimate

$$0 < e - s_n < \frac{1}{n!n},$$

where $s_n = \Sigma_{k=0}^{n} 1/(k!)$. From this estimate it is easy to deduce that e is irrational. Indeed, suppose that e were a rational number, i.e., $e = p/q$, where p and q are positive integers. By the estimate above,

$$0 < q!(e - s_q) < \frac{1}{q}.$$

By our assumption, $q!e$ is an integer. Since

$$q!s_q = q!\left(1 + 1 + \frac{1}{2!} + \cdots + \frac{1}{q!}\right)$$

is an integer, we see that $q!(e - s_q)$ is an integer. Since $q \geq 1$, $0 < q!(e - s_q) < 1/q$ implies the existence of an integer between 0 and 1. This however is a contradiction because it violates the corollary of Proposition 1.

N.B.: In Proposition 36 of Chapter 3 we shall show that the number e is transcendental. It is not known whether Euler's constant is irrational, let alone transcendental.

Exercises

1. Prove the *Binomial Theorem*: For any real numbers a and b and any positive integer n, we have

$$(a + b)^n = \sum_{k=0}^{n} \binom{n}{k} a^{n-k} b^k,$$

where $\binom{n}{0} = 1$ and, for $k \geq 1$,

$$\binom{n}{k} = \frac{n!}{k!(n - k)!}.$$

Hint Use the fact that

$$\binom{n}{k} + \binom{n}{k + 1} = \binom{n + 1}{k + 1}$$

and proceed by induction.

2. Using induction, verify the following two identities:

(a) $$\sum_{k=0}^{n} \binom{m+k}{k} = \binom{m+1+n}{n};$$

(b) $$\sum_{k=0}^{n} \binom{m}{k}\binom{j}{n-k} = \binom{m+j}{n}.$$

3. Given any positive integer n, show that the number of odd binomial coefficients, i.e., the number of odd integers in the collection

$$\binom{n}{0}, \binom{n}{1}, \binom{n}{2}, \ldots, \binom{n}{n}$$

is a power of 2.

4. The *Archimedean property* of real numbers asserts: If a and b are positive real numbers, then there is a positive integer n such that $na > b$. Deduce the following corollaries:

(a) If x is any real number, there exists a positive integer n such that $n > x$.

(b) If ε is any positive real number, there exists a positive integer n such that $1/n < \varepsilon$.

(c) If x is any real number, there exist integers m and n such that $m < x < n$.

5. Prove Propositions 19 and 23.

6. Show that for any $a > 0$, we have $a + 1/a \geq 2$ with equality only if $a = 1$.

7. For $n \geq 1$, verify the following claims:

(a) $$(a_1 + a_2 + \cdots + a_n)^2 \leq n(a_1^2 + a_2^2 + \cdots + a_n^2);$$

(b) $$\frac{1}{1\cdot 2} + \frac{1}{2\cdot 3} + \cdots + \frac{1}{n(n+1)} < 1;$$

(c) $$1 + 2^{-2} + 3^{-2} + \cdots + n^{-2} < 2;$$

(d) $$\left(1 + \frac{1}{2} + \cdots + \frac{1}{n}\right)^2 < 2n;$$

(e) $$\left(\frac{1}{n+1} + \frac{1}{n+2} + \cdots + \frac{1}{2n}\right)^2 < \frac{1}{2};$$

(f) $\frac{1}{2} n^2 < 1 + 2 + 3 + \cdots + n < \frac{1}{2}(n + 1)^2$;

(g) $\frac{1}{3} n^3 < 1^2 + 2^2 + 3^2 + \cdots + n^2 < \frac{1}{3}(n + 1)^3$;

(h) $\frac{1}{p + 1} n^{p+1} < 1^p + 2^p + 3^p + \cdots + n^p < \frac{1}{p + 1}(n + 1)^{p+1}$.

8. Answer the following:

(a) Is $n^{n+1} < (n + 1)^n$ or $n^{n+1} > (n + 1)^n$?

(b) When is $2^n < n^3$, when $2^n > n^3$?

(c) When is $3^n < n^4$, when $3^n > n^4$?

(d) For what n is $(n/3)^n < n! < (n/2)^n$?

9. Establish the formulas

(a) $1 + 2 + \cdots + n = \frac{1}{2} n(n + 1)$;

(b) $1^2 + 2^2 + \cdots + n^2 = \frac{1}{6} n(n + 1)(2n + 1)$;

(c) $1^3 + 2^3 + \cdots + n^3 = \frac{1}{4} n^2(n + 1)^2 = (1 + 2 + \cdots + n)^2$;

(d) $1^4 + 2^4 + \cdots + n^4 = \frac{1}{30} n(n + 1)(2n + 1)(3n^2 + 3n - 1)$;

(e) $1^2 + 3^2 + \cdots + (2n - 1)^2 = \frac{1}{3} (4n^3 - n)$;

(f) $1^3 + 3^3 + \cdots + (2n - 1)^3 = 2n^4 - n^2$;

(g) $1^4 - 2^4 + 3^4 - \cdots + (-1)^{n-1}n^4 = (-1)^{n-1} \frac{1}{2}(n^4 + 2n^3 - n)$;

(h) $\sum_{k=1}^{n-1} k^2(n - k)^2 = \frac{1}{30} (n^5 - n)$;

(i) $1 - \frac{1}{2} + \frac{1}{3} - \frac{1}{4} + \cdots - \frac{1}{2n} = \frac{1}{n + 1} + \frac{1}{n + 2} + \cdots + \frac{1}{2n}$;

(j) $\frac{1}{m} - \frac{n}{m + p} + \frac{n(n - 1)}{2!} \frac{1}{m + 2p} - \cdots + (-1)^n \frac{1}{m + np}$

$$= \frac{n!p^n}{m(m + p) \cdots (m + np)}.$$

10. Let

$$A_n = \frac{a_1 + a_2 + \cdots + a_n}{n}$$

with $a_j > 0$ for $j = 1, 2, \ldots, n$. Show that, for $p > 1$,

$$\sum_{n=1}^{m} A_n^p \le \frac{p}{p-1} \sum_{n=1}^{m} A_n^{p-1} a_n.$$

Solution By Proposition 47 we get

$$A_n^{p-1} A_{n-1} \le \frac{(p-1)A_n^p + A_{n-1}^p}{p}.$$

But

$$A_n^p - \frac{p}{p-1} A_n^{p-1} a_n = A_n^p - \frac{p}{p-1} [nA_n - (n-1)A_{n-1}]A_n^{p-1}$$

$$= A_n^p\left(1 - \frac{np}{p-1}\right) + \frac{(n-1)p}{p-1} A_n^{p-1} A_{n-1}$$

$$\le A_n^p\left(1 - \frac{np}{p-1}\right) + \frac{n-1}{p-1}\left((p-1)A_n^p + A_{n-1}^p\right)$$

$$= \frac{1}{p-1}\left((n-1)A_{n-1}^p - nA_n^p\right).$$ □

11. Prove the *Inequality of Hardy and Landau*: For $p > 1$ and $a_1, a_2, \ldots, a_n$ positive,

$$\sum_{n=1}^{m} \left(\frac{a_1 + a_2 + \cdots + a_n}{n}\right)^p < \left(\frac{p}{p-1}\right)^p \sum_{n=1}^{m} a_n^p.$$

Solution Using Hölder's Inequality (see Proposition 48) we get from the result in Exercise 10 the following:

$$\sum_{n=1}^{m} A_n^p < \frac{p}{p-1} \sum_{n=1}^{m} A_n^{p-1} a_n \le \frac{p}{p-1} \left(\sum_{n=1}^{m} a_n^p\right)^{1/p} \left(\sum_{n=1}^{m} A_n^p\right)^{1/p'},$$

where $p' = p/(p-1)$. Dividing by the last factor on the right (which is certainly positive) and raising the result to the power p, we get the desired inequality. □

12. Prove the *Inequality of Carleman*: If $a_n > 0$ for all $n \in N$, then

$$\sum_{n=1}^{\infty} (a_1 a_2 \cdots a_n)^{1/n} \le e \sum_{n=1}^{\infty} a_n$$

provided that $\Sigma_{n=1}^{\infty} a_n$ converges.

Solution From Exercise 11 we see that

$$\sum_{n=1}^{\infty} \left(\frac{a_1^{1/p} + a_2^{1/p} + \cdots + a_n^{1/p}}{n} \right)^p \le \left(\frac{p}{p-1} \right)^p \sum_{n=1}^{\infty} a_n$$

provided that $\Sigma_{n=1}^{\infty} a_n$ converges; the "<" turned into "≤" because of the limit process involved (as $m \to \infty$). If we now let $p \to \infty$, and note that

$$(a_1 a_2 \cdots a_n)^{1/n} \le \frac{a_1 + a_2 + \cdots + a_n}{n}$$

(see Proposition 44), we obtain the desired result since

$$\left(1 + \frac{1}{p-1}\right)^p \to e \quad \text{as} \quad p \to \infty$$

by Example 1 following Proposition 55.

N.B.: It is worthwhile to note that strict inequality holds in the Inequality of Carleman as all $a_n > 0$; the reader might want to pursue this point.

13. Show that Hölder's Inequality (see Proposition 48) and Minkowski's Inequality (see Proposition 49) remain intact for infinite sums, i.e., when $n \to \infty$, provided that the series on the right-hand side of these inequalities converge; the convergence of the series on the left-hand side of these inequalities follows as a bargain.

14. Verify the following:

(a) If $p > 0$, then $\lim_{n \to \infty} n^{-p} = 0$.

(b) If $p > 0$, then $\lim_{n \to \infty} (p)^{1/n} = 1$.

(c) $\lim\limits_{n \to \infty} (n)^{1/n} = 1$.

(d) If $p > 0$ and s is a fixed real number, then

$$\lim_{n\to\infty} \frac{n^s}{(1+p)^n} = 0.$$

(e) If $|x| < 1$, then $\lim_{n\to\infty} x^n = 0$.

Solution

(a) Take $n > (1/\varepsilon)^{1/p}$.

(b) If $p > 1$, put $x_n = \sqrt[n]{p} - 1$. Then $x_n > 0$, and, by the Binomial Theorem (see Exercise 1),

$$1 + nx_n \le (1 + x_n)^n = p$$

so that

$$0 < x_n \le \frac{p-1}{n}.$$

Hence $x_n \to 0$. If $p = 1$, (b) is trivial; and if $0 < p < 1$, the result is obtained by considering reciprocals.

(c) Put $x_n = \sqrt[n]{n} - 1$. Then $x_n \ge 0$, and, by the Binomial Theorem,

$$n = (1 + x_n)^n \ge \frac{n(n-1)}{2} x_n^2.$$

Hence, for $n \ge 2$,

$$0 \le x_n \le \left(\frac{2}{n-1}\right)^{1/2}.$$

(d) Let k be an integer such that $k > s$, $k > 0$. For $n > 2k$,

$$(1+p)^n > \binom{n}{k} p^k = \frac{n(n-1)\cdots(n-k-1)}{n!} p^k > \frac{n^k p^k}{2^k k!}.$$

Hence, for $n > 2^k$,

$$0 < \frac{n^s}{(1+p)^n} < \frac{2^k k!}{p^k} n^{s-k}.$$

Since $s - k < 0$, $n^{s-k} \to 0$, by part (a).

(e) Take $s = 0$ in part (d). □

15. Prove the following:

(a) *Abel's Lemma:* If $(a_n)_{n=1}^{\infty}$ is a sequence of real numbers whose partial sums $s_n = a_1 + \cdots + a_n$ satisfy, for all $n \in N$, $m \le s_n \le M$ for some real numbers m and M, and if $(b_n)_{n=1}^{\infty}$ is such that $b_1 \ge b_2 \ge b_3 \ge \cdots \ge 0$, then, for all $n \in N$,

$$mb_1 \le \sum_{k=1}^{n} a_k b_k \le Mb_1.$$

Proof An easy calculation shows that

$$\sum_{k=1}^{n} a_k b_k = \sum_{k=1}^{n} s_k(b_k - b_{k+1}) + s_n b_{n+1}.$$

Since $b_k - b_{k+1} \ge 0$ and $s_k \le M$, we get

$$\sum_{k=1}^{n} a_k b_k \le M \sum_{k=1}^{n} (b_k - b_{k+1}) + Mb_{n+1} = Mb_1.$$

The left-hand side inequality is verified in a similar fashion. □

(b) *Dirichlet's Test:* Let the sequences $(a_n)_{n=1}^{\infty}$ and $(b_n)_{n=1}^{\infty}$ be as in part (a) and suppose, moreover, that $b_n \to 0$ as $n \to \infty$. Then $\Sigma_{k=1}^{\infty} a_k b_k$ converges.

Proof By hypothesis, there is an $M > 0$ such that $|s_n| \le M$ for all $n \in N$. Hence, for any $j, n \in N$ and $j \le n$,

$$\left|\sum_{k=j}^{n} a_k\right| = |s_n - s_{j-1}| \le |s_n| + |s_{j-1}| \le 2M.$$

By part (a), applied to $(a_k)_{k=j}^{\infty}$ and $(b_k)_{k=j}^{\infty}$, this implies, for $n \ge j$,

$$\left|\sum_{k=j}^{n} a_k b_k\right| \le 2Mb_j$$

By assumption there is an integer n_0 such that $b_n < \varepsilon/2M$ for all $n \ge n_0$. Hence $2Mb_j < \varepsilon$ for $j \ge n_0$; thus, for $n \ge j \ge n_0$,

$$\left|\sum_{k=j}^{n} a_k b_k\right| < \varepsilon.$$

The claim now follows from Proposition 64. □

(c) *Abel's Test:* If $\Sigma_{n=1}^{\infty} a_n$ is a convergent series and if the sequence $(b_n)_{n=1}$ is monotonic and bounded, then $\Sigma_{n=1}^{\infty} a_n b_n$ is also a convergent series.

Proof Suppose $(b_n)_{n=1}$ is monotonically increasing and let $c_n = b - b_n$, where $b = \lim_{n \to \infty} b_n$. By part (b) the series $\Sigma_{n=1}^{\infty} a_n c_n$ converges. Clearly, $\Sigma_{n=1}^{\infty} ba_n$ also converges. But $a_n b_n = ba_n - a_n c_n$; thus

$$\sum_{n=1}^{\infty} a_n b_n = \sum_{n=1}^{\infty} (ba_n - a_n c_n)$$

converges. □

16. Let $a_n > 0$ for all $n \in N$ and $a_n \to a$ with $0 < a < \infty$. Show that $\log a_n \to \log a$.

Solution Note that

$$\log a_n - \log a = \log \frac{a_n}{a} = \log \left(1 + \frac{a_n - a}{a}\right).$$

Let $x_n = (a_n - a)/a > -1$. We verify that $\log (1 + x_n) \to 0$ as $x_n \to 0$. Let $\varepsilon > 0$ and put $e^{\varepsilon} - 1 = \varepsilon_1$ and $1 - e^{-\varepsilon} = \varepsilon_2$. Then $\varepsilon' = \min\{\varepsilon_1, \varepsilon_2\} > 0$ and for this ε' there exists an n_0 such that $|x_n| < \varepsilon'$ for all $n \geq n_0$. Thus, for $n \geq n_0$, $-\varepsilon_2 < x_n < \varepsilon_1$, or $e^{-\varepsilon} < 1 + x_n < e^{\varepsilon}$, or $|\log (1 + x_n)| < \varepsilon$. □

17. Let a_n and a be as in Exercise 16. Show:

(a) $$\lim_{n \to \infty} \log (a_1 \cdots a_n)^{1/n} = \lim_{n \to \infty} \frac{\log a_1 + \cdots + \log a_n}{n} = \log a,$$

that is,

$$(a_1 \cdots a_n)^{1/n} \to a \quad \text{as} \quad n \to \infty.$$

(b) The sequence

$$a_1, \frac{a_2}{a_1}, \frac{a_3}{a_2}, \ldots, \frac{a_n}{a_{n-1}}, \frac{a_{n+1}}{a_n}, \ldots$$

satisfies

$$\lim_{n\to\infty} (a_n)^{1/n} = \lim_{n\to\infty} \frac{a_{n+1}}{a_n}.$$

(c) $\lim_{n\to\infty} (n!)^{1/n} = \frac{1}{e}.$

18. Consider the harmonic series $\Sigma_{n=1}^{\infty} 1/n$. Since

$$\frac{1}{n+1} + \frac{1}{n+2} + \cdots + \frac{1}{2n} > n\,\frac{1}{2n} = \frac{1}{2},$$

we see that

$$\frac{1}{2^{k-1}+1} + \cdots + \frac{1}{2^k} > \frac{1}{2}$$

and so the sums

$$\frac{1}{3} + \frac{1}{4}, \quad \frac{1}{5} + \cdots + \frac{1}{8}, \quad \frac{1}{9} + \cdots + \frac{1}{16}, \quad \ldots$$

are all larger than 1/2 and therefore, letting $H_n = \Sigma_{k=1}^{n} 1/k$,

$$H_{2^k} > k\,\frac{1}{2}.$$

The sums H_n grow very slowly, however. Euler calculated for example that

$$H_{1000} = 7.48 \ldots, \quad H_{1000000} = 14.39 \ldots, \quad \text{etc.}$$

(a) Using the inequality

$$0 < \frac{1}{n} - \log\frac{n+1}{n} < \frac{1}{n(n+1)} < \frac{1}{n^2}$$

(in this connection see Example 2 following Proposition 55), show

$$\sum_{n=1}^{\infty}\left(\frac{1}{n} - \log\frac{n+1}{n}\right) = C,$$

where C is Euler's constant.

(b) Show that if we delete in the harmonic series all terms that have the digit 9 in their denominators, then the resulting series converges and that its sum is less than 80.

19. Show that the series

$$L = \sum_{n=1}^{\infty} n^{-s}$$

converges for $s > 1$ by first noting that

$$\frac{1}{(n+1)^s} + \frac{1}{(n+2)^s} + \cdots + \frac{1}{(2n)^s} < n\,\frac{1}{n^s} = \frac{1}{n^t},$$

where $s = 1 + t$, $t > 0$. But

$$\frac{1}{(2^{k-1}+1)^s} + \cdots + \frac{1}{(2^k)^s} < \frac{1}{(2^{k-1})^t} = \frac{1}{(2^t)^{k-1}}$$

and so

$$L < \left(1 + \frac{1}{2^s}\right) + \frac{2^{-t}}{1 - 2^{-t}}.$$

20. Let $[t]$ denote the largest integer contained in t. Show that the series

$$\sum_{n=1}^{\infty} \frac{(-1)^{[\sqrt{n}]}}{n}$$

converges.

Hint The first three terms of the series are negative, the next five terms are positive, etc. If we group together the terms of equal sign, we get the alternating series

$$\sum_{k=1}^{\infty} (-1)^k \left(\frac{1}{k^2} + \frac{1}{k^2+1} + \cdots + \frac{1}{(k+1)^2 - 1}\right).$$

But

$$\frac{2}{k+1} < \frac{1}{k^2} + \frac{1}{k^2+1} + \cdots + \frac{1}{(k+1)^2 - 1} < \frac{2}{k};$$

note, for example, that the sum of the first k terms is less than

$$k\,\frac{1}{k^2} = \frac{1}{k}$$

and the sum of the remaining k + 1 terms is less than

$$(k+1)\,\frac{1}{k^2+k} = \frac{1}{k}$$

and so the sum of all terms is smaller than 2/k. From this we may conclude that the terms of the new series which we formed tend to zero because the absolute values of the terms are monotonically decreasing. Thus, by Proposition 70, the new series which we formed converges. This, however, implies that the series under investigation converges because of the following *Associative Law for Convergent Infinite Series:* The insertion of parentheses in convergent series is permissible without restriction. More precisely: Let $\Sigma_{n=1}^{\infty} a_n = s$ be a convergent series. If $(n_k)_{k=0}^{\infty}$, with $0 = n_0 < n_1 < n_2 < \cdots$, is an arbitrary sequence of integers, and if we set

$$a_{n_k+1} + \cdots + a_{n_{k+1}} = A_k,$$

then the series $\Sigma_{k=0}^{\infty} A_k$ is also convergent and has the value s. (To see that the Associative Law for Convergent Infinite Series is true we merely need to observe that the sequence of partial sums of $\Sigma_{k=0}^{\infty} A_k$ is a subsequence of the sequence of partial sums of $\Sigma_{n=1}^{\infty} a_n$; but the latter sequence of partial sums converges by hypothesis and so any of its subsequences converges to the same limit s.)

21. Show that the series

$$\sum_{n=1}^{\infty} \frac{(-1)^{[\sqrt{n}]}}{n^s}$$

converges for $s > 1/2$ and diverges for $s < 1/2$.

22. Describe the region in the x, y plane for which the series

$$\sum_{k=0}^{\infty} (|x| + |y|)^k$$

converges.

23. Let

$$\sum_{n=0}^{\infty} \frac{n^m}{n!} = ep_m.$$

Show that the integers p_m satisfy the relation

$$p_{m+1} = 1 + mp_1 + \frac{m(m - 1)}{2!} p_2 + \cdots + p_m.$$

24. Show that

$$\sum_{n=1}^{\infty} \frac{1}{n(n + m)} = \frac{1}{m}\left(1 + \frac{1}{2} + \cdots + \frac{1}{m}\right).$$

25. Show that the series

$$\sum_{n=1}^{\infty} (-1)^n \sin \frac{x}{n} \quad \text{for arbitrary } x \neq 0$$

converges conditionally.

Hint Note that, for sufficiently large n, sin x/n has the same sign as x. Also observe that

$$\sin \frac{x}{n} \Big/ \frac{1}{n} \to x \quad \text{as} \quad n \to \infty.$$

26. (a) Let the entries t_{nm} $(1 \leq m \leq n)$ of an infinite triangular matrix

$$\begin{array}{lll} t_{11} & & \\ t_{21} & t_{22} & \\ t_{31} & t_{32} & t_{33} \\ \cdots & \cdots & \cdots \end{array}$$

satisfy the following two conditions:

(i) The elements of each column tend to zero, i.e.,

$t_{nm} \to 0$ as $n \to \infty$ (m fixed);

(ii) the sums of the absolute values of all elements of any row do not exceed a constant K, i.e., for all $n \in N$,

$|t_{n1}| + |t_{n2}| + \cdots + |t_{nn}| \leq K < \infty.$

If $(x_n)_{n=1}^{\infty}$ is a sequence that converges to zero, show that the sequence

$$x'_n = t_{n1}x_1 + t_{n2}x_2 + \cdots + t_{nn}x_n$$

converges to zero also.

(b) Suppose that the coefficients t_{nm} satisfy in addition to conditions (i) and (ii) of part (a) also the condition

(iii) $T_n = t_{n1} + t_{n2} + \cdots + t_{nn} \to 1$ as $n \to \infty$.

If $(y_n)_{n=1}^{\infty}$ is a sequence that converges to a (finite) real number a, show that

$$y'_n = t_{n1}y_1 + t_{n2}y_2 + \cdots + t_{nn}y_n \to a \quad \text{as} \quad n \to \infty.$$

Proof

(a) Let $\varepsilon > 0$. Then there exists an integer m such that $|x_n| < \varepsilon/2K$ for $n \geq m$; for these n we have, by condition (ii),

$$|x'_n| < |t_{n1}x_1 + \cdots + t_{nm}x_m| + \frac{\varepsilon}{2}.$$

Since m is kept fixed, there exists by condition (i) an $n_0 \geq m$ such that for $n \geq n_0$ we have $|t_{n1}x_1 + \cdots + t_{nm}x_m| < \varepsilon/2$. Thus $|x'_n| < \varepsilon$.

(b) Since

$$y'_n = t_{n1}(y_1 - a) + t_{n2}(y_2 - a) + \cdots + t_{nn}(y_n - a) + T_na,$$

we may apply part (a) to the sequence $(y_n - a)_{n=1}^{\infty}$. □

27. Putting $t_{n1} = t_{n2} = \cdots = t_{nn} = 1/n$, we see that conditions (i), (ii), and (iii) of Exercise 26 are satisfied. Deduce now the

corollary of Proposition 59.

28. Consider Proposition 59 and for simplicity assume that the sequence $(y_n)_{n=1}^{\infty}$ is strictly increasing and suppose that

$$\frac{x_n - x_{n-1}}{y_n - y_{n-1}} \qquad (n = 1, 2, \ldots; x_0 = y_0 = 0)$$

tends to a; putting

$$t_{nm} = \frac{y_m - y_{m-1}}{y_n},$$

verify that conditions (i), (ii), and (iii) of Exercise 26 are satisfied. Conclude that

$$\frac{x_n}{y_n} = \sum_{m=1}^{n} t_{nm} \frac{x_m - x_{m-1}}{y_m - y_{m-1}}$$

converges to a, thereby having another proof for Proposition 59.

29. Let $(x_n)_{n=1}^{\infty}$ and $(y_n)_{n=1}^{\infty}$ both tend to zero and

$$|y_1| + |y_2| + \cdots + |y_n| \le K \quad (n = 1, 2, \ldots; K \text{ constant}).$$

Show that

$$z_n = x_1 y_n + x_2 y_{n-1} + \cdots + x_n y_1$$

tends to zero also.

Hint Set $t_{nm} = y_{n-m+1}$ and apply part (a) of Exercise 26.

30. Let $x_n \to a$ and $y_n \to b$. Show that

$$z_n = \frac{x_1 y_n + x_2 y_{n-1} + \cdots + x_n y_1}{n} \to ab.$$

Proof Assume first that a = 0. If we replace y_n by y_n/n in Exercise 29, we get $z_n \to 0$. The condition imposed on y_n in Exercise 29 is not violated by this since the y_n's are bounded, $|y_n| \le K$. Turning now to the general case, we note that

$$z_n = \frac{(x_1 - a)y_n + \cdots + (x_n - a)y_1}{n} + a\,\frac{y_1 + \cdots + y_n}{n}.$$

But we have already shown that

$$\frac{(x_1 - a)y_n + \cdots + (x_n - a)y_1}{n} \to 0$$

and, by Exercise 27,

$$a\,\frac{y_1 + \cdots + y_n}{n} \to ab.$$

<u>31</u>. Let $x_n \to a$. Show that

$$x_n' = \frac{1\cdot x_0 + \binom{n}{1}x_1 + \cdots + \binom{n}{n}x_n}{2^n} \to a.$$

Hint We set

$$t_{nm} = \frac{\binom{n}{m}}{2^n}$$

and apply part (b) of Exercise 26. Since $\binom{n}{m} < n^m$ and $(n^m)/(2^n) \to 0$ (see part (d) of Exercise 14) we see that condition (i) of Exercise 26 is satisfied. Since

$$\sum_{m=0}^{n} \binom{n}{m} = (1 + 1)^n = 2^n,$$

conditions (ii) and (iii) of Exercise 26 are satisfied too.

<u>32</u>. Let $x_n \to a$ and $z > 0$. Show that

$$x_n' = \frac{1\cdot x_0 + \binom{n}{1}zx_1 + \cdots + \binom{n}{n}z^n x_n}{(1 + z)^n}$$

and

$$x_n'' = \frac{z^n x_0 + \binom{n}{1}z^{n-1}x_1 + \cdots + \binom{n}{n}x_n}{(1 + z)^n}$$

tend to a as well.

<u>33</u>. Let $a > -1$, $b > -1$, and $a \neq b$. Show that

$$\sum_{n=1}^{\infty} \frac{1}{(n + a)(n + b)} = \frac{1}{b - a} \int_0^1 \frac{x^a - x^b}{1 - x}\, dx.$$

34. If $x_n > 0$, $x_n - x_{n+1} > 0$, and $x_n - 2x_{n+1} + x_{n+2} > 0$, show that

$$\frac{x_0}{2} < x_0 - x_1 + x_2 - x_3 + x_4 - \cdots < x_0 - \frac{1}{2} x_1.$$

35. Consider the two convergent series

$$a_1 + a_2 + \cdots + a_n + \cdots \tag{A}$$

and

$$b_1 + b_2 + \cdots + b_n + \cdots \tag{B}$$

and suppose that the series (A) converges to A and the series (B) converges to B. The *Cauchy product* (or *product*, for short) of these two series is defined to be the series

$$a_1b_1 + (a_1b_2 + a_2b_1) + (a_1b_3 + a_2b_2 + a_3b_1) + \cdots \tag{C}$$

Prove the following two results:

(a) *Theorem of Mertens:* If one of the two convergent series (A) and (B) converges absolutely, then the series (C) converges to AB.

(b) *Theorem of Abel:* If the Cauchy product (C) of the two convergent series (A) and (B) converges to C, then C is necessarily equal to AB.

Proof We first look at the theorem of Mertens. Without loss of generality we assume that the series (A) is absolutely convergent, that is, we suppose that the series

$$|a_1| + |a_2| + \cdots + |a_n| + \cdots \tag{A*}$$

converges to A*. Let

$$A_n = a_1 + a_2 + \cdots + a_n,$$

$$B_n = b_1 + b_2 + \cdots + b_n,$$

$$c_n = a_1b_n + a_2b_{n-1} + \cdots + a_{n-1}b_2 + a_nb_1,$$

and

$$C_n = c_1 + c_2 + \cdots + c_n.$$

We have to show that C_n tends to AB.

We observe that

$$C_n = a_1B_n + a_2B_{n-1} + \cdots + a_{n-1}B_2 + a_nB_1.$$

Setting

$$B_m = B - \beta_m,$$

where β_m tends to zero as $m \to \infty$, we can write

$$C_n = A_nB - \gamma_n,$$

where

$$\gamma_n = a_1\beta_n + a_2\beta_{n-1} + \cdots + a_{n-1}\beta_2 + a_n\beta_1.$$

Since A_n tends to A we only need to show that γ_n tends to 0. But the latter is a consequence of Exercise 29 (for $x_n = \beta_n$ and $y_n = a_n$) because

$$|a_1| + |a_2| + \cdots + |a_n| \le A^*.$$

To verify the theorem of Abel, we note that

$$C_1 + C_2 + \cdots + C_n = A_1B_n + A_2B_{n-1} + \cdots + A_nB_1.$$

But $C_n \to C$ and so, by the corollary of Proposition 59,

$$\frac{C_1 + C_2 + \cdots + C_n}{n} \to C.$$

By Exercise 30 (for $x_n = A_n$ and $y_n = B_n$) we obtain, however, that

$$\frac{A_1B_n + A_2B_{n-1} + \cdots + A_nB_1}{n} \to AB.$$

Thus C = AB. □

36. Given the two infinite series

$$\sum_{n=1}^{\infty} a_n \quad \text{and} \quad \sum_{n=1}^{\infty} b_n,$$

we can form the infinite matrix

$$\begin{array}{cccc} a_1b_1 & a_2b_1 & a_3b_1 & \cdots \\ a_1b_2 & a_2b_2 & a_3b_2 & \cdots \\ a_1b_3 & a_2b_3 & a_3b_3 & \cdots \\ \vdots & \vdots & \vdots & \end{array}$$

The Cauchy product of the two series is obtained by grouping together the entries on successive diagonals:

$$a_1b_1; \ a_1b_2, a_2b_1; \ a_1b_3, a_2b_2, a_3b_1; \ \ldots .$$

Another product of the two series is obtained by the rectangular grouping

$$a_1b_1; \ a_1b_2, a_2b_2, a_2b_1; \ a_1b_3, a_2b_3, a_3b_3, a_3b_2, a_3b_1; \ \ldots .$$

The first grouping gives rise to the series

$$a_1b_1 + (a_1b_2 + a_2b_1) + (a_1b_3 + a_2b_2 + a_3b_1) + \cdots,$$

while the second grouping gives rise to the series

$$a_1b_1 + (a_1b_2 + a_2b_2 + a_2b_1)$$
$$+ (a_1b_3 + a_2b_3 + a_3b_3 + a_3b_2 + a_3b_1) + \cdots .$$

In other words, different groupings give rise to different products of the two given infinite series.

We denote the series $\Sigma_{n=1}^{\infty} a_n$ by (A), the series $\Sigma_{n=1}^{\infty} b_n$ by (B), the series $\Sigma_{n=1}^{\infty} |a_n|$ by (A*), the series $\Sigma_{n=1}^{\infty} |b_n|$ by (B*); if all of these series converge, we denote their respective limits by A, B, A*, and B*.

Prove the following result:

Theorem of Cauchy: If the series (A) and (B) are both absolutely convergent, then any of their products converges; moreover, all products converge to AB.

<u>*Proof*</u> Let us enumerate the entries in the infinite matrix in whatever fashion we please; from the resulting sequence we form the series

$$\sum_{s=1}^{\infty} a_{i_s} b_{k_s} = a_{i_1} b_{k_1} + a_{i_2} b_{k_2} + \cdots + a_{i_s} b_{k_s} + \cdots.$$

To see that the series

$$\sum_{s=1}^{\infty} |a_{i_s} b_{k_s}|$$

converges, we note that

$$\begin{aligned} &|a_{i_1} b_{k_1}| + \cdots + |a_{i_s} b_{k_s}| \\ &\quad \le (|a_1| + \cdots + |a_j|)(|b_1| + \cdots + |b_j|) \\ &\quad \le A^*B^*, \end{aligned}$$

where j denotes the largest of the indices $i_1, k_1, \ldots, i_s, k_s$. By Proposition 66, the series $\Sigma_{s=1}^{\infty} |a_{i_s} b_{k_s}|$ converges; by Proposition 74 therefore $\Sigma_{s=1}^{\infty} a_{i_s} b_{k_s}$ converges as well.

We now only need to determine the sum of the series $\Sigma_{s=1}^{\infty} a_{i_s} b_{k_s}$. Since it is absolutely convergent, all rearrangements converge (and to the same sum). But the Cauchy product converges to AB by Exercise 35. □

<u>37</u>. With an infinite matrix

$$\begin{matrix} a_{11} & a_{12} & a_{13} & \cdots \\ a_{21} & a_{22} & a_{23} & \cdots \\ a_{31} & a_{32} & a_{33} & \cdots \\ \vdots & \vdots & \vdots & \end{matrix}$$

of real numbers we can associate a *double series*

$$\sum_{\mu,\nu=1}^{\infty} a_{\mu\nu}.$$

If $A: (\mu,\nu) \to \lambda$ is an enumeration of the pairs (μ,ν) of natural numbers, we put $c_\lambda(A) = a_{\mu\nu}$ and call the series

$$\sum_{\lambda=1}^{\infty} c_\lambda(A)$$

the *ordering (relative to* A*) of* $\Sigma_{\mu,\nu=1}^{\infty} a_{\mu\nu}$ *into a single series.*

Prove the following result:

Main Rearrangement Theorem: Let $\Sigma_{\mu,\nu=1}^{\infty} a_{\mu\nu}$ be a given double series. Suppose there exists a number M such that

$$\sum_{\mu,\nu=1}^{m} |a_{\mu\nu}| \le M < \infty \quad \text{for all m.}$$

Then the following statements hold:

(a) Any ordering of the double series into a single series is absolutely convergent and all single series have the same value s.

(b) The series $\Sigma_{\nu=1}^{\infty} a_{\mu\nu}$ (column series) are absolutely convergent ($\mu = 1, 2, 3, \ldots$), and so are the series $\Sigma_{\mu=1}^{\infty} a_{\mu\nu}$ (row series) ($\nu = 1, 2, 3, \ldots$).

(c) The two series

$$\sum_{\mu=1}^{\infty}\left(\sum_{\nu=1}^{\infty} a_{\mu\nu}\right) \quad \text{and} \quad \sum_{\nu=1}^{\infty}\left(\sum_{\mu=1}^{\infty} a_{\mu\nu}\right)$$

are absolutely convergent and their values are equal (sum of the row series = sum of the column series), namely, equal to s.

<u>Proof</u>

(a) Let $n \in N$. We consider the pairs (μ,ν) which correspond to the numbers $\lambda = 1, \ldots, n$ under the enumeration A and let m be the largest of the μ and ν. Then

$$\sum_{\lambda=1}^{n} |c_\lambda(A)| \le \sum_{\mu,\nu=1}^{m} |a_{\mu\nu}| \le M.$$

By Proposition 66 we see that $\Sigma_{\lambda=1}^{\infty} c_\lambda(A)$ converges absolutely. If A^* is a second enumeration, then $\Sigma_{\lambda=1}^{\infty} c_\lambda(A^*)$ is a rearrangement of $\Sigma_{\lambda=1}^{\infty} c_\lambda(A)$; by the absolute convergence both series must have the same value.

(b) If we put $m = \max\{\mu,n\}$ resp. $m = \max\{\nu,n\}$, then

$$\sum_{\nu=1}^{n} |a_{\mu\nu}| \le \sum_{\mu,\nu=1}^{m} |a_{\mu\nu}| \le M$$

$$\text{resp.} \quad \sum_{\mu=1}^{n} |a_{\mu\nu}| \le \sum_{\mu,\nu=1}^{m} |a_{\mu\nu}| \le M;$$

thus, by Proposition 66, this point is also settled.

(c) For $m \le n$ we have

$$\sum_{\mu=1}^{m} \left|\sum_{\nu=1}^{n} a_{\mu\nu}\right| \le \sum_{\mu=1}^{m} \sum_{\nu=1}^{n} |a_{\mu\nu}| \le \sum_{\mu,\nu=1}^{m} |a_{\mu\nu}| \le M,$$

and passage to the limit as $n \to \infty$ gives

$$\sum_{\mu=1}^{m} \left|\sum_{\nu=1}^{\infty} a_{\mu\nu}\right| \le M,$$

showing the absolute convergence of the series formed by the column series. In the same way we get the absolute convergence of the series formed by the row series.

Let $\Sigma_{\lambda=1}^{\infty} c_\lambda$ be some ordering of the double series into a single series. For any $\varepsilon > 0$ there exists a k_0 such that

$$\sum_{\lambda=k+1}^{\infty} |c_\lambda| < \varepsilon \quad \text{for all } k \ge k_0.$$

Corresponding to k we can find some n_0 such that the $c_1, c_2, \ldots, c_k$ are among the $a_{\mu\nu}$ for $1 \le \mu \le n_0$, $1 \le \nu \le n_0$. For $m \ge n_0$, $n \ge n_0$ we then have

$$\left| \sum_{\mu=1}^{m} \sum_{\nu=1}^{n} a_{\mu\nu} - \sum_{\lambda=1}^{k} c_\lambda \right| \le \sum_{\lambda=k+1}^{\infty} |c_\lambda| < \varepsilon.$$

Upon passage to the limit as $n \to \infty$, $m \to \infty$ (resp. in the reverse order) and then followed by passage to the limit as $k \to \infty$, we get

$$\left| \sum_{\mu=1}^{\infty} \sum_{\nu=1}^{\infty} a_{\mu\nu} - \sum_{\lambda=1}^{\infty} c_\lambda \right| \le \varepsilon$$

$$\text{resp.} \quad \left| \sum_{\nu=1}^{\infty} \sum_{\mu=1}^{\infty} a_{\mu\nu} - \sum_{\lambda=1}^{\infty} c_\lambda \right| \le \varepsilon.$$

But $\varepsilon > 0$ is arbitrarily small. □

38. Prove the following *Theorem of Goldbach:* If a_n are all those natural numbers that have the form p^q, where p and q are integers larger than 1, then

$$G = \sum_{n=1}^{\infty} \frac{1}{a_n - 1} = 1.$$

Hint Let m run through all natural numbers larger than 1 which are not powers; then

$$G = \sum_m \frac{1}{m^2 - 1} + \sum_m \frac{1}{m^3 - 1} + \cdots$$

$$= \sum_m \left(\frac{1}{m^2 - 1} + \frac{1}{m^3 - 1} + \cdots \right)$$

$$= \sum_m \left[\left(\frac{1}{m^2} + \frac{1}{m^4} + \frac{1}{m^6} + \cdots \right) + \left(\frac{1}{m^3} + \frac{1}{m^6} + \frac{1}{m^9} + \cdots \right) + \cdots \right]$$

$$= \sum_m \left[\left(\frac{1}{m^2} + \frac{1}{m^3} + \cdots \right) + \left(\frac{1}{m^4} + \frac{1}{m^6} + \cdots \right) + \left(\frac{1}{m^6} + \frac{1}{m^9} + \cdots \right) + \cdots \right]$$

$$= \sum_m \left(\frac{1}{m(m-1)} + \frac{1}{m^2(m^2-1)} + \frac{1}{m^3(m^3-1)} + \cdots \right) = \sum_{n=2}^{\infty} \frac{1}{n(n-1)}.$$

39. The series

$$\sum_{n=0}^{\infty} \frac{x^n}{n!}$$

is absolutely convergent for all real numbers x. Denoting this series by "exp x" or e^x, show that

$$\exp(x + y) = (\exp x)\cdot(\exp y) \quad \text{for all } x, y \in R^1.$$

Hint For any $x, y \in R^1$ we have

$$\sum_{n=0}^{\infty} \frac{x^n}{n!} \cdot \sum_{k=0}^{\infty} \frac{y^k}{k!} = \sum_{j=0}^{\infty} \left(\sum_{n=0}^{j} \frac{x^n}{n!} \frac{y^{j-n}}{(j-n)!} \right)$$

$$= \sum_{j=0}^{\infty} \frac{1}{j!} \sum_{n=0}^{j} \binom{j}{n} x^n y^{j-n} = \sum_{j=0}^{\infty} \frac{1}{j!} (x + y)^j.$$

40. This exercise deals with an elementary proof of the formula

$$\sum_{k=1}^{\infty} \frac{1}{k^2} = \frac{\pi^2}{6}.$$

Proof We start with the inequality $\sin x < x < \tan x$ for $0 < x < \pi/2$, take reciprocals, and square each member to obtain

$$\cot^2 x < 1/x^2 < 1 + \cot^2 x.$$

Now we put $x = k\pi/(2m + 1)$ where k and m are integers, $1 \le k \le m$, and sum on k to obtain

$$\sum_{k=1}^{m} \cot^2 \frac{k\pi}{2m + 1} < \frac{(2m + 1)^2}{\pi^2} \sum_{k=1}^{m} \frac{1}{k^2} < m + \sum_{k=1}^{m} \cot^2 \frac{k\pi}{2m + 1}. \qquad \text{(E.1)}$$

But

$$\sum_{k=1}^{m} \cot^2 \frac{k\pi}{2m + 1} = \frac{m(2m - 1)}{3} \qquad \text{(E.2)}$$

(as we shall show further below), and so relation (E.1) gives

$$\frac{m(2m - 1)}{3} < \frac{(2m + 1)^2}{\pi^2} \sum_{k=1}^{m} \frac{1}{k^2} < m + \frac{m(2m - 1)}{3}.$$

Multiplying by $\pi^2/(4m^2)$ and letting $m \to \infty$, we obtain

$$\lim_{m\to\infty} \sum_{k=1}^{m} \frac{1}{k^2} = \frac{\pi^2}{6}.$$

Finally, we show the validity of (E.2). By equating imaginary parts in the formula

$$\cos n\theta + i \sin n\theta = (\cos\theta + i \sin\theta)^n = \sin^n\theta\,(\cot\theta + i)^n$$

$$= \sin^n\theta \sum_{k=0}^{n} \binom{n}{k} i^k \cot^{n-k}\theta,$$

we get the trigonometric identity

$$\sin n\theta = \sin^n\theta \left\{ \binom{n}{1}\cot^{n-1}\theta - \binom{n}{3}\cot^{n-3}\theta + \binom{n}{5}\cot^{n-5}\theta - \cdots \right\}.$$

Taking $n = 2m + 1$, we write this identity in the form

$$\sin(2m + 1)\theta = \sin^{2m+1}\theta\, P_m(\cot^2\theta) \tag{E.3}$$

with $0 < \theta < \pi/2$, where P_m is the polynomial of degree m given by

$$P_m(x) = \binom{2m+1}{1}x^m - \binom{2m+1}{3}x^{m-1} + \binom{2m+1}{5}x^{m-2} - \cdots.$$

Since $\sin\theta \neq 0$ for $0 < \theta < \pi/2$, equation (E.3) shows that $P_m(\cot^2\theta) = 0$ if and only if $(2m + 1)\theta = k\pi$ for some integer k. Therefore $P_m(x)$ vanishes at the m distinct points

$$x_k = \cot^2 \frac{\pi k}{2m + 1}$$

for $k = 1, 2, \ldots, m$. These are all the zeros of $P_m(x)$ and their sum is

$$\sum_{k=1}^{m} \cot^2 \frac{\pi k}{2m + 1} = \frac{\binom{2m+1}{3}}{\binom{2m+1}{1}} = \frac{m(2m - 1)}{3},$$

which proves (E.2). □

41. Consider an infinite series whose n-th term is $\pm(1/n)$, the $\pm$ signs being determined according to a pattern that repeats periodically in blocks of eight.

(a) Show that a sufficient condition for the series to be conditionally convergent is that there be four "+" signs and four "-" signs in the block of eight.

(b) Is this sufficient condition also necessary?

42. For $n \in N$, let $s_1, \ldots, s_n$ be arbitrary real numbers and $t_1, \ldots, t_n$ be any real numbers such that $t_1 + \cdots + t_n = 0$. Show that

$$\sum_{k=1}^{n} \sum_{j=1}^{n} t_k t_j |s_k - s_j| \le 0.$$

Hint When $n = 2$, we can use the identity

$$(t_1 + t_2)^2 - (t_1 - t_2)^2 = 4t_1t_2;$$

but $t_1 + t_2 = 0$.

When $n = 3$, we have

$$0 = (t_1 + t_2 + t_3)^2 = t_1^2 + t_2^2 + t_3^2 + 2t_1t_2 + 2t_2t_3 + 2t_3t_1.$$

Thus

$$\begin{aligned}
&2t_1t_2|s_1 - s_2| + 2t_2t_3|s_1 - s_2| + 2t_3t_1|s_1 - s_2| \\
&+ 2t_1t_2|s_2 - s_3| + 2t_2t_3|s_2 - s_3| + 2t_3t_1|s_2 - s_3| \\
&+ 2t_1t_2|s_3 - s_1| + 2t_2t_3|s_3 - s_1| + 2t_3t_1|s_3 - s_1| \\
&= -(t_1^2 + t_2^2 + t_3^2)(|s_1 - s_2| + |s_2 - s_3| + |s_3 - s_1|) \\
&\quad + 2t_3(t_2 + t_1)|s_2 - s_1| \\
&\quad + 2t_1(t_2 + t_3)|s_2 - s_3| \\
&\quad + 2t_2(t_1 + t_3)|s_1 - s_3|
\end{aligned}$$

$$= -(t_1^2 + t_2^2 + t_3^2)(|s_1 - s_2| + |s_2 - s_3| + |s_3 - s_1|)$$
$$+ 2t_3^2|s_2 - s_1| + 2t_1^2|s_2 - s_3| + 2t_2^2|s_3 - s_1|$$
$$= -t_1^2(|s_2 - s_1| + |s_1 - s_3|) + t_1^2|s_2 - s_3|$$
$$-t_2^2(|s_1 - s_2| + |s_2 - s_3|) + t_2^2|s_1 - s_3|$$
$$-t_3^2(|s_2 - s_3| + |s_3 - s_1|) + t_3^2|s_2 - s_1|.$$

But $|A - B| \le |A - C| + |C - B|$ for any three real numbers A, B, and C.

43. Let $\Sigma_{k=1}^{\infty} u_k$ be a series and set $s_n = u_1 + \cdots + u_n$ and

$$\sigma_n = \frac{s_1 + \cdots + s_n}{n}$$

Without using Proposition 59, show the following: If $\Sigma_{k=1}^{\infty} u_k$ converges and its sum is s, then the sequence of arithmetic means σ_n also converges to s; if the series properly diverges to $+\infty$, then so does its sequence of arithmetic means.

Solution

(a) First, we assume that $s_n \to 0$ as $n \to \infty$, and we prove that $\sigma_n \to 0$ as $n \to \infty$. Take an arbitrary positive number ε. Then there is a positive integer n_0, such that

$$|s_n| < \frac{\varepsilon}{2} \quad \text{when} \quad n \ge n_0.$$

Also

$$|\sigma_n| \le \frac{|s_1 + \cdots + s_{n_0}|}{n} + \frac{|s_{n_0+1}| + \cdots + |s_n|}{n}.$$

But we can choose $k > n_0$, so that

$$\frac{|s_1 + \cdots + s_{n_0}|}{n} < \frac{\varepsilon}{2} \quad \text{when} \quad n \ge k.$$

Therefore

$$|\sigma_n| < \frac{\varepsilon}{2} + \frac{1}{2}\left(1 - \frac{n_0}{n}\right)\varepsilon \quad \text{when} \quad n \geq k.$$

Thus $\sigma_n \to 0$ as $n \to \infty$.

(b) Let $s_n \to s \neq 0$ as $n \to \infty$. Then

$$0 = (u_1 - s) + u_2 + u_3 + \cdots.$$

The arithmetic mean for n terms of this series is equal to

$$\frac{1}{n}\left(\sum_{j=1}^{n} s_n - ns\right) = \sigma_n - s.$$

By (a), however, the limit of this sequence of arithmetic means is zero. Therefore $\sigma_n \to s$ as $n \to \infty$.

Finally, in the case of proper divergence to $+\infty$, however large K may be, we know that there is a positive integer n_0 such that $s_n > K$, when $n \geq n_0$. But

$$\sigma_n = \frac{s_1 + \cdots + s_{n_0}}{n} + \frac{s_{n_0+1} + \cdots + s_n}{n}.$$

The first summand on the right-hand side tends to zero as $n \to \infty$, and the second summand is greater than

$$\left(1 - \frac{n_0}{n}\right)K,$$

which tends to K as $n \to \infty$. □

<u>44</u>. Prove the following theorem due to *Hardy* and *Landau*: Let the sequence of arithmetic means σ_n of the series $u_1 + u_2 + \cdots$ converge to σ and either $n(s_n - s_{n+1}) < K$ or $n(s_{n+1} - s_n) < K$, where K is some positive constant and the s_n denote partial sums of the given series. Then $s_n \to \sigma$ as $n \to \infty$.

Proof We may without loss of generality take

$$\sigma = 0, \quad K = 1, \quad \text{and} \quad n(s_n - s_{n+1}) < 1.$$

For if we put

$$U_1 = \frac{u_1 - \sigma}{K}, \quad U_2 = \frac{u_2}{K},$$

etc., and

$$S_n = U_1 + \cdots + U_n,$$

we have

$$S_n = \frac{s_n - \sigma}{K}, \quad \text{and} \quad \frac{1}{n}(S_1 + \cdots + S_n) = \frac{\sigma_n - \sigma}{K}.$$

Thus

$$n(S_n - S_{n+1}) < 1, \quad \text{if} \quad n(s_n - s_{n+1}) < K.$$

In the other case we put

$$U_1 = \frac{\sigma - u_1}{K}, \quad U_2 = -\frac{u_2}{K}, \quad \text{etc.}$$

Suppose then we are given $\sigma_n \to 0$ and $n(s_n - s_{n+1}) < 1$. However, it is clear that $\lim_{n\to\infty} s_n$ is not equal to $+\infty$ (or $-\infty$), because if it were we would have $\lim_{n\to\infty} \sigma_n = +\infty$ (or $-\infty$).

(a) If possible, let

$$\overline{\lim_{n\to\infty}} s_n = \Lambda,$$

where Λ is $+\infty$, or a finite positive number. Then, if A is any positive number less than Λ,

$$s_n > A, \quad \text{for infinitely many of the values of } n,$$

say $M_1, M_2, \ldots$. But to the arbitrary positive number ε, there corresponds a positive integer μ, such that

$$|\sigma_n| < \varepsilon, \quad \text{when} \quad n \geq \mu.$$

Let M be the first of the sequence $M_1, M_2, \ldots$ which is greater than μ, and such that $MA \geq$ an even positive integer.

Let 2p be the largest even positive integer not greater than MA.

Then $2p \le MA < 4p$. Also

$$s_{M+1} > s_M - \frac{1}{M},$$

$$s_{M+2} > s_{M+1} - \frac{1}{M+1} > s_M - \left(\frac{1}{M} + \frac{1}{M+1}\right) > s_M - \frac{2}{M},$$

$$s_{M+3} > \cdots\cdots\cdots\cdots\cdots\cdots > s_M - \frac{3}{M},$$

$$\cdots\cdots\cdots\cdots\cdots\cdots\cdots\cdots\cdots\cdots\cdots\cdots$$

$$s_{M+p} > \cdots\cdots\cdots\cdots\cdots\cdots > s_M - \frac{p}{M},$$

and each of these is greater than $A - p/M$. But

$$\sigma_{M+p} = \frac{M\sigma_M + (s_{M+1} + s_{M+2} + \cdots + s_{M+p})}{M + p}$$

$$> \frac{M}{M+p}\sigma_M + \frac{p}{M+p}\left(A - \frac{p}{M}\right)$$

$$> \frac{a}{a+1}(A - a) - \varepsilon, \quad \text{where } a = \frac{p}{M} \le \frac{A}{2}.$$

Now $A - a \ge A/2$ and

$$\frac{a}{a+1} = \frac{A/4}{A/4 + 1},$$

since $p/M = a > A/4$. Therefore

$$\frac{a}{a+1}(A - a) > \frac{A^2}{2(A+4)}, \quad \text{and} \quad \sigma_{M+p} > \frac{A^2}{2(A+4)} - \varepsilon.$$

If

$$\varepsilon = \frac{A^2}{4(A+4)},$$

we have $\sigma_{M+p} > \varepsilon$, which is impossible. Thus

$$\overline{\lim_{n\to\infty}}\ s_n < 0.$$

(b) We shall now show that

$$\varliminf_{n\to\infty} s_n \ge 0.$$

For, if possible, let

$$\varliminf_{n\to\infty} s_n = \lambda < 0.$$

Take B any positive number less than $-\lambda$. Then

$$s_n < -B, \quad \text{for an infinite number of values of } n,$$

say $H_1, H_2, \ldots,$ and

$$|\sigma_n| < \varepsilon, \quad \text{when } n \ge \nu.$$

Let H be the first of the sequence $H_1, H_2, \ldots$ such that

$$\frac{H}{1 + \frac{1}{2}B} > \nu$$

and such that between

$$\frac{H}{1 + \frac{1}{2}B} \quad \text{and} \quad \frac{H}{1 + \frac{1}{4}B}$$

there shall be at least one positive integer. Let the integer next above $H/(1 + \frac{1}{2}B)$ be q, and write $H/q = 1 + b$. Then since

$$\frac{H}{1 + \frac{1}{2}B} < q < \frac{H}{1 + \frac{1}{4}B},$$

we have $\frac{1}{4}B < b < \frac{1}{2}B$. Also as before

$$\sigma_H = \frac{q\sigma_q + s_{q+1} + \cdots + s_H}{H}.$$

But

$$s_{H-1} < s_H + \frac{1}{H - 1},$$

$$s_{H-2} < s_{H-1} + \frac{1}{H - 2} < s_H + \left(\frac{1}{H - 1} + \frac{1}{H - 2}\right),$$

....

$$s_{q+1} < \cdots < s_H + \left(\frac{1}{H-1} + \frac{1}{H-2} + \cdots + \frac{1}{q+1} + \frac{1}{q}\right).$$

Therefore each of these is less than $-B + (H - q)/q < -B + b$. Also

$$\sigma_H < \frac{q}{H}\sigma_q + \frac{H-q}{H}(-B + b) < -\frac{b}{1+b}(B - b) + \varepsilon.$$

But

$$B - b > \frac{1}{2}B \quad \text{and} \quad \frac{b}{1+b} > \frac{B}{B+4}.$$

Therefore

$$\frac{b}{1+b}(B - b) < \frac{B^2}{2(B+4)} \quad \text{and} \quad \sigma_H < -\frac{B^2}{2(B+4)} + \varepsilon.$$

If we take

$$\varepsilon = \frac{B^2}{4(B+4)},$$

we have $\sigma_H < -\varepsilon$, which is impossible. Thus

$$\underline{\lim}_{n\to\infty} s_n \geq 0.$$

But we have seen that

$$\overline{\lim}_{n\to\infty} s_n \leq 0.$$

It follows that $\lim_{n\to\infty} s_n = 0$. □

45. Using the result in Exercise 44, we easily see that the following statement is valid: Let the sequence of arithmetic means σ_n for the series $u_1 + u_2 + \cdots$ converge to σ. If a positive integer n_0 exists such that

$$|u_n| < \frac{K}{n}, \quad \text{when} \quad n \geq n_0,$$

where K is a positive number independent of n, then the series $\Sigma_{n=1}^{\infty} u_n$ converges and its sum is σ.

46. For every integer $p \geq 2$ there are exactly $p - 1$ numbers of the form $1/k + 1/m$ for which the sum of the natural numbers k and m equals p. For $p = 2, 3, 4, \ldots$, consider these numbers enumerated. Then we obtain the sequence

$$2, \frac{3}{2}, \frac{3}{2}, \frac{4}{3}, 1, \frac{4}{3}, \frac{5}{4}, \frac{5}{6}, \frac{5}{6}, \ldots.$$

Verify that this sequence has the accumulation points

$$0, 1, \frac{1}{2}, \frac{1}{3}, \frac{1}{4}, \ldots$$

and no other accumulation points.

47. Let m be a positive integer and $|x| < 1$. Show that

$$1 - |x| < \sqrt[m]{1 + x} < 1 + |x|.$$

48. Show that, for $n = 2, 3, \ldots$ and $1 + h > 0$, we have

$$\sqrt[n]{1 + h} \leq 1 + \frac{h}{n}$$

(with equality holding if and only if $h = 0$).

49. Determine all polynomials of the form

$$x^n + a_1 x^{n-1} + \cdots + a_n$$

with all $a_k = \pm 1$ and which have all roots real.

Hint Let the roots be $x_1, \ldots, x_n$. Since the roots are real, the numbers $x_1^2, \ldots, x_n^2$ are positive, and we know that their sum is

$$x_1^2 + \cdots + x_n^2 = a_1^2 - 2a_2 = 1 \pm 2 = 3 \quad \text{(since } -1 \text{ is negative)}.$$

The product

$$x_1^2 \cdots x_n^2 = a_n^2 = 1.$$

Hence, by Proposition 44, we have $1 \leq 3/n$, that is, $n \leq 3$, with equality only if $x_1^2 = x_2^2 = x_3^2 = 1$. In case $n = 3$, all roots are ± 1 and $a_2 = -1$. The rest is trivial.

50. Let k be a positive integer. Using Proposition 59, verify that

(a) $$z_n = \frac{1^k + 2^k + \cdots + n^k}{n^{k+1}} \to \frac{1}{k+1}$$

and

(b) $$u_n = n\left(z_n - \frac{1}{k+1}\right) \to \frac{1}{2}$$

as $n \to \infty$.

51. Let $a > 1$. Show that

(a) $$\frac{\log_a n}{n} \to 0$$

and

(b) $$\frac{a^n}{n} \to \infty$$

as $n \to \infty$.

52. Let a and b be two given real numbers. We set $x_0 = a$ and $x_1 = b$ and define, for $n \geq 2$,

$$x_n = \frac{x_{n-2} + x_{n-1}}{2}.$$

For $n \geq 2$, we therefore have that

$$x_n - x_{n-1} = -\frac{1}{2}(x_{n-1} - x_{n-2}).$$

Show that

$$\lim_{n \to \infty} x_n = \frac{a + 2b}{3}.$$

Hint Note that

$$\lim_{n \to \infty} (x_n - a) = \frac{b - a}{1 - (-\frac{1}{2})}.$$

53. Show that

$$\lim_{n \to \infty} \sqrt{n}(\sqrt{n+1} - \sqrt{n}) = \frac{1}{2}.$$

54. Let $a_1, \ldots, a_m$ be m positive real numbers and let A denote the largest of these numbers. Show that, for $n \in N$,

$$\lim_{n\to\infty} (a_1^n + \cdots + a_m^n)^{1/n} = A.$$

Hint Note that

$$A \le (a_1^n + \cdots + a_m^n)^{1/n} \le A \sqrt[n]{m}.$$

55. Let

$$z_n = \frac{1}{(n^2+1)^{1/2}} + \frac{1}{(n^2+2)^{1/2}} + \cdots + \frac{1}{(n^2+n)^{1/2}}.$$

Show that $z_n \to 1$ as $n \to \infty$.

Hint Note that

$$\frac{n}{(n^2+n)^{1/2}} < z_n < \frac{n}{(n^2+1)^{1/2}}.$$

56. Let a be a positive rational number such that $a^2 < 2$. Show that it is possible to find a positive integer n such that

$$\left(a + \frac{1}{n}\right)^2 < 2.$$

Hint The foregoing inequality is equivalent to

$$\frac{2a}{n} + \frac{1}{n^2} < 2 - a^2$$

and the latter inequality certainly holds provided that n satisfies

$$n > \frac{2a+1}{2-a^2};$$

but there exists such an n by the Archimedean property.

57. Let A and B be given real numbers and suppose that $A \le B + \varepsilon$ for any $\varepsilon > 0$. Show that $A \le B$.

Hint Does $A > B$ imply $A > B + \varepsilon$ for some $\varepsilon > 0$?

58. Give the prime factorization of 104060401.

Hint Observe that $104060401 = x^4 + 4x^3 + 6x^2 + 4x + 1$ for $x = 100$.

59. Discover integer solutions of the equation

$$\frac{n(n+1)}{2} = k^2.$$

Solution Setting $x = 2n + 1$ and $y = 2k$, the foregoing equation becomes

$$x^2 - 2y^2 = 1. \qquad \text{(P)}$$

If $x = u$ and $y = v$ are integers which satisfy the equation (P), we say, for simplicity, that the number $u + v\sqrt{2}$ is a *solution of* (P); two solutions $u + v\sqrt{2}$ and $u' + v'\sqrt{2}$ are *equal* if $u = u'$ and $v = v'$ and the first solution is *greater than* the second if

$$u + v\sqrt{2} > u' + v'\sqrt{2}.$$

Let us consider all solutions $x + y\sqrt{2}$ of the equation (P) with positive x and y. Among these there is a least solution $x_1 + y_1\sqrt{2}$, in which x_1 and y_1 have their least (positive) values. The number $x_1 + y_1\sqrt{2}$ is called the *fundamental solution of* the equation (P). In fact, $x_1 + y_1\sqrt{2} = 3 + 2\sqrt{2}$. Putting

$$x_j + y_j\sqrt{2} = (x_1 + y_1\sqrt{2})^j \quad \text{for } j = 2, 3, \ldots,$$

we see that $x_j + y_j\sqrt{2}$ are also solutions of the equation (P); actually, all solutions of (P) can be obtained in this way. We leave it to the reader to check out that $(n,k) = (1,1)$, $(8,6)$, $(49,35)$, and $(288,204)$ are the first four solutions of

$$\frac{n(n+1)}{2} = k^2.$$

The equation $x^2 - Dy^2 = 1$ with D being a natural number which is not a perfect square, is known as *Pell's Equation* in Diophantine analysis.

60. Suppose that $\Sigma_{i=1}^{\infty} x_i$ is a convergent series of positive terms which monotonically decrease (that is, $x_1 \geq x_2 \geq x_3 \geq \cdots$). Let P

denote the set of all numbers which are sums of some (finite or infinite) subseries of $\Sigma_{i=1}^{\infty} x_i$. Show that P is an interval if and only if, for every integer n,

$$x_n \leq \sum_{i=n+1}^{\infty} x_i.$$

<u>*Solution*</u> If $x_k = 2p + \Sigma_{i=k+1}^{\infty} x_i$ with $p > 0$ and if

$$x_k - p = p + \sum_{i=k+1}^{\infty} x_i = \sum_{i=1}^{\infty} x_i \theta_i$$

with each θ_i either zero or one, then $\theta_i = 0$, $i = 1, 2, \ldots, k$, by the monotoneity of the terms, but then

$$\sum_{i=1}^{\infty} x_i \theta_i \leq \sum_{i=k+1}^{\infty} x_i = x_k - 2p$$

gives a contradiction. Thus the condition is necessary.

If $x_n \leq \Sigma_{i=n+1}^{\infty} x_i$, $n = 1, 2, \ldots$, and if $0 < p < \Sigma_{i=1}^{\infty} x_i$, take i_1 as the least positive integer n for which $x_n \leq p$. Assuming inequality occurs, take j_1 as the largest positive integer n such that $S(i_1, j_1) \leq p < S(i_1, j_1 + 1)$, where by $S(m,n)$ we mean the sum $\Sigma_{i=m}^{n} x_i$. Such a choice of j_1 is possible since $S(i_1, \infty) \geq x_{i_1 - 1} > p$ if $i_1 > 1$ and $S(1, \infty) > p$ if $i_1 = 1$. If k blocks of consecutive terms have been chosen such that $i_h < j_h < i_{h+1} < j_{h+1}$, $h = 1, 2, \ldots, k - 1$, and $0 \leq p - \Sigma_{r=1}^{k} S(i_r, j_r) < x_{i_k + 1}$ and if equality does not obtain in the last formula then take i_{k+1} as the smallest positive integer n such that $x_n \leq p - \Sigma_{r=1}^{k} S(i_r, j_r)$ and j_{k+1} as the largest positive integer n such that $0 \leq \Sigma_{r=1}^{k+1} S(i_r, j_r) < x_{j_{k+1}+1}$. The existence of i_{k+1} and j_{k+1} with the required properties follows as before. The process either terminates if an inequality occurs or produces an infinite subseries each partial sum of which is less than p. Since any partial sum which terminates in the k-th group of terms exceeds $\Sigma_{r=1}^{k-1} S(i_r, j_r) > p - x_{j_{k-1}+1}$, the series converges to p. □

61. Prove that every positive rational number is the sum of a finite number of distinct terms of the harmonic series

$$1 + \frac{1}{2} + \frac{1}{3} + \cdots + \frac{1}{n} + \cdots.$$

Proof Let A and B be positive integers. Then by the divergence of the series under consideration there is a unique nonnegative integer n_0 such that

$$\sum_{j=0}^{n_0} \frac{1}{j} < \frac{A}{B} \le \sum_{j=0}^{n_0+1} \frac{1}{j}.$$

(The $\Sigma_{j=0}^{1}$ 1/j is taken as 0.) If equality holds the desired representation is at hand and so we assume that

$$\frac{A}{B} < \sum_{j=0}^{n_0+1} \frac{1}{j}.$$

Then $A/B - \Sigma_{j=0}^{n_0} 1/j = C/D < 1/(n_0 + 1)$. Take n_1 as the unique positive integer such that $1/(n_1 + 1) \le C/D < 1/n_1$. We again suppose inequality as the problem is otherwise solved, and put

$$\frac{C}{D} - \frac{1}{n_1 + 1} = \frac{E}{F} > 0.$$

But $E/F = [C(n_1 + 1) - D]/D(n_1 + 1)$ and $C(n_1 + 1) - D < C$ so that with E/F in lowest terms we must have $E < C$. Then

$$\frac{E}{F} < \frac{1}{n_1(n_1 + 1)}$$

and so the unique integer n_2 such that $1/(n_2 + 1) \le E/F < 1/n_2$ satisfies $n_2 > n_1$. In a finite number of steps we must obtain the desired representation, since if the equality does not occur before, it must occur when the numerator of the reduced fraction has become 1. □

62. We say that a sequence of points in an interval is *dense* when every subinterval contains at least one point of the sequence. With this agreement in mind, let $P_1, P_2, \ldots$ be a sequence of distinct points which is dense in the interval (0,1). The points

$$P_1, P_2, \ldots, P_{n-1}$$

decompose the interval into n parts, and P_n decomposes one of these into two parts. Let a_n and b_n be the lengths of these two intervals. Prove that

$$\sum_{n=1}^{\infty} a_n b_n (a_n + b_n) = \frac{1}{3}.$$

Solution The identity to be proved can be rewritten in the equivalent form

$$-(a_1 + b_1)^3 + \sum_{n=1}^{\infty} \left[(a_n + b_n)^3 - a_n^3 - b_n^3 \right] = 0.$$

Consider the partial sum

$$-a_1^3 - b_1^3 + (a_2 + b_2)^3 - a_2^3 - b_2^3 + (a_3 + b_3)^3 - a_3^3 - b_3^3 + \cdots$$
$$+ (a_n + b_n)^3 - a_n^3 - b_n^3.$$

Each of the positive terms cancels one of the preceding negative terms. Because the sequence $P_1, P_2, \ldots$ is dense, every negative term eventually gets cancelled out. If we choose n sufficiently large, the only terms that are not cancelled out in the above sum are the cubes of the lengths of disjoint intervals, each of which is of length ε at most. The sum of the terms that are not cancelled out is therefore majorized by ε. It follows that the above sum is at most ε in absolute value for n large. Now let $n \to \infty$. □

63. Let $a_1, a_2, \ldots$ be a sequence of positive real numbers; define

$$s_n = \frac{a_1 + a_2 + \cdots + a_n}{n} \quad \text{and} \quad r_n = \frac{a_1^{-1} + a_2^{-1} + \cdots + a_n^{-1}}{n}.$$

Given that the limits $\lim_{n\to\infty} s_n$ and $\lim_{n\to\infty} r_n$ exist, prove that the product of these two limits is not less than 1.

Hint In Section 3 we showed that

harmonic mean $\le$ geometric mean $\le$ arithmetic mean

for the same set of positive numbers.

64. Any positive integer may be written in the form

$$n = 2^k(2j + 1).$$

Let $a_n = e^{-k}$ and $b_n = a_1 a_2 \cdots a_n$. Prove that $\Sigma_{n=1}^{\infty} b_n$ converges.

Hint The problem means that every positive integer can be written uniquely in the form $2^k(2j + 1)$, where k and j are nonnegative integers. Thus k = 0 when n is odd, and k is a positive integer when n is even. Hence $a_n = 1$ when n is odd, and $a_n \le e^{-1}$ when n is even. Therefore $b_{2n} = b_{2n+1}$ and the series is equivalent to

$$b_1 + 2b_2 + 2b_4 + \cdots.$$

But since $b_{2n+2}/b_{2n} = a_{2n+1}a_{2n+2} \le e^{-1}$, the latter series converges. Since the terms approach zero, the grouping of the given series does not affect convergence and the given series converges.

65. Let $(x_n)_{n=1}^{\infty}$ be a given sequence and let $y_n = x_{n-1} + 2x_n$ for n = 2, 3, ... Suppose that the sequence $(y_n)_{n=2}^{\infty}$ converges. Prove that the sequence $(x_n)_{n=1}^{\infty}$ also converges.

Solution Let $\bar{y} = \lim_{n\to\infty} y_n$ and set $\bar{x} = \bar{y}/3$. We will show that $\bar{x} = \lim_{n\to\infty} x_n$. For $\varepsilon > 0$ there is an n_0 such that for all n larger than n_0, $|y_n - \bar{y}| < \varepsilon/2$.

$$\varepsilon/2 > |y_n - \bar{y}| = |x_{n-1} + 2x_n - 3\bar{x}| = |2(x_n - \bar{x}) + (x_{n-1} - \bar{x})|$$

$$\ge 2|x_n - \bar{x}| - |x_{n-1} - \bar{x}|.$$

This may be rewritten as $|x_n - \bar{x}| < \varepsilon/4 + \frac{1}{2}|x_{n-1} - \bar{x}|$, which can be iterated to give

$$|x_{n+m} - \bar{x}| < \frac{\varepsilon}{4}\left(\sum_{i=1}^{m} 2^{-i}\right) + 2^{-(m+1)}|x_{n-1} - \bar{x}|$$

$$< \frac{\varepsilon}{2} + 2^{-(m+1)}|x_{n-1} - \bar{x}|.$$

By taking m large enough,

$$2^{-(m+1)}|x_{n-1} - \bar{x}| < \frac{\varepsilon}{2}.$$

Thus for all sufficiently large k, $|x_k - \bar{x}| < \varepsilon$. □

66. Given a sequence $(a_n)_{n=1}^{\infty}$ of nonnegative real numbers such that $a_{n+m} \le a_n a_m$ for all pairs of positive integers, m and n, prove that the sequence $(\sqrt[n]{a_n})_{n=1}^{\infty}$ converges.

Solution Since $a_{n+1} \le a_n a_1$, it follows by induction that $a_n \le a_1^n$ for all n, and hence that $0 \le a_n^{1/n} \le a_1$ for all n. The sequence $(a_n^{1/n})_{n=1}^{\infty}$ is thus bounded. Let

$$L = \overline{\lim_{n\to\infty}} \sqrt[n]{a_n}.$$

We have $0 \le L \le a_1$. We shall show that $a_n^{1/n} \ge L$ for all n. Once this is established, it will follow that

$$\underline{\lim_{n\to\infty}} \sqrt[n]{a_n} \ge L = \overline{\lim_{n\to\infty}} \sqrt[n]{a_n},$$

which implies the existence of $\lim_{n\to\infty} a_n^{1/n}$.

Let $m_1 < m_2 < m_3 < \cdots$ be such that

$$\lim_{k\to\infty} \sqrt[m_k]{a_{m_k}} = L.$$

Let n be any integer, and let the rational number m_k/n be given by $q_k + d_k$ where q_k is an integer and $0 \le d_k < 1$. We observe that $q_k/m_k = (1/n) - d_k/m_k$, so that as $k \to \infty$, $q_k/m_k \to 1/n$, since $d_k/m_k < 1/m_k \to 0$. Now by the given relation

$$a_{m_k} = a_{nq_k+nd_k} \le a_n^{q_k} a_1^{nd_k}.$$

Hence

$$a_{m_k}^{1/m_k} \le a_n^{q_k/m_k} a_1^{nd_k/m_k},$$

and letting $k \to \infty$ we obtain

$$L \le a_n^{1/n},$$

since $q_k/m_k \to 1/n$ and $nd_k/m_k \to 0$. It follows that $\lim_{n\to\infty} a_n^{1/n}$ exists.

The sequence $(a_n^{1/n})_{n=1}^{\infty}$ need not be monotonic. As an example let

$$a_n = e^{-n} \quad \text{if } n \text{ is even,}$$
$$= e^{-n+1} \quad \text{if } n \text{ is odd.}$$

For this sequence $(a_n^{1/n})_{n=1}^{\infty}$ is an oscillating sequence converging to e^{-1}.

67. Establish a one-to-one correspondence between the set of all x such that $0 < x \le 1$ and the set of all y such that $0 < y < 1$.

Solution Let

$$y = \frac{3}{2} - x \quad \text{for} \quad \frac{1}{2} < x \le 1; \quad \text{then} \quad \frac{1}{2} \le y < 1 \quad \text{holds.}$$

Next, let

$$y = \frac{3}{4} - x \quad \text{for} \quad \frac{1}{4} < x \le \frac{1}{2}; \quad \text{then} \quad \frac{1}{4} \le y < \frac{1}{2} \quad \text{holds.}$$

At the n-th step, let

$$y = \frac{1}{2^n} - x \quad \text{for} \quad \frac{1}{2^n} < x \le \frac{1}{2^{n-1}}; \quad \text{then} \quad \frac{1}{2^n} \le y < \frac{1}{2^{n-1}} \quad \text{holds.}$$

Do this for all $n \in N$. □

68. Given a sequence $(x_n)_{n=1}^{\infty}$ such that $x_n - x_{n-2} \to 0$ as $n \to \infty$. Prove that

$$\lim_{n\to\infty} \frac{x_n - x_{n-1}}{n} = 0.$$

Solution For $\varepsilon > 0$, let n_0 be sufficiently large so that $|x_n - x_{n-2}| < \varepsilon$ for all $n \ge n_0$. Observe that for $n > n_0$,

$$x_n - x_{n-1} = (x_n - x_{n-2}) - (x_{n-1} - x_{n-3}) + (x_{n-2} - x_{n-4}) - \cdots$$
$$\pm \Big((x_{n_0+1} - x_{n_0-1}) - (x_{n_0} - x_{n_0-1})\Big).$$

Thus $|x_n - x_{n-1}| \le (n - n_0)\varepsilon + |x_{n_0} - x_{n_0-1}|$ and so $(x_n - x_{n-1})/n$ tends to zero as $n \to \infty$. □

69. Let $(a_n)_{n=1}^{\infty}$ be a sequence of positive real numbers. Show that

$$\overline{\lim_{n\to\infty}}\ n\left(\frac{1 + a_{n+1}}{a_n} - 1\right) \ge 1.$$

Solution Assume that the conclusion is false. Then there is a positive integer m such that for $n \ge m$,

$$n\left(\frac{1 + a_{n+1}}{a_n} - 1\right) < 1.$$

This inequality is equivalent to

$$\frac{1}{m + 1} < \frac{a_n}{n} - \frac{a_{n+1}}{n + 1}.$$

Replacing n by m, m + 1, ..., m + k - 1, in turn, and adding the results, we obtain

$$\frac{1}{m + 1} + \frac{1}{m + 2} + \cdots + \frac{1}{m + k} < \frac{a_m}{m} - \frac{a_{m+k}}{m + k} < \frac{a_m}{m},$$

in contradiction to the divergence of the harmonic series $\Sigma_{j=1}^{\infty} 1/j$.

To see that 1 cannot be replaced by a larger number, let $a_n = kn$, $n = 1, 2, \ldots$. Then

$$n\left(\frac{1 + a_{n+1}}{a_n} - 1\right) = \frac{1 + k}{k},$$

which is arbitrarily near 1 for large k. Alternately, if

$$a_n = n \log n,$$

we get

$$\lim_{n\to\infty} n\left(\frac{1 + a_{n+1}}{a_n}\right) = 1.$$ □

70. Suppose that $u_0, u_1, u_2, \ldots$ is a sequence of real numbers such that

$$u_n = \sum_{k=1}^{\infty} u_{n+k}^2$$

for $n = 0, 1, 2, \ldots$. Prove that if $\Sigma_{n=0}^{\infty} u_n$ converges then $u_k = 0$ for all k.

Solution Clearly $0 \le u_{n+1} \le u_n$ for $n = 1, 2, \ldots$. If $\Sigma_{j=0}^{\infty} u_j$ converges, we take $k \ge 1$ so that $\Sigma_{j=k+1}^{\infty} u_j < 1$. Then

$$u_{k+1} \le u_k = \sum_{j=k+1}^{\infty} u_j^2 \le u_{k+1}\left(\sum_{j=k+1}^{\infty} u_j\right) \le u_{k+1}.$$

Hence $u_k = u_{k+1}$ and so $u_{k+1} = 0$, implying $u_j = 0$ for $j > k$ since

$$u_{k+1} = \sum_{j=k+1}^{\infty} u_j^2.$$

By induction $u_j = 0$ for $j < k + 1$. □

71. Let p and q be positive numbers. Show that

$$\int_0^1 \frac{x^{p-1}}{1 + x^q}\, dx = \frac{1}{p} - \frac{1}{p+q} + \frac{1}{p+2q} - \frac{1}{p+3q} + \cdots.$$

72. Let $(p_n)_{n=1}^{\infty}$ be a sequence of positive real numbers. Show that if the series $\Sigma_{n=1}^{\infty} p_n^{-1}$ converges, then so does the series

$$\sum_{n=1}^{\infty} \frac{n^2}{(p_1 + p_2 + \cdots + p_n)^2}\, p_n.$$

Solution Put $q_n = p_1 + \cdots + p_n$ and $q_0 = 0$. We are led to the estimate

$$S_m = \sum_{n=1}^{m} \left(\frac{n}{q_n}\right)^2 (q_n - q_{n-1})$$

in terms of $T = \Sigma_{n=1}^{\infty} 1/p_n$. We observe that

$$S_m \le \frac{1}{p_1} + \sum_{n=2}^{m} \frac{n^2}{q_n q_{n-1}}(q_n - q_{n-1}) = \frac{1}{p_1} + \sum_{n=2}^{m} \frac{n^2}{q_{n-1}} - \sum_{n=2}^{m} \frac{n^2}{q_n}$$

$$= \frac{1}{p_1} + \sum_{n=1}^{m-1} \frac{(n+1)^2}{q_n} - \sum_{n=2}^{m} \frac{n^2}{q_n} \le \frac{5}{p_1} + 2\sum_{n=2}^{m} \frac{n}{q_n} + \sum_{n=2}^{m} \frac{1}{q_n}.$$

By the Cauchy-Schwarz Inequality (see Proposition 48),

$$\left(\sum_{n=2}^{m} \frac{n}{q_n}\right)^2 \le \left(\sum_{n=2}^{m} \frac{n^2}{q_n^2} p_n\right)\left(\sum_{n=1}^{m} \frac{1}{p_n}\right)$$

and thus $S_m \le 5/p_1 + 2\sqrt{S_m T} + T$. But this inequality implies that $\sqrt{S_m} \le \sqrt{T} + \sqrt{2T + 5/p_1}$. □

73. Let $(u_k)_{k=1}^{\infty}$ be a sequence of integers, and let V_n be the number of those which are less than or equal to n. Show that if $\Sigma_{k=1}^{\infty} 1/u_k < \infty$, then $V_n/n \to 0$ as $n \to \infty$.

Hint For any $j < n$ we have $(V_n - j)/n \le \Sigma_{k=j+1}^{V_n} 1/u_k$. Thus

$$\overline{\lim_{n\to\infty}} \frac{V_n}{n} \le \sum_{k=j+1}^{\infty} \frac{1}{u_k}.$$

74. Consider the following enumeration of proper fractions

$$\frac{0}{1}, \frac{1}{1}, \frac{0}{2}, \frac{1}{2}, \frac{2}{2}, \frac{0}{3}, \frac{1}{3}, \frac{2}{3}, \ldots.$$

Here we have ordered the proper fractions into groups according to increasing denominators and, within each group, according to increasing numerators and we have also listed fractions in their unreduced form. It is clear that in this enumeration the fraction p/q has index

$$k = k(p,q) = p + \tfrac{1}{2}q(q+1).$$

For each k, we now cover the k-th term in this enumeration of rational numbers between 0 and 1 by an interval of length 2^{-k} in such a way that the k-th term is at the center of its covering interval. The length of the interval cover is $\Sigma_{k=1}^{\infty} 2^{-k} = 1$.

Show that the point $\sqrt{2}/2$ belongs to none of the covering intervals.

Solution Suppose $t = \sqrt{2}/2$ were in one of the covering intervals. Then there would be integers p and q ($q \geq p \geq 0$, $q \geq 1$) such that

$$\left|\frac{p}{q} - t\right| < \frac{1}{2}\,\frac{1}{2^{p+\frac{1}{2}q(q+1)}} \leq \frac{1}{2^{\frac{1}{2}q(q+1)+1}}.$$

But for every p and q with $q \geq 1$, we have, since $\sqrt{2}$ is an irrational number, $|2p^2 - q^2| \geq 1$ and therefore

$$\left|\frac{p}{q} - \frac{\sqrt{2}}{2}\right| = \frac{|(p/q)^2 - (1/2)|}{(p/q) + (\sqrt{2}/2)} = \frac{|2p^2 - q^2|}{2pq + q^2\sqrt{2}} > \frac{1}{4q^2}$$

$$\geq \frac{1}{2^{\frac{1}{2}q(q+1)+1}}.$$ □

75. Prove that if x and y are positive irrational numbers such that $1/x + 1/y = 1$, then the sequences $[x]$, $[2x]$, ..., $[nx]$, ... and $[y]$, $[2y]$, ..., $[ny]$, ... together include every positive integer exactly once. (The notation $[x]$ means the largest integer contained in x.)

Solution Since x and y are both positive, $1/x$ and $1/y$ are both less than 1, so that both x and y are greater than 1. Thus no two multiples of x have the same integral part, and no two multiples of y have the same integral part. Therefore no integer appears more than once in either of the two sequences. Suppose that an integer K appeared in both sequences. Then we could find integers p and q such that $K < px < K + 1$ and $K < qy < K + 1$; no equality is possible because x and y are irrational numbers. Solving these inequalities for $1/x$ and $1/y$, we find

$$\frac{p}{K+1} < \frac{1}{x} < \frac{p}{K}, \quad \frac{q}{K+1} < \frac{1}{y} < \frac{q}{K}.$$

Adding we find

$$\frac{p+q}{K+1} < 1 < \frac{p+q}{K} \quad \text{or} \quad K < p + q < K + 1,$$

which is impossible since K, p, and q are all integers. Finally, suppose that an integer M is missing from both sequences. Then we can find integers p and q such that

$$px < M, \quad (p + 1)x > M + 1, \quad (q + 1)y > M + 1.$$

Solving for $1/x$ and $1/y$ as before, we are led to $M - 1 < p + q < M$, which is again impossible. Therefore every integer is present in one sequence or the other, and each integer occurs exactly once. □

76. Show that

$$3 = \lim_{n \to \infty} \sqrt{1 + 2\sqrt{1 + 3\sqrt{1 + \cdots \sqrt{1 + (- 1)\sqrt{1 + n}}}}}.$$

Solution We see that

$$3 = \sqrt{1 + 2 \cdot 4} = \sqrt{1 + 2\sqrt{16}} = \sqrt{1 + 2\sqrt{1 + 3\sqrt{25}}}$$

$$= \sqrt{1 + 2\sqrt{1 + 3\sqrt{1 + 4\sqrt{36}}}}.$$

This leads to conjecturing the relation

$$3 = \sqrt{1 + 2\sqrt{1 + 3\sqrt{1 + \cdots \sqrt{1 + n\sqrt{(n + 2)^2}}}}}$$

for all $n \geq 1$. Proceeding by induction we verify that

$$(n + 2)^2 = 1 + (n + 1)(n + 3) = 1 + (n + 1)\sqrt{(n + 3)^2}.$$

This given, we must have

$$3 \geq \sqrt{1 + 2\sqrt{1 + 3\sqrt{1 + \cdots \sqrt{1 + (n - 1)\sqrt{1 + n}}}}}.$$

To get an inequality in the other direction going, observe that for any $\alpha > 1$

$$\sqrt{1 + n\alpha} \leq \sqrt{\alpha}\sqrt{1 + n}.$$

A repetition of this inequality gives then

$$3 \le (n+2)^a \sqrt{1 + 2\sqrt{1 + 3\sqrt{1 + \cdots \sqrt{1 + (n-1)\sqrt{1+n}}}}},$$

where $a = 2^{1-n}$. □

77. Investigate the convergence of the sequences

(a) $a_n = \sqrt{1 + \sqrt{1 + \cdots + \sqrt{1}}}$ (n roots)

and

(b) $b_n = \sqrt{1 + \sqrt{2 + \cdots + \sqrt{n}}}$ (n roots).

78. Let $0 < x_1 < 1$ and $x_{n+1} = x_n(1 - x_n)$ for $n = 1, 2, 3, \ldots$. Show that $nx_n \to 1$ as $n \to \infty$.

Solution Observe that

$$\begin{aligned}(n+1)x_{n+1} &= nx_n + x_n - (n+1)x_n^2 \\ &= nx_n + x_n\{1 - (n+1)x_n\}.\end{aligned} \tag{E.4}$$

To see that nx_n is increasing, we need to show that $1 - (n+1)x_n \ge 0$. From the graph of $x(1 - x)$ we note that $x_2 \le 1/4$ and that $x_n \le a \le 1/2$ implies $x_{n+1} \le a(1 - a)$. So by induction

$$(n+1)x_n \le (n+1)\frac{1}{n}\left(1 - \frac{1}{n}\right) = 1 - \frac{1}{n^2} \le 1.$$

Furthermore, $nx_n < (n+1)x_n \le 1$ and so nx_n is bounded above by 1. Thus nx_n converges to a limit α with $0 < nx_n < \alpha \le 1$. Now summing (E.4) from 2 to n we obtain

$$\begin{aligned}1 &\ge (n+1)x_{n+1} \\ &= 2x_2 + x_2(1 - 3x_2) + x_3(1 - 4x_3) + \cdots + x_n\{1 - (n+1)x_n\}.\end{aligned} \tag{E.5}$$

If $\alpha \ne 1$ then $\{1 - (n+1)x_n\} \ge (1 - \alpha)/2$ for all large n and thus

(E.5) would show that $\Sigma_{n=1}^{\infty} x_n$ is convergent. However $nx_n \geq x_1$ and so $\Sigma_{n=1}^{\infty} x_n \geq x_1 \Sigma_{n=1}^{\infty} 1/n$. □

79. The following result gives a scheme to calculate on a computer the reciprocal of a number c.

(a) Let $0 < x_0 < 1$ and define

$$x_{n+1} = x_n(2 - x_n) \quad \text{for} \quad n = 0, 1, 2, \ldots.$$

Show that $0 < x_n < x_{n+1} < 1$ and $x_n \to 1$ as $n \to \infty$.

(b) Let c be any positive number. Set $x_n = cy_n$ in the recursion formula of part (a); this gives $y_{n+1} = y_n(2 - cy_n)$. Assuming that $0 < y_0 < 1/c$, we can see that $y_n \to 1/c$ as $n \to \infty$.

80. Let

$$b_n = \sum_{k=0}^{n} \binom{n}{k}^{-1}$$

for $n \geq 1$. Show that $\lim_{n\to\infty} b_n = 2$.

Hint Note that

$$b_n = \frac{n + 1}{2n} b_{n-1} + 1$$

for $n \geq 2$.

81. Show that

$$\sum_{k=1}^{n} \binom{n}{k} k^2 = n(n + 1)2^{n-2}.$$

82. Consider

$$\frac{m}{(m + 1)(m + 2)} + \frac{m}{(m + 2)(m + 3)} + \cdots = \frac{m}{m + 1}.$$

Letting $m \to \infty$, observe that the sum of the limits need not equal the limit of the sum in case the sum is infinite.

Chapter 2

CONTINUITY

1. *Point Set Theory*

In the sequel we shall denote by R^1 the real number system (or the real number line) and by R_e^1 the extended real number system. The simplest subsets of R^1 are the intervals. The *open interval* (a,b) is the set of all real numbers x such that $a < x < b$, and the *closed interval* [a,b] is the set of all real numbers x such that $a \le x \le b$. The *half-open intervals* (a,b] and [a,b) are defined as the sets $\{x \in R^1: a < x \le b\}$ and $\{x \in R^1: a \le x < b\}$, respectively.

Definition A set G of real numbers is said to be *open* if for each $x \in G$ there is a $\delta > 0$ such that each y with $|x - y| < \delta$ belongs to the set G.

In other words, a set G is said to be open if for every x in G there is an open interval $I = (x - \delta, x + \delta)$ with $\delta > 0$ such that $I \subset G$. The open intervals are examples of open sets, and both the empty set $\emptyset$ and the set R^1 of all real numbers are open. (The fact that the empty set $\emptyset$ is open can be seen as follows: if $G = \emptyset$, then the situation $x \in G$ never arises and hence no corresponding δ need be found.)

Proposition 1 The intersection $G_1 \cap G_2$ of two open sets G_1 and G_2 is open.

Proof Let $x \in G_1 \cap G_2$. Since $x \in G_1$ and G_1 is open, there is a $\delta_1 > 0$ such that all y with $|x - y| < \delta_1$ belong to G_1. Similarly, there is a $\delta_2 > 0$ such that all y with $|x - y| < \delta_2$ belong to G_2. Take $\delta = \min\{\delta_1, \delta_2\}$; then $\delta > 0$, and if $|x - y| < \delta$, then y belongs to both G_1 and G_2, that is, y belongs to $G_1 \cap G_2$. □

Remark The intersection of any finite collection of open sets is open; this is a simple consequence of Proposition 1. It is not true, however, that the intersection of any collection of open sets is open. Take for example $G_n = (-1/n, 1/n)$ for $n \in N$. It is clear that only the point $x = 0$ is common to all these open intervals G_n; but a set of real numbers containing a single point only is certainly not open.

Proposition 2 The union of any collection $\mathcal{G}$ of open sets is open.

Proof Let U be the union of the collection $\mathcal{G}$, and $x \in U$. Then there is a set $G \in \mathcal{G}$ with $x \in G$. Since G is open, there is a $\delta > 0$ such that all y with $|x - y| < \delta$ belong to G and hence to U, since $G \subset U$. Thus U is open. □

Proposition 3 Every open set of real numbers is the union of a countable collection of disjoint open intervals. This representation is unique up to the order in which the terms appear in the union.

Proof Since G is open, for each $x \in G$ there is a $y > x$ such that $(x,y) \subset G$. Let

$$a = \inf\{z: (z,x) \subset G\} \quad \text{and} \quad b = \sup\{y: (x,y) \subset G\}.$$

Then $a < x < b$, and $I_x = (a,b)$ is an open interval containing x. But $I_x \subset G$ because, if $w \in I_x$, say $x < w < b$, we have by the definition of the number b a number $y > w$ such that $(x,y) \subset G$, and so $w \in G$. Moreover, $b \notin G$, since if $b \in G$, then for some $\delta > 0$ we have $(b - \delta, b + \delta) \subset G$, whence $(x, b + \delta) \subset G$, contradicting the definition of b. In the same way we see that $a \notin G$.

We now consider the collection of all open intervals $\{I_x: x \in G\}$. Since each x in G is contained in I_x and each I_x is contained in G, we have $G = \cup \{I_x: x \in G\}$. Let (a,b) and (c,d) be two intervals in this collection with a point in common. Then we must have $c < b$ and $a < d$. Since $c \notin G$, we have $c \notin (a,b)$ and so $c \leq a$. Since $a \notin G$, we have $a \notin (c,d)$ and so $a \leq c$. Thus $a = c$. Similarly, $b = d$ and so $(a,b) = (c,d)$. Thus two different intervals in this collection $\{I_x: x \in G\}$ must be disjoint. Therefore G is the union of the disjoint collection $\{I_x: x \in G\}$ of open intervals and we must show next that this collection is countable. But each open interval contains some rational point by Proposition 25 of Chapter 1. Since we have a collection of disjoint open intervals, each open interval contains a different rational number, and the collection can be put in one-to-one correspondence with a subset of the set Q of all rational numbers. Thus the collection $\{I_x: x \in G\}$ is countable.

Finally, if there were two distinct representations of G in term terms of countable unions of disjoint open intervals, then there would exist a point $x \in G$ which would belong to a component interval I in the first representation and to a component interval $J \neq I$ in the second representation. But then one of these component intervals, e.g., J, would extend beyond the other; it would follow from this that one of the endpoints of I belongs to J, which is impossible, inasmuch as the endpoints of I do not belong to G. □

N.B.: In connection with the foregoing proof have a look at

Exercise 45 at the end of this chapter.

Lemma The union of a countably infinite collection of countably infinite sets is a countably infinite set.

Proof For the case in which there are no elements common to any of the sets we can make the result clear as follows: Let the sets be $E_1, E_2, \ldots$, and let the elements of E_i be $e_{i1}, e_{i2}, \ldots$. Then the elements of the union $\cup_{i=1}^{\infty} E_i$ can be enumerated in the order indicated in the following display.

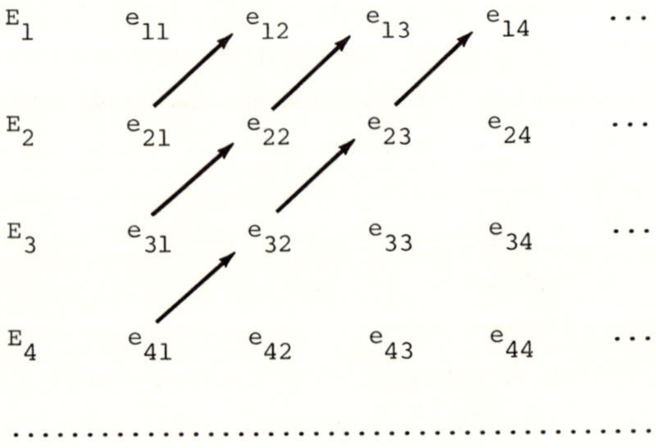

That is, first we take e_{11}, then those e_{ij}'s such that $i + j = 3$, then those for which $i + j = 4$, and so forth.

If any two of the sets E_i have elements in common, these will of course appear more than once in the enumeration

$$e_{11};\quad e_{21}, e_{12};\quad e_{31}, e_{22}, e_{13};\quad e_{41}, e_{32}, e_{23}, e_{14};\quad \ldots$$

But, since E_1 for example is known to be infinite and $E_1 \subset \cup_{i=1}^{\infty} E_i$, it is clear that $\cup_{i=1}^{\infty} E_i$ must be countably infinite. □

Proposition 4 (Theorem of Lindelöf) Let $\mathcal{G}$ be an infinite collection of open sets of real numbers. Then there is a countable subcollection $(G_n)_{n=1}^{\infty}$ of $\mathcal{G}$ such that $\bigcup \{G \in \mathcal{G}\} = \bigcup \{G_n : n \geq 1\}$.

Proof Let $U = \bigcup\{G: G \in \mathcal{G}\}$, and $x \in U$. Then there is a $G \in \mathcal{G}$ with $x \in G$. Since G is open, there is an open interval I_x such that $x \in I_x \subset G$. By Proposition 25 of Chapter 1, we can find an open interval J_x with rational endpoints such that $x \in J_x \subset I_x$. Since the collection of all open intervals with rational endpoints is countable by the foregoing lemma, the collection $\{J_x: x \in U\}$ is countable, and $U = \bigcup\{J_x: x \in U\}$. For each interval in $\{J_x: x \in U\}$ choose a set G in $\mathcal{G}$ with contains it. This gives a countable subcollection $(G_n)_{n=1}^{\infty}$ of $\mathcal{G}$, and $U = \cup_{n=1}^{\infty} G_n$. □

Definition A real number x is called a *point of closure of a set* E (of real numbers) if for every $\delta > 0$ there is a y in E such that $|x - y| < \delta$. We shall write $\overline{E}$ to denote the points of closure of E and call $\overline{E}$ the *closure of* E.

The real number x is therefore a point of closure of E if every open interval $(x - \delta, x + \delta)$ with $\delta > 0$ also contains a point of E. Evidently, each point of E is a point of closure of E; thus $E \subset \overline{E}$.

Proposition 5 If $A \subset B$, then $\overline{A} \subset \overline{B}$. Moreover, $\overline{A \cup B} = \overline{A} \cup \overline{B}$.

Proof The first part of the proposition is trivial. To verify the second part we note that $\overline{A} \subset \overline{A \cup B}$ and $\overline{B} \subset \overline{A \cup B}$ because $A \subset A \cup B$ and $B \subset A \cup B$. Thus $\overline{A} \cup \overline{B} \subset \overline{A \cup B}$. Suppose that $x \notin \overline{A} \cup \overline{B}$. Then there is a $\delta_1 > 0$ such that there exists no $y \in A$ with $|x - y| < \delta_1$, and there is a $\delta_2 > 0$ such that there exists no $y \in B$ with $|x - y| < \delta_2$. For $\delta = \min\{\delta_1, \delta_2\}$, there exists no $y \in A \cup B$ with $|x - y| < \delta$. This means that $x \notin \overline{A \cup B}$, and we have $\overline{A \cup B} \subset \overline{A} \cup \overline{B}$. □

Definition A set F of real numbers is said to be *closed* if $F = \overline{F}$.

Proposition 6 For any set E of real numbers the set $\overline{E}$ is closed; that is, $\overline{\overline{E}} = \overline{E}$.

Proof Let x be a point of closure of $\overline{E}$. Then, given $\delta > 0$, there is a point $y \in \overline{E}$ with $|x - y| < \delta/2$. Since $y \in \overline{E}$, there is a $z \in E$ with $|y - z| < \delta/2$. Thus $|x - z| < \delta$, and we see that x is a point of closure of E; hence $\overline{\overline{E}} \subset E$. The inclusion $E \subset \overline{\overline{E}}$ follows from $E \subset \overline{E}$ by Proposition 5. □

Proposition 7 If F_1 and F_2 are closed sets, then $F_1 \cup F_2$ is a closed set.

Proof By Proposition 5, $\overline{F_1 \cup F_2} = \overline{F}_1 \cup \overline{F}_2 = F_1 \cup F_2$. □

Proposition 8 The finite union of closed sets is closed.

Proof This follows by induction from Proposition 7. □

Corollary The intersection of any collection of closed sets is closed.

Proof Let x be a point of closure of $\bigcap\{F: F \in \mathcal{F}\}$. Then, given $\delta > 0$, there is a $y \in \bigcap\{F: F \in \mathcal{F}\}$ such that $|x - y| < \delta$. Since such a y belongs to each $F \in \mathcal{F}$, we see that x is a point of closure of each $F \in \mathcal{F}$. Since each $F \in \mathcal{F}$ is closed, we have $x \in F$ for each $F \in \mathcal{F}$. This means that $x \in \bigcap\{F: F \in \mathcal{F}\}$. □

Definition Let E be a set of real numbers. The *complement of* E (denoted by E^c) is the set of all real numbers x such that $x \notin E$. On occasion we denote the complement of E by the symbol $R^1 - E$. If A and B are subsets of R^1 we define the *difference* B - A, or *relative complement of* A *in* B, as the set of elements in B which are not in A.

Proposition 9 The complement of an open set is closed and the complement of a closed set is open.

Proof Let G be an open set. We show that $R^1 - G$ is a closed set. If $x \in G$, there is a $\delta > 0$ such that, if $|x - y| < \delta$, then $y \in G$. Hence x cannot be a point of closure of $R^1 - G$, since there is no $y \in R^1 - G$ with $|x - y| < \delta$. Thus $R^1 - G$ contains all of its points of closure and is therefore closed.

On the other hand, let F be a closed set and $x \in R^1 - F$. Then, since x is not a point of closure of F, there is a $\delta > 0$ such that there exists no $y \in F$ with $|x - y| < \delta$. Thus, if $|x - y| < \delta$, then $y \in R^1 - F$ and so $R^1 - F$ is an open set. □

Remark Two special laws relating complements to unions and intersections are *De Morgan's laws:* If A and B are subsets of R^1, then

$$(A \cup B)^c = A^c \cap B^c \quad \text{and} \quad (A \cap B)^c = A^c \cup B^c.$$

De Morgan's laws hold also for arbitrary unions and intersections; if $\{E_t: t \in T\}$ is any collection of sets of real numbers, then

$$\Big(\bigcup_{t\in T} E_t\Big)^c = \bigcap_{t\in T} (E_t^c) \quad \text{and} \quad \Big(\bigcap_{t\in T} E_t\Big)^c = \bigcup_{t\in T} (E_t^c).$$

We shall quickly prove one of these equations and leave the verification of the other equation to the reader. Let

$$A = \Big(\bigcup_{t\in T} E_t\Big)^c \quad \text{and} \quad B = \bigcap_{t\in T} (E_t^c);$$

we want to show that A = B.

If $x \in A$, then $x \notin \cup_{t\in T} E_t$, hence $x \notin E_t$ for any $t \in T$, hence $x \in E_t^c$ for every $t \in T$, so that $x \in \cap_{t\in T} (E_t^c)$. Thus $A \subset B$. Conversely, if $x \in B$, then $x \in E_t^c$ for every $t \in T$, hence $x \notin E_t$ for any $t \in T$, hence $x \notin \cup_{t\in T} E_t$, so that $x \in (\cup_{t\in T} E_t)^c$. Thus $B \subset A$. It follows that A = B.

Definition A collection $\mathcal{G}$ of sets is said to *cover a set* E if $E \subset \bigcup\{G: G \in \mathcal{G}\}$. In this case the collection $\mathcal{G}$ is called a *covering of* E. If each $G \in \mathcal{G}$ is an open set, we call $\mathcal{G}$ an *open covering of* E.

If $\mathcal{G}$ contains only a finite number of sets, we call $\mathcal{G}$ a *finite covering of* E.

Proposition 10 (Heine-Borel Theorem) Let E be a closed and bounded set of real numbers. Then each open covering of E has a finite subcovering. That is, if $\mathcal{G}$ is a collection of open sets such that $E \subset \bigcup \{G: G \in \mathcal{G}\}$, then there is a finite collection $\{G_1, G_2, \ldots, G_n\}$ of sets in $\mathcal{G}$ such that

$$E \subset \bigcup_{k=1}^{n} G_k.$$

Proof We first consider the case where E is the closed interval $[a,b]$ with $-\infty < a < b < \infty$. Let X be the set of real numbers $x \le b$ with the property that the interval $[a,x]$ can be covered by a finite number of the sets of $\mathcal{G}$. The set X is certainly nonempty and is bounded above by b; thus X has a supremum, say c. Since $c \in [a,b]$, there is a $G \in \mathcal{G}$ which contains c. Since G is open, there is a $\delta > 0$ such that the open interval $(c - \delta, c + \delta)$ is contained in G. But $c - \delta$ is not an upper bound for X, and so there must be an $x \in X$ with $x > c - \delta$. Since $x \in X$, there is a finite collection $\{G_1, G_2, \ldots, G_j\}$ of sets in $\mathcal{G}$ which covers $[a, c + \delta)$. Thus each point of $[c, c + \delta)$ would be in X if it were less than or equal to b. Since no point of $[c, c + \delta)$ except c can belong to X, we must have $c = b$ and $b \in X$, proving our special case.

Now let E be any closed and bounded set of real numbers and $\mathcal{G}$ an open covering of E. Since E is bounded, it is contained in some closed, bounded interval $[a,b]$. Let $\mathcal{G}^*$ be the collection obtained by adjoining $R^1 - E$, that is, E^c, to $\mathcal{G}$. Thus

$$\mathcal{G}^* = \mathcal{G} \cup \{E^c\}.$$

Since E is closed, E^c is open, and $\mathcal{G}^*$ is a collection of open sets. By assumption $E \subset \bigcup \{G: G \in \mathcal{G}\}$, and so

$$R^1 = E \cup E^c \subset E^c \cup \bigcup \{G: G \in \mathcal{G}\} = \bigcup \{G: G \in \mathcal{G}^*\}.$$

Thus $\mathcal{G}^*$ is an open covering of R^1 and therefore of $[a,b]$. By our previous case there is a finite subcollection of $\mathcal{G}^*$ which covers $[a,b]$ and hence E. If this finite subcollection does not contain E^c, it is a subcollection of $\mathcal{G}$ and the claim of our proposition holds. If the subcollection contains E^c, we shall denote it by $\{G_1, G_2, \ldots, G_n, E^c\}$. Then $E \subset E^c \cup G_1 \cup G_2 \cup \ldots \cup G_n$. Since no point of E is contained in E^c, we have

$$E \subset G_1 \cup G_2 \cup \ldots \cup G_n$$

and the collection $\{G_1, G_2, \ldots, G_n\}$ is a finite subcollection of $\mathcal{G}$ which covers E. □

Definition A set E is said to be *compact* if every open covering $\mathcal{G}$ of E has a finite subcovering.

By Proposition 10, any closed and bounded set of real numbers is compact.

Definition Let $\varepsilon > 0$. By a *neighborhood* $V(x)$ of a point x we mean the open interval $(x - \varepsilon, x + \varepsilon)$.

A set G is seen to be open if and only if for each point $x \in G$, some neighborhood $V(x)$ is contained in G.

Definition A point x is said to be a *limit point of a set* E if every neighborhood of x contains a point $y \neq x$ such that $y \in E$. If $z \in E$ and z is not a limit point of E, then z is called an *isolated point of* E.

A point x is a limit point of a set E if and only if it is a point of closure of $E - \{x\}$, where $\{x\}$ denotes the set containing the single element x. A set F is closed if and only if every limit point of F is a point of F. The closure $\bar{E}$ of E consists of all points of E together with all limit points of E. If E is the set of values of a sequence $(x_n)_{n=1}^{\infty}$, then an accumulation point of $(x_n)_{n=1}^{\infty}$ need not be a limit point of the set E.

Proposition 11 A point x is a limit point of E if and only if there is a sequence $(x_n)_{n=1}^{\infty}$ of distinct points in E convergent to x.

Proof Let x be a limit point of E. In the interval (x - 1, x + 1) we pick a point $x_1 \in E$ with $x_1 \neq x$. Then we pick in the interval $(x - \frac{1}{2}, x + \frac{1}{2})$ a point $x_2 \in E$ such that $x_2 \neq x_1$ and $x_2 \neq x$. At the n-th step of this process of selection we pick a point $x_n \in E$ from the interval (x - 1/n, x + 1/n) which is different from the points x, $x_1, x_2, \ldots, x_{n-1}$. In this way we get the desired sequence $(x_n)_{n=1}^{\infty}$ of distinct points in E that converges to x.

The implication in the other direction is obvious. □

Proposition 12 If S is an infinite subset of a compact set E, then S has a limit point in E.

Proof If no point of E were a limit point of S, then each $y \in E$ would have a neighborhood V(x) which contains at most one point of S (namely, y, if $y \in S$). Hence no finite subcollection of $\{V(y): y \in E\}$ can cover S; and the same is true of E, since $S \subset E$. This contradicts the compactness of E. □

Proposition 13 If x is a limit point of a set E of real numbers, then every neighborhood of x contains infinitely many points of E.

Proof Suppose there is a neighborhood V(x) of x which contains only a finite number of points of E. Let $y_1, \ldots, y_n$ be those points of $V(x) \cap E$, which are distinct from x, and put

$$r = \min\{|x - y_m|: 1 \le m \le n\}.$$

Then $r > 0$ and (x - r, x + r) contains no point y of E such that $y \neq x$. Thus x is not a limit point of E which is a contradiction. This proves the proposition. □

Corollary A finite point set has no limit points.

Proposition 14 If every infinite subset of E has a limit point in E, then E is closed and bounded.

Proof If E is not bounded, then E contains points x_n such that $|x_n| > n$ for $n = 1, 2, \ldots$. The set S consisting of these points x_n is infinite and clearly has no limit point in R^1 (the set of all finite real numbers), hence none in E. Thus E has to be bounded.

If E is not closed, then there is a point $x_0 \in R^1$ which is a limit point of E but not a point of E. For $n = 1,2, \ldots,$ there are points $x_n \in E$ such that $|x_n - x_0| < 1/n$. Let S be the set of these points x_n. Then S is infinite (otherwise $|x_n - x_0|$ would have a constant positive value, for infinitely many n). S has x_0 as a limit point, and S has no other limit point in R^1. For if $y \in R^1$, $y \neq x_0$, then

$$|x_n - y| \geq |x_0 - y| - |x_n - x_0| \geq |x_0 - y| - \frac{1}{n} \geq \frac{1}{2}|x_0 - y|$$

for all but finitely many n; this shows that y is not a limit point of S (by Proposition 13). Thus S has no limit point in E. This shows that E has to be closed whenever every infinite subset of E has a limit point in E. □

Summing up the content of Propositions 10, 12, and 14, we have:

Proposition 15 Let E be a set of real numbers. Then the following three statements are equivalent:

(i) E is closed and bounded.

(ii) E is compact.

(iii) Every infinite subset of E has a limit point in E.

Proposition 16 (Bolzano-Weierstrass Theorem) Every bounded infinite subset of R^1 has a limit point.

Proof Being bounded, the set E is a subset of a closed

bounded interval [a,b]. But [a,b] is compact and so E has a limit point in [a,b], by Proposition 12. □

Proposition 17 (Cantor's Intersection Theorem) Let $(S_n)_{n=1}^{\infty}$ be a sequence of nonempty compact sets in R^1 such that $S_{n+1} \subset S_n$ for each $n \in N$, that is, the sequence of sets is assumed to be nested. Then the intersection of all the S_n's is not empty.

Proof We form a sequence $(x_n)_{n=1}^{\infty}$ so that $x_n \in S_n$ for each $n \in N$. Then all of the points x_n are in S_1, which is a bounded set. Hence the sequence $(x_n)_{n=1}^{\infty}$ is bounded. It therefore contains a convergent subsequence with some limit x (by Proposition 56 of Chapter 1). We assert that $x \in S_n$ for each $n \in N$. Indeed, if the convergent subsequence is $x_{n_1}, x_{n_2}, \ldots$, and if n is any positive integer, we certainly have $n_j > n$ if j is large enough. But then, for such sufficiently large values of j, we have $x_{n_j} \in S_{n_j} \subset S_n$ because the sets S_n are nested, that is, $S_{n+1} \subset S_n$ for every $n \in N$. But S_n is a closed set. Hence it contains the limit x of the convergent subsequence because all but a finite number of terms of the subsequence belong to S_n. We have used here the fact that if the terms $(y_n)_{n=1}^{\infty}$ of a convergent sequence are in a set Y, then the limit y of the sequence belongs to $\overline{Y}$, that is, y is a point of closure of Y. □

Definition We say that a set is *countable* if it is either finite or countably infinite; in other words, a set is countable if it can be put in one-to-one correspondence with a subset of N (the set of all natural numbers). A set which is not countable is called an *uncountable set;* in other words, an uncountable set is an infinite set which cannot be put in one-to-one correspondence with the set N.

The empty set, any finite set and any countably infinite set are countable; the set of all real numbers, the set of all points of an interval are examples of uncountable sets.

Definition A point x is called a *point of condensation of a set* E of real numbers if every neighborhood of x contains uncountably many points of E.

Proposition 18 Every uncountable set E of real numbers has at least one point of condensation in E.

Proof Suppose by way of contradiction that no point of E is a condensation point of E. Then for each $x \in E$, there is an open interval I_x, containing x, such that $I_x \cap E$ is countable. Let J_x be an open interval contained in I_x, and containing x, but having rational endpoints; indeed $J_x \cap E$ is again countable. Moreover, the collection of all such intervals J_x is countable, by the lemma preceding Proposition 4, and we may enumerate it as follows

$$J_1, J_2, \ldots, J_n, \ldots.$$

Since each point of E is in some J_k, we may conclude that

$$E = \bigcup_{k=1}^{\infty} (J_k \cap E).$$

The countable union of countable sets being countable by the lemma preceding Proposition 4, we have the desired contradiction. □

Remark Proposition 18 is analogous to Proposition 16, but stronger in the sense that the condition of boundedness is not required. Moreover, Proposition 16 assures only the existence of a limit point; Proposition 18 assures that some point of condensation is actually contained in the set.

Definition The set of all points of condensation of a set E is called the *condensed set of* E; we will denote the condensed set of E by E_c.

Proposition 19 The set $E - E_c$ is countable.

Proof Let $x \in E - E_c$. Then x is not a point of condensation of E; as such, there exists a neighborhood of x which contains only countably many points of E and, even more so, countably many points of $E - E_c$. Thus, no point of $E - E_c$ is a point of condensation of $E - E_c$. Thus $E - E_c$ must be countable by the foregoing proposition. □

Definition A set E of real numbers is called *perfect* if E is closed and every point of E is a limit point of E.

Proposition 20 For any set E of real numbers, the set E_c is perfect.

Proof If E_c is empty, then it is perfect. If E_c is not empty, then E is uncountable. Let $x \in E_c$, and let I_x be any neighborhood of x. Then I_x contains uncountably many points of E. Thus, I_x contains uncountably many points of E_c, for otherwise I_x would contain uncountably many points of $E - E_c$. This shows that x is a point of condensation of E_c and, even more so, a limit point of E_c.

Now let y be a point of closure of E_c. Then any neighborhood I_y of y contains at least one point t ($\neq y$) of E_c. But I_y is also a neighborhood of t and must contain uncountably many points of E. This implies that $y \in E_c$. Hence E_c is closed.

Thus E_c is perfect. □

Proposition 21 (Theorem of Cantor and Bendixsohn) Every closed set of real numbers is the union of a perfect set and a countable set.

Proof Let F be a closed set and let F_c be the condensed set of F. By the foregoing proposition F_c is perfect. Moreover, by Proposition 19, $F - F_c$ is countable. Let F' denote the set of all limit points of F. Since F is closed, $F_c \subset F' \subset F$. Thus

$$F = F_c \cup (F - F_c).$$ □

<u>*Remark*</u> Let E be any set of real numbers and let E' denote the set of all limit points of E. The set E' is called the *derived set of* E. To show that E' is closed, we must verify that if p is a limit point of E', then p is also a limit point of E, that is, $p \in E'$. We shall do this now. Let G_p be an open set containing p. Since p is a limit point of E', G_p contains at least one point $q \in E'$ different from p. But G_p is an open set containing $q \in E'$; hence G_p contains (infinitely many) points of E. So, there exists an $x \in E$ such that $x \neq p$, $x \neq q$, and $x \in G_p$. Thus, each open set containing p contains points of E other than p; so $p \in E'$.

Having noted that E' is closed for any set E of real numbers, we therefore see that E is perfect if and only if E = E', that is, if the set is equal to its derived set.

<u>*Examples of perfect sets*</u> Any closed, bounded interval [a,b] and the Cantor (ternary) set are perfect sets. Cantor's (ternary) set was introduced as Application 3 of Proposition 80 in Chapter 1.

To see that Cantor's (ternary) set is closed we only have to note that it is the complement of the union of the removed open intervals and the open intervals $(-\infty, 0)$ and $(1, +\infty)$. To verify that every point of the Cantor set is a limit point of the Cantor set, we proceed as follows. Let x_0 be a point of the Cantor set. If we write x_0 in ternary expansion, say

$$x_0 = 0.a_1a_2a_3 \ldots a_n \ldots,$$

then $a_n = 0$ or 2. We construct a sequence of points in the following manner:

$$x_1 = 0.a_1'a_2a_3a_4 \ldots a_n \ldots$$

$$x_2 = 0.a_1a_2'a_3a_4 \ldots a_n \ldots$$

$$x_3 = 0.a_1a_2a_3'a_4 \ldots a_n \ldots$$

$$\vdots$$

$$x_n = 0.a_1a_2a_3a_4 \ldots a_n' \ldots$$

$$\vdots$$

where $a_n' = 0$ if $a_n = 2$ and $a_n' = 2$ if $a_n = 0$. In this way we get a sequence of distinct points $(x_n)_{n=1}^{\infty}$ all belonging to the Cantor set such that x_n differs from x_0 in the n-th place in the ternary expansion. But $x_n \to x_0$ as $n \to \infty$ and so x_0 is seen to be a limit point of the Cantor set.

We also note an obvious one-to-one correspondence between the Cantor set and the set of all points of the interval [0,1]. Let $x \in [0,1]$ and express x as p-adic fraction to the base p = 2, that is, let

$$x = 0.b_1 b_2 b_3 \ldots b_n \ldots$$

with $b_n = 0$ or 1. Make this x correspond to that point y in the Cantor set which is of the form

$$y = 0.a_1 a_2 a_3 \ldots a_n \ldots$$

with $a_n = 2$ when $b_n = 1$ and $a_n = 0$ when $b_n = 0$. In other words, this one-to-one correspondence $x \leftrightarrow y$ merely interchanges the digits 1 and 2.

Another interesting property of the Cantor set C is: Every subinterval [a,b] of [0,1] contains an interval (a',b') free of points of C such that its length satisfies the inequality

$$b' - a' \geq \frac{1}{5}(b - a).$$

Indeed. If [a,b] is free of points of C, then the claim is trivial. If not, then let V be one of those closed intervals of length 3^{-n} that remained at the n-th step of the removal of the open mid-third such that $V \subset [a,b]$ but with n being as small as possible. If V = [a,b], then the claim is true, for at the (n + 1)-th step we remove the open mid-third of V. In the other case, that is, in the case when V is proper subset of [a,b], we adjoin to V (possibly on the left and on the right) intervals of the same length as V which are free of points of C. If one of these (two) intervals has an interval of length $\geq (1/2)(b - a)$ in common with (a,b), the claim is fulfilled. If this, however, is not the case, then the length of V is $\geq (3/5)(b - a)$. Since the mid-third (a',b') of V is free of points of C and $b' - a' \geq (1/5)(b - a)$ holds, the proof is complete.

Definition A set E of real numbers is said to be *connected* if there do not exist two disjoint open sets A and B such that A intersects E, B intersects E, and $E \subset A \cup B$.

Proposition 22 A subset E of R^1 is connected if and only if E has the following property:

If $x \in E$, $y \in E$, and $x < z < y$, then $z \in E$.

Proof Suppose that the property fails for some numbers x, y, z, that is, $x \in E$, $y \in E$, $x < z < y$, but $z \notin E$. If A is the set of all $\alpha < z$ and B is the set of all $\beta > z$, then E is not connected.

Conversely, suppose that E is not connected. Then there exist points $x \in E$, $y \in E$, with $x < y$, and there exist disjoint open sets A, B in R^1 such that $x \in A$, $y \in B$, $E \subset A \cup B$. Let

$$S = A \cap [x,y],$$

and let $z = \sup S$.

Since $y \in B$ and B is open, we have $z < y$. Thus if we had $z \in A$, the fact that A is open would show that z is not an upper bound of S. Hence $z \notin A$.

Since $x \in A$ and A is open, we have $x < z$. Thus if we had $z \in B$, the fact that B is open would show that z is not the least upper bound of S. Hence $z \notin B$.

Since $E \subset A \cup B$, it follows that $z \notin E$, which completes the proof. □

Corollary 1 Let $a \leq b$ with $a, b \in R^1$. A set E is connected if and only if it is one of the following intervals: $(-\infty,b)$, $(-\infty,b]$, (a,∞), $[a,\infty)$, $(-\infty,\infty)$, (a,b), $[a,b)$, $(a,b]$, $[a,b]$.

Corollary 2 The only subsets of R^1 which are both open and closed are the empty set and R^1 itself.

Proof If S is a subset of R^1 which is both open and closed, its complement S^c is also both open and closed. If we then assume

$S \neq \emptyset$ and $S \neq R^1$, it follows that S and S^c are separated sets with $R^1 = S \cup S^c$, in contradiction to the fact that R^1 is connected. □

Proposition 23 Let E be closed and bounded above and suppose that $y = \sup E$. Then $y \in E$.

Proof Suppose $y \notin E$. For every $h > 0$, there is a point $x \in E$ such that $y - h \le x \le y$ for otherwise $y - h$ would be an upper bound of E. Thus every neighborhood of y contains points x of E, and $x \neq y$, since $y \notin E$. It follows that y is a limit point of E which is not a point of E, so that E is not closed. But this contradicts the hypothesis. □

Remark It is easy to see that a closed bounded set of real numbers contains both its supremum and infimum.

2. Continuous Functions

Let x and y be objects. The *ordered pair* (x,y) is the set $\{\{x\}, \{x, y\}\}$ (that is, a set of two elements, of which one is the set whose only element is x, and the other is the set of two elements x and y).

Let X, Y be sets. Then $X \times Y = \{(x,y): x \in X \text{ and } y \in Y\}$ is called the *Cartesian product* of X and Y. R is called a *relation from* X *to* Y if R is a subset of $X \times Y$. A *function* mapping a set X into a set Y is by definition a relation from X to Y with the property that, for any element x of X, there is one and only one element y of Y such that (x,y) is a pair of the relation; if Y is a set of (finite) real numbers and f is a function mapping X into Y, we say that f is a *real-valued function*. If $E \subset X$, $f(E)$ denotes the set of all elements $f(x)$ for $x \in E$. We call $f(E)$ the *image of* E *under* f. In this notation, $f(X)$ is the *range of* f; the set X is called the *domain of definition of* f. It is clear that $f(X) \subset Y$. If $f(X) = Y$, we say that f maps X *onto* Y. If $S \subset Y$, $f^{-1}(S)$ denotes the set of all $x \in X$ such that $f(x) \in S$. We call $f^{-1}(S)$ the *inverse image of* S *under* f. If $y \in Y$, $f^{-1}(y)$ is the set of all $x \in X$ such that $f(x) = y$. If for each $y \in Y$, $f^{-1}(y)$ consists of at most one element of X, then f is called a *one-to-one mapping of* X into Y, that is, $f(x_1) \neq f(x_2)$ whenever $x_1 \neq x_2$, $x_1 \in X$, $x_2 \in X$. A one-to-one mapping of X onto Y is called a *one-to-one correspondence*. In this case there is a function f^{-1} defined on Y by $f^{-1}[f(x)] = x$ for all $x \in X$; f^{-1} is called the *inverse function of* f and f is called *invertible*.

Definition Let a and L be in R^1 and suppose that f is a real-valued function defined on some interval $(a - h, a + h)$ with $h > 0$ except possibly at the point a itself. We say that f(x) tends to L as x tends to a, and write

$$\lim_{x \to a} f(x) = L, \quad \text{or} \quad f(x) \to L \quad \text{as} \quad x \to a,$$

if, given any $\varepsilon > 0$, there exists $\delta > 0$, such that, for $0 < |x - a| < \delta$, we have $|f(x) - L| < \varepsilon$. More generally, if f is a real-valued function defined on a subset E of R^1 and a is a limit point of E, we say that f(x) *tends to* L *as* x *tends to* a if there is a point L in R^1 with the following property: For any $\varepsilon > 0$ there exists some $\delta > 0$ such that $|f(x) - L| < \varepsilon$ for all $x \in E$ for which $0 < |x - a| < \delta$. L is said to be the *limit of* f(x) *as* x *tends to* a.

Examples

1. From elementary calculus we know that

$$\lim_{x \to 0} \frac{\sin x}{x} = 1$$

but $(\sin x)/x$ is not defined for $x = 0$.

2. $\lim_{x \to 0} \sin \frac{1}{x}$ does not exist.

Indeed. Assume there exists $L \in R^1$ such that

$$\lim_{x \to 0} \sin \frac{1}{x} = L.$$

Then for $\varepsilon = 1$ there would exist $\delta > 0$ such that, for $0 < |x| < \delta$,

$$\left|\sin \frac{1}{x} - L\right| < 1. \tag{1}$$

But $\sin (1/x) = 1$ for $x = 2/\pi(4n + 1)$ with $n \in N$; since $2/\pi(4n + 1)$ tends to 0 as $n \to \infty$, we see that $\sin (1/x) = 1$ for some $x \in (0,\delta)$. For this x (1) implies $|1 - L| < 1$. Similarly, $\sin (1/x) = -1$ for $x = (2n\pi + 3\pi/2)^{-1}$ with $n \in N$. There will thus exist an $x \in (0,\delta)$ for which $\sin (1/x) = -1$. By (1) again, $|-1 - L| < 1$. But this is a contradiction, for $|1 - L| < 1$ means that $L \in (0,2)$ and $|-1 - L| < 1$ means that $L \in (-2,0)$; the open intervals $(0,2)$ and $(-2,0)$ are disjoint and thus cannot have a point L in common.

Proposition 24 If $\lim_{x \to a} f(x) = L$ and $L > A$, then for all $x \in E$ sufficiently close to a we have $f(x) > A$.

Proof Pick $\varepsilon < L - A$; then $L - \varepsilon > A$. By the definition of limit for this ε there is a δ such that for all $x \in E$ with $0 < |x - a| < \delta$,

$$L - \varepsilon < f(x) < L + \varepsilon;$$

for these x we clearly have $f(x) > A$. □

Remark An analogous result can be obtained when we take $L < A$.

Proposition 25 From the definition of limit of a function we obtain

$$\lim_{x \to a} f(x) = L$$

if and only if

$$\lim_{n \to \infty} f(x_n) = L$$

for every sequence $(x_n)_{n=1}^{\infty}$ in E such that $x_n \neq a$, $\lim_{n \to \infty} x_n = a$.

Proof Suppose $\lim_{x \to a} f(x) = L$. We choose a sequence $(x_n)_{n=1}^{\infty}$ in the set E such that $x_n \neq a$ and $x_n \to a$ as $n \to \infty$. Let $\varepsilon > 0$ be given. Then there is a $\delta > 0$ such that $|f(x) - L| < \varepsilon$ whenever x belongs to E and $0 < |x - a| < \delta$. Also, there exists an n_0 such that $n \geq n_0$ implies $0 < |x_n - a| < \delta$. Thus, for all $n \geq n_0$, we have $|f(x_n) - L| < \varepsilon$ and we see that $\lim_{n \to \infty} f(x_n) = L$ holds.

Conversely, suppose $\lim_{x \to a} f(x) = L$ fails. Then there exists some $\varepsilon > 0$ such that for every $\delta > 0$ there is a point x in the set E (depending on δ), for which $|f(x) - L| \geq \varepsilon$ but $0 < |x - a| < \delta$. Taking $\delta_n = 1/n$ with $n \in N$, we thus find a sequence in the set E satisfying $x_n \neq a$, $x_n \to a$ as $n \to \infty$, for which $\lim_{n \to \infty} x_n = L$ is false. □

Corollary 1 If the limit of f(x) as x tends to a exists, then it is unique.

Corollary 2 If $\lim_{x \to a} f(x) = A$ and $\lim_{x \to a} g(x) = B$, then

(i) $\lim_{x \to a} (f(x) + g(x)) = A + B$,

(ii) $\lim_{x \to a} (f(x)\, g(x)) = AB$,

(iii) $\lim_{x \to a} \dfrac{f(x)}{g(x)} = \dfrac{A}{B}$, if $B \neq 0$.

Remark The preceding two corollaries are consequence of the foregoing Proposition 25 and Propositions 51 and 53 of Chapter 1.

Definition Let f be a real-valued function whose domain of definition is a set E of real numbers. We say that f is *continuous at a point* x of E if, given $\varepsilon > 0$, there exists a $\delta > 0$ such that for all y in E with $|x - y| < \delta$ we have $|f(x) - f(y)| < \varepsilon$ [in other words, if for any $\varepsilon > 0$ there is a neighborhood V(x) of x such that $f(x) - \varepsilon < f(y) < f(x) + \varepsilon$ for any y in $E \cap V(x)$]. The function f is said to be *continuous on a subset* A of E if it is continuous at each point of A. If we say that f is *continuous* we mean that f is continuous on its domain of definition.

It should be noted that f has to be defined at the point x in order to be continuous at x. If x is an *isolated point of* E, that is, if x is in E and x is not a limit point of E, then our definition of continuity at x implies that every function f with domain of definition E is continuous at x. For, no matter which $\varepsilon > 0$ we pick, we can find $\delta > 0$ so that the only point $y \in E$ for which $|x - y| < \delta$ is $y = x$; then $|f(x) - f(y)| = 0 < \varepsilon$.

The negation of continuity reads as follows: A function f is *discontinuous at a point* x of its domain of definition E if there is a positive number ε having the property that corresponding to an arbitrary positive number δ (however small), there is a point y in E such that $|x - y| < \delta$ and $|f(x) - f(y)| \geq \varepsilon$. [Observe that the negation of the statement "for any $\varepsilon > 0$ and some $\delta > 0$ we have $|f(x) - f(y)| < \varepsilon$ for any y in E such that $|x - y| < \delta$" is the statement "for some $\varepsilon > 0$ and any $\delta > 0$ we have $|f(x) - f(y)| \geq \varepsilon$ for some y in E such that $|x - y| < \delta$."]

Proposition 26 Let f be a real-valued function defined on a set E of real numbers and suppose that x is a limit point of E. Then f is continuous at x if and only if $\lim_{t \to x} f(t) = f(x)$.

Proof The proposition is a direct consequence of the definitions of continuity and limit of a function. □

Proposition 27 Let f and g be continuous functions with the same domain of definition E. Then the functions

(i) $(f + g)(x) = f(x) + g(x)$,

(ii) $(fg)(x) = f(x)\, g(x)$,

(iii) $(f/g)(x) = f(x)/g(x)$ [if $g(x) \neq 0$ for all $x \in E$],

(iv) $|f|(x) = |f(x)|$

are continuous on E.

Proof At isolated points of E there is nothing to prove; at limit points, the statements (i), (ii), and (iii) follow from Proposition 25. In order to verify statement (iv), we observe that for any real numbers A and B the inequality $||A| - |B|| \leq |A - B|$ holds; thus the continuity of f implies the continuity of $|f|$ because

$$||f(x)| - |f(y)|| \leq |f(x) - f(y)|.$$ □

Remark If f and g are as in Proposition 27, then the functions

$$(f \vee g)(x) = \max\{f(x), g(x)\} \quad \text{and} \quad (f \wedge g)(x) = \min\{f(x), g(x)\}$$

are continuous on E as well. Indeed,

$$(f \vee g)(x) = \frac{f(x) + g(x) + |f(x) - g(x)|}{2}$$

and

$$(f \wedge g)(x) = \frac{f(x) + g(x) - |f(x) - g(x)|}{2};$$

Thus the claim follows from Proposition 27. □

Proposition 28 Let X, Y, Z be sets of real numbers. Suppose that the function f maps X into Y, the function g maps the range of f, f(X), into Z, and h is the function mapping X into Z defined by

$$h(x) = g[f(x)]$$

with $x \in X$. If f is continuous at a point $a \in X$ and if g is continuous at a point f(a), then h is continuous at a.

Proof Let $\varepsilon > 0$ be given. Since g is continuous at f(a), there is $\gamma > 0$ such that $|g(y) - g[f(a)]| < \varepsilon$ if $|y - f(a)| < \gamma$ and $y \in f(X)$. Since f is continuous at a, there is a $\delta > 0$ such that $|f(x) - f(a)| < \gamma$ if $|x - a| < \delta$ and $x \in X$. It follows that

$$|h(x) - h(a)| = |g[f(x)] - g[f(a)]| < \varepsilon$$

if $|x - a| < \delta$ and $x \in X$. Thus h is continuous at a. □

Proposition 29 Let f be a continuous real-valued function on a compact set E of real numbers. Then the range of f, that is, f(E), is compact and f assumes on E its largest and its smallest value.

Proof We first show that f is a bounded function on E, that is, f(E) is a bounded set of real numbers. Since f is continuous on E, for each $x \in E$ there is an open interval I_x containing x such that $|f(y) - f(x)| < 1$ for $y \in I_x \cap E$; this follows from the continuity of f at x and having chosen $\varepsilon = 1$. Thus, for $y \in I_x \cap E$, we have $|f(y)| \leq |f(x)| + 1$ and so f is bounded on I_x. The collection of sets $\{I_x\colon x \in E\}$ is a collection of open intervals that covers E; since E is compact, there is a finite subcollection $\{I_{x_1}, \ldots, I_{x_n}\}$ which covers E. Let $M = 1 + \max\{|f(x_1)|, \ldots, |f(x_n)|\}$. Then each y in E belongs to some interval I_{x_k} with $1 \leq k \leq n$, and hence $|f(y)| < 1 + |f(x_k)| \leq M$. This shows that f is bounded on E by M.

Next we show that the range f(E) is closed. Suppose that f(x) is a limit point of the range of f. Then there exists a sequence $x_n \in E$ such that $f(x_n) \to L$. The bounded sequence $(x_n)_{n=1}^{\infty}$ contains a convergent subsequence $(x_{n_j})_{j=1}^{\infty}$ whose limit t belongs to E (since E is compact and therefore closed). Since $f(x_{n_j}) \to L$, the continuity of f at t implies that $f(t) = L$.

Having seen that f(E) is both bounded and closed, we infer that it is compact.

To see that f assumes its largest value on E, let $m^* = \sup f(E)$. Since f is bounded, m^* is finite; our purpose is to show that there is an $x_1 \in E$ such that $f(x_1) = m^*$. Suppose not. Then $f(x) < m^*$ for each $x \in E$, and by continuity there is an open interval I_x containing x such that $f(y) < \frac{1}{2}[f(x) + m^*]$ for all $y \in I_x \cap E$. But E is compact and so there exists a finite number of these intervals $\{I_{x_1}, \ldots, I_{x_n}\}$ which already cover E. We let $a = \max\{f(x_1), \ldots, f(x_n)\}$. But each $y \in E$ belongs to some interval I_{x_k} with $1 \leq k \leq n$ and

$$f(y) < \frac{1}{2}[f(x_k) + m^*] \leq \tfrac{1}{2}(a + m^*).$$

Thus $\frac{1}{2}(a + m^*)$ is a bound for f on E. This is impossible, however, because $\frac{1}{2}(a + m^*) < m^*$. Thus, there must be an $x_1 \in E$ such that $f(x_1) = m^*$. Similarly, there exists an $x_2 \in E$ at which f assumes its smallest value. In this connection note that $\inf\{f(x): x \in E\} = -\sup\{-f(x): x \in E\}$ and since f is continuous on E, so is -f. □

Remark The last part of the foregoing proof can also be rendered as follows. Since f(E) is bounded,

$$m_* = \inf f(E) \quad \text{and} \quad m^* = \sup f(E)$$

exist; since f(E) is closed, m_* and m^* will belong to f(E) in view of the remark following Proposition 23.

Proposition 30 Let f be a real-valued function defined on R^1. Then f is continuous on R^1 if and only if for each open set G of R^1 the set $f^{-1}(G) = \{x \in R^1: f(x) \in G\}$ is open

Proof Let $f^{-1}(G)$ be open for each open set G and let x be any real number. Then, given $\varepsilon > 0$, the interval $I = (f(x) - \varepsilon, f(x) + \varepsilon)$ is an open set, and so its inverse image $f^{-1}(I)$ must be open. Since $x \in f^{-1}(I)$, there must be some $\delta > 0$ such that $(x - \delta, x + \delta) \subset f^{-1}(I)$. But this means that if $|y - x| < \delta$, then f(y) is in $(f(x) - \varepsilon, f(x) + \varepsilon)$; that is, $|f(x) - f(y)| < \varepsilon$. Hence f is continuous at x. Since x is arbitrary, f is continuous on R^1.

Conversely, let f be continuous on R^1 and G be an open set. Suppose that $x \in f^{-1}(G)$. Then $f(x) \in G$, and there exists an $\varepsilon > 0$ such that $(f(x) - \varepsilon, f(x) + \varepsilon) \subset G$. Since f is continuous at x, there is a $\delta > 0$ such that $|f(x) - f(y)| < \varepsilon$ for $|x - y| < \delta$. Hence, for every $y \in (x - \delta, x + \delta)$, we have $f(y) \in (f(x) - \varepsilon, f(x) + \varepsilon) \subset G$. Thus $(x - \delta, x + \delta) \subset f^{-1}(G)$, and so $f^{-1}(G)$ is open. □

Proposition 31 Let f be a continuous real-valued function on a closed bounded interval [a,b] and suppose that $f(a) \neq f(b)$. Then for every real number A between f(a) and f(b) there exists a point t such that $f(t) = A$.

Proof We suppose that $f(a) < f(b)$. Starting with the points $x_0 = a$ and $y_0 = b$, we construct by successive bisections a sequence of nested intervals $([x_n, y_n])_{n=1}^{\infty}$ such that $f(x_n) \le A \le f(y_n)$ for all $n \in N$. To this end we only have to set

$$x_{n+1} = \frac{x_n + y_n}{2} \quad \text{and} \quad y_{n+1} = y_n \quad \text{in case} \quad f\left(\frac{x_n + y_n}{2}\right) \le A$$

and

$$x_{n+1} = x_n \quad \text{and} \quad y_{n+1} = \frac{x_n + y_n}{2} \quad \text{in case} \quad f\left(\frac{x_n + y_n}{2}\right) > A.$$

Let t be the point determined by this sequence of nested intervals, that is, let t be the point common to all these nested intervals. Then $x_n \to t$ and $y_n \to t$. Since $f(x_n) \le A \le f(y_n)$ and f is continuous at t, we have that $f(t) = A$.

The case $f(b) < f(a)$ reduces to the case already discussed when we consider the function $-f$. □

Corollary Let f be as in Proposition 31. If $f(x) \ne B$ for all x in $[a,b]$, then either $f(x) > B$ for all $x \in [a,b]$ or $f(x) < B$ for all $x \in [a,b]$.

Remark The foregoing corollary shows that every polynomial of odd degree over the field of real numbers has at least one real root. Indeed. For $a_1, \ldots, a_n$ in R^1, let

$$f(x) = x^n + a_1 x^{n-1} + \ldots + a_n$$

and let

$$M = \max\left\{1, |a_1| + \cdots + |a_n|\right\}.$$

Then

$$f(t) > 0 \quad \text{for} \quad t > M$$

and

$$(-1)^n f(t) > 0 \quad \text{for} \quad t < -M.$$

Thus, if f has roots in R^1, then these roots will be in the interval $-M \le t \le M$.

We leave the details as an exercise for the reader.

Examples

1. Consider the function

$$f(x) = 0 \qquad \text{for irrational } x > 0,$$
$$= \frac{1}{p+q} \quad \text{for } x = \frac{p}{q}$$

with p and q being natural numbers such that the largest common divisor of p and q is 1, that is, p and q are *relatively prime*. It is clear that f(x) is defined for all $x > 0$. The function f is discontinuous for rational values t of the variable and is continuous for irrational t. Indeed, let $\varepsilon > 0$ be given. Then there are only a finite number of positive integers p, q such that $p + q \le 1/\varepsilon$. Thus there is a $\delta > 0$ such that for these finitely many positive integers p, q we have $|(p/q) - t| \ge \delta$. For all rational x with $|x - t| < \delta$ we therefore have

$$|f(x) - f(t)| = |f(x)| = \frac{1}{p+q} < \varepsilon.$$

For irrational x we have in any case that $|f(x) - f(t)| = 0 < \varepsilon$.

2. Let g and h be the functions defined by

$$g(x) = 1 \quad \text{for rational } x,$$
$$= -1 \quad \text{for irrational } x$$

and

$$h(x) = x \quad \text{for rational } x,$$
$$= -x \quad \text{for irrational } x.$$

Both g and h are defined on all of R^1; g is everywhere discontinuous and h is continuous only at $x = 0$.

Definition A real-valued function f defined on a set E of real numbers is said to be *uniformly continuous on* E if, given any $\varepsilon > 0$, there exists a $\delta > 0$ such that for all x and y in E with $|x - y| < \delta$ we have

$$|f(x) - f(y)| < \varepsilon.$$

The negation of uniform continuity reads as follows: A function defined on a set E fails to be uniformly continuous on E if there exists a positive real number ε having the property that for any positive real number δ there exist points x and y in E such that

$$|x - y| < \delta \quad \text{and} \quad |f(x) - f(y)| \geq \varepsilon.$$

Remark We emphasize the distinction between the two statements "f is continuous at each point of E" and "f is uniformly continuous on E." The The first statement is expressed by saying that if a point x in E and a positive number ε are given, there is some positive number δ (and by implication it may depend on both x and ε) such that $|f(x) - f(y)| < \varepsilon$ if y is in E and $|x - y| < \delta$. In the definition of uniform continuity no point of E is specified in advance. Instead, merely some positive ε is given; the corresponding δ then depends on ε, but not on any particular point of E. Moreover, to speak of a given function as being uniformly continuous at a certain point is meaningless.

Proposition 32 If a real-valued function f is continuous on a compact set E of real numbers, then it is uniformly continuous on E.

Proof Given any $\varepsilon > 0$ and x in E, there exists a $\delta_x > 0$ such that $|x - y| < \delta_x$ implies $|f(x) - f(y)| < \varepsilon/2$. Let G_x be the interval

$$\left(x - \frac{1}{2}\delta_x, x + \frac{1}{2}\delta_x\right).$$

Then $\{G_x : x \in E\}$ is an open covering of E. Since E is compact, there is a finite subcollection

$$\left\{G_{x_1}, \ldots, G_{x_n}\right\}$$

which covers E. Let

$$\delta = \frac{1}{2}\min\left\{\delta_{x_1}, \ldots, \delta_{x_n}\right\}.$$

Then $\delta > 0$. Given two points y and z in E such that $|y - z| < \delta$, the point y must belong to some G_{x_k} with $1 \le k \le n$ and hence there is a k such that

$$|y - x_k| < \frac{1}{2}\delta_{x_k}.$$

Thus

$$|z - x_k| \le |z - y| + |y - x_k| < \frac{1}{2}\delta_{x_k} + \delta \le \delta_{x_k}.$$

We therefore have

$$|f(y) - f(x_k)| < \frac{\varepsilon}{2} \quad \text{and} \quad |f(z) - f(x_k)| < \frac{\varepsilon}{2}.$$

But

$$|f(z) - f(y)| \le |f(z) - f(x_k)| + |f(x_k) - f(y)|$$

and so

$$|f(z) - f(y)| < \varepsilon,$$

showing that f is uniformly continuous on E. □

Definition Let f be a bounded real-valued function defined on an interval J; we define

$$\omega(f;J) = \sup f(J) - \inf f(J)$$

and call $\omega(f;J)$ the *oscillation of* f *over* J.

It is clear that $\omega(f;J) \ge 0$ and that $\omega(f;I) \le \omega(f;J)$ for any subinterval I of J.

Definition If [a,b] is a closed interval and $x_0, x_1, \ldots, x_n$ are finitely many points such that $a = x_0 < x_1 < \cdots < x_n = b$, then the points $x_0, x_1, \ldots, x_n$ are said to form a *partition of* the interval [a,b] into the subintervals $[a,x_1], \ldots, [x_{n-1},b]$.

Proposition 32 has the following obvious but very useful corollary.

Corollary Let f be a continuous real-valued function on a closed bounded interval [a,b]. Then for any $\varepsilon > 0$ there is a $\delta > 0$ with the following property: There exists a partition of [a,b] into a finite number of subintervals with each subinterval having length less than δ such that the oscillation of f on each subinterval is less than ε.

Proposition 33 If a real-valued function f is uniformly continuous on a set E of real numbers and if $(x_n)_{n=1}^{\infty}$ is any Cauchy sequence of elements in E, then $\{f(x_n)\}_{n=1}^{\infty}$ is also a Cauchy sequence. Conversely, if a real-valued function f, defined on a bounded set E of real numbers transforms Cauchy sequences of elements of E into Cauchy sequences, then f is uniformly continuous on E.

Proof Let f be uniformly continuous on E and let $(x_n)_{n=1}^{\infty}$ be a Cauchy sequence of elements in E. Given $\varepsilon > 0$, there exists $\delta > 0$ such that if x' and x" are in E and $|x' - x"| < \delta$ then

$$|f(x') - f(x")| < \varepsilon.$$

But, associated with the number $\delta > 0$, there exists an index n_0 such that if m, n $\geq n_0$ then $|x_m - x_n| < \delta$; therefore, $|f(x_n) - f(x_m)| < \varepsilon$, which shows that $\{f(x_n)\}_{n=1}^{\infty}$ is a Cauchy sequence.

On the other hand, assume that the function f is not uniformly continuous on a bounded set E. The negation of uniform continuity may be thus expressed: there exists an $\varepsilon > 0$ such that for any $\delta > 0$ there exist points x_n', x_n'' in E such that $|x_n' - x_n''| < \delta$, however $|f(x_n') - f(x_n'')| \geq \varepsilon$.

Since E is bounded, it is possible to extract a convergent subsequence of $(x_n')_{n=1}^{\infty}$, namely $(x_{n_k}')_{k=1}^{\infty}$; it is not implied, however, that the limit x_0 of $(x_{n_k}')_{k=1}^{\infty}$ belongs to E; at any rate, $(x_{n_k}')_{k=1}^{\infty}$ is a Cauchy sequence. In the same way, there exists a Cauchy sequence $(x_{n_k}'')_{k=1}^{\infty}$ having the same limit x_0, as it follows from the inequality

$$|x''_{n_k} - x_0| \le |x''_{n_k} - x'_{n_k}| + |x'_{n_k} - x_0|$$

since both $|x''_{n_k} - x'_{n_k}|$ and $|x'_{n_k} - x_0|$ may be arbitrarily small.

Therefore, the sequence $(x_k)_{k=1}^{\infty}$, where $x_1 = x'_{n_1}$, $x_2 = x''_{n_1}$, ..., $x_{2k-1} = x'_{n_k}$, $x_{2k} = x''_{n_k}$, ..., obtained by "mixing" the two convergent subsequences $(x'_{n_k})_{k=1}^{\infty}$ and $(x''_{n_k})_{k=1}^{\infty}$, is again a Cauchy sequence.

By hypothesis, we deduce that $\{f(x_k)\}_{k=1}^{\infty}$ is also a Cauchy sequence; thus, given any $\varepsilon > 0$, there exists an index k_0 such that if $k, k' \ge k_0$, then $|f(x_k) - f(x_{k'})| < \varepsilon$. However, this is not the case, since for every k_0 we have

$$|f(x_{2k_0-1}) - f(x_{2k_0})| = |f(x'_{n_{k_0}}) - f(x''_{n_{k_0}})| \ge \varepsilon.$$

This shows that f must be uniformly continuous on the bounded set E.

□

Proposition 34 (Principle of Extension by Continuity) Let E be a set of real numbers and E' be a subset of E such that every element of E is a limit point of E' (in other words, we assume that the closure of E' is E). Let f' be a function uniformly continuous on E'. Then there exists one and only one continuous function f on the set E such that $f'(x) = f(x)$ for every $x \in E'$. Moreover, if E is bounded, then f is uniformly continuous on E.

Proof First, we observe that if there is at all a continuous function f on E, with the property stated, then, for every element $x \in E$ and for every sequence $(x'_n)_{n=1}^{\infty}$ of elements of E' such that $\lim_{n\to\infty} x'_n = x$ (which exists because x is a limit point of E'), we must have $f(x) = \lim_{n\to\infty} f(x'_n) = \lim_{n\to\infty} f'(x'_n)$ since $x'_n \in E'$. Thus, f(x) must be the unique limit of the convergent sequence $\{f'(x'_n)\}_{n=1}^{\infty}$ and, consequently, only one continuous function f can exist with the required property.

We have also found out how we should define the value of f at every point $x \in E$. In fact, if $x \in E$, it is a limit point of E', hence there exists a sequence $(x'_n)_{n=1}^{\infty}$ of elements of E' such that $\lim_{n\to\infty} x'_n = x$; thus $(x'_n)_{n=1}^{\infty}$ is a Cauchy sequence. But f' is uniformly continuous on E' and so, by Proposition 33, $\{f'(x'_n)\}_{n=1}^{\infty}$ is a Cauchy sequence. Hence there exists a real number $y = \lim_{n\to\infty} f'(x'_n)$. We define $f(x) = y$.

Now, we justify this definition by showing that if $(z'_n)_{n=1}^{\infty}$ is another Cauchy sequence in E' such that $\lim_{n\to\infty} z'_n = x$, then we have $\lim_{n\to\infty} f'(z'_n) = \lim_{n\to\infty} f'(x'_n)$. This, however, follows from the uniform continuity of f' on E'. Given any $\varepsilon > 0$, there exists a $\delta > 0$ such that if $x', z' \in E'$ and $|x' - z'| < \delta$ then $|f'(x') - f'(z')|$ is less than ε. Associated with this number δ, there exists an index n_0 such that both $|x'_n - x| < \delta/2$ and $|z'_n - x| < \delta/2$ for every index $n \geq n_0$; hence, for any index $n \geq n_0$,

$$|x'_n - z'_n| \leq |x'_n - x| + |x - z'_n| < \delta.$$

Thus, for $n \geq n_0$,

$$|f'(x'_n) - f'(z'_n)| < \varepsilon,$$

which shows that $\{f'(x'_n)\}_{n=1}^{\infty}$ and $\{f'(z'_n)\}_{n=1}^{\infty}$ are Cauchy sequences converging to the same limit.

Finally, we must show that f is continuous on E. Since every point x of E is in some bounded subset T of the set E, it is enough to prove that if T is any bounded subset of E, then f is uniformly continuous on T. This implies, indeed, that f is continuous on T, and, therefore, at the arbitrary point x of E. At the same time, since we have not excluded that T = E whenever E is assumed to be bounded, this will also establish that f is uniformly continuous on E when E is bounded.

Thus we assume that T is a bounded subset of E. By Proposition 33, it is sufficient to show that if $(x_n)_{n=1}^{\infty}$ is any Cauchy sequence of elements of T, then $\{f(x_n)\}_{n=1}^{\infty}$ is also a Cauchy sequence.

Given any $\varepsilon > 0$, since f' is uniformly continuous on E', there exists $\delta > 0$ such that $x', z' \in E'$, $|x' - z'| < \delta$, then

$$|f'(x') - f'(z')| < \frac{\varepsilon}{3}.$$

Since $(x_n)_{n=1}^{\infty}$ is a Cauchy sequence, associated with this δ, there exists an index n_0 such that if m, $n \geq n_0$ then $|x_m - x_n| < \delta/3$.

Consider any two indices m, $n \geq n_0$. Since

$$x_m = \lim_{k \to \infty} x'_{m,k}, \quad x_n = \lim_{k \to \infty} x'_{n,k},$$

there exists a sufficiently large index k_0 (dependent on m, n) such that, for $k \geq k_0$,

$$|x_m - x'_{m,k}| < \frac{\delta}{3} \text{ and also } |x_n - x'_{n,k}| < \frac{\delta}{3};$$

therefore, if m, $n \geq n_0$, and taking any index $k \geq k_0$, we have

$$x'_{m,k} - x'_{n,k} = (x'_{m,k} - x_m) + (x_m - x_n) + (x_n - x'_{n,k}),$$

so

$$|x'_{m,k} - x'_{n,k}| < \delta.$$

Hence, by the uniform continuity of f', $|f'(x'_{m,k}) - f'(x'_{n,k})| < \varepsilon$ provided that m, $n \geq n_0$ and $k \geq k_0$.

On the other hand, considering any two natural numbers m, $n \geq n_0$, since

$$f(x_m) = \lim_{k \to \infty} f'(x'_{m,k}) \text{ and } f(x_n) = \lim_{k \to \infty} f'(x'_{n,k}),$$

given the number $\varepsilon > 0$, there is a sufficiently large index k_1, which we may take such that $k_1 \geq k_0$, with the property: if $k \geq k_1$, then $|f(x_m) - f'(x'_{m,j})| < \varepsilon/3$ and $|f(x_n) - f'(x'_{m,j})| < \varepsilon/3$. Combining these inequalities, if m, $n \geq n_0$ and using $k \geq k_1$, it follows from

$$f(x_m) - f(x_n) = \{f(x_m) - f'(x'_{m,k})\} + \{f'(x'_{m,k}) - f'(x'_{n,k})\} + \{f'(x'_{n,k}) - f(x_n)\}$$

that

$$|f(x_m) - f(x_n)| \leq \frac{\varepsilon}{3} + \frac{\varepsilon}{3} + \frac{\varepsilon}{3} = \varepsilon.$$

This shows that $\{f(x_n)\}_{n=1}^{\infty}$ is a Cauchy sequence. □

Remark The foregoing proposition shows that if a function f' is uniformly continuous on the set of all rational points of an interval I, then there exists a unique function f, defined also for the irrational points of I and which is continuous on I; if I is bounded, f will even be uniformly continuous on I. The function f is called the *continuous extension of* f'.

As a simple application of Proposition 34 we consider the definition of the exponential function with basis $a > 0$, that is,

$$f(x) = a^x \quad \text{with} \quad x \in R^1.$$

It is easy to check that the functions $f_n(r) = a^r$ with rational r are uniformly continuous on the set of all rational points of the interval $[-n,n]$. By Proposition 34 we may define a function f, on the whole real line, as follows: if x is any real number, if n is a positive integer, $|x| \le n$, we put $f(x) = f_n(x)$; this definition is independent of the integer n.

Definition Let f be a real-valued function on the interval [a,b]. Then f is said to be *monotonically increasing on* [a,b] if

$$a \le x < y \le b \quad \text{implies} \quad f(x) \le f(y);$$

f is said to be *strictly increasing on* [a,b] if

$$a \le x < y \le b \quad \text{implies} \quad f(x) < f(y).$$

Analogously, f is said to be *monotonically decreasing on* [a,b] if

$$a \le x < y \le b \quad \text{implies} \quad f(x) \ge f(y)$$

and f is said to be *strictly decreasing on* [a,b] if

$$a \le x < y \le b \quad \text{implies} \quad f(x) > f(y).$$

The class of *monotonic functions* consists of both the increasing and the decreasing functions; the class of *strictly monotonic functions* consists of the strictly increasing and the strictly decreasing functions. Monotonically increasing functions we sometimes also call *non-decreasing functions;* monotonically decreasing functions we shall on occasion also refer to as *nonincreasing functions*.

Proposition 35 Let f be a real-valued continuous function on a bounded closed interval $I = [a,b]$. Then f is invertible if and only if f is strictly monotonic. Trivially, f^{-1} is increasing or decreasing depending on whether f is increasing or decreasing. Moreover, the inverse function f^{-1} is continuous on its interval of definition I^*.

Proof It is clear that strict monotonicity is a sufficient condition for the invertibility of f. We only need to show that in the case of continuity it is also a necessary condition. For f to be invertible we need to have that $f(x_1) \neq f(x_2)$ for any two distinct points x_1 and x_2 in I. Since f is assumed to be continuous, we have for any x between x_1 and x_2 that

$$\{f(x) - f(x_1)\}\{f(x) - f(x_2)\} < 0. \tag{2}$$

To see this, we suppose that x_1 and x_2 are so indexed that $f(x_1)$ is less than $f(x_2)$. Suppose now that $f(x) > f(x_2)$. By Proposition 31, the value $f(x_2)$ is attained between x_1 and x; however this contradicts the invertibility of f since x_2 is not situated between x_1 and x. In the same manner the assumption that f(x) is less than $f(x_1)$ leads to a contradiction. Thus $f(x_1) < f(x) < f(x_2)$ and we get (2).

Suppose now that f is not monotonic on I. Then there are two pairs of points y_1, y_2 and y_3, y_4 in I such that $y_1 < y_2$ and $y_3 < y_4$ satisfying

$$f(y_1) < f(y_2), \quad f(y_3) > f(y_4). \tag{3}$$

Without loss of generality we may assume that none of the points y_1, y_2, y_3, and y_4 is an endpoint of the interval I. (Observe: if f is continuous at t and $f(t) > c$ [resp. $f(t) < c$] for a given real number c, then there is a neighborhood U of t such that $f(x) > c$ [resp. $f(x) < c$] for all x belonging to U and x being in I, the domain of definition of I. Indeed, we only need to let $\varepsilon = |f(t) - c|$. Then for all $x \in I$ such that $|x - t| < \delta$ we have

$$-|f(t) - c| < f(x) - f(t) < |f(t) - c|$$

and everything is clear.) Then there are points α, $\beta \in I$ such that for $k = 1, 2, 3, 4$,

$$\alpha < y_k < \beta.$$

By (2) we have

$$\{f(y_1) - f(\alpha)\}\{f(y_1) - f(y_2)\} < 0$$

and

$$\{f(y_2) - f(y_1)\}\{f(y_2) - f(\beta)\} < 0;$$

adding the foregoing two inequalities we obtain

$$\{f(y_1) - f(y_2)\}^2 - \{f(\alpha) - f(\beta)\}\{f(y_1) - f(y_2)\} < 0.$$

Thus

$$\{f(\alpha) - f(\beta)\}\{f(y_1) - f(y_2)\} > 0.$$

In the same way we obtain

$$\{f(\alpha) - f(\beta)\}\{f(y_3) - f(y_4)\} > 0.$$

By multiplication of the last two inequalities, we see that

$$\{f(y_1) - f(y_2)\}\{f(y_3) - f(y_4)\} > 0.$$

But this violates the inequality (3).

We have therefore shown that f is strictly monotonic. To complete the proof, we must show that f^{-1} is continuous on the range of f.

Let r be in the range of f and write $r = f(s)$ with $s \in I$. We want to prove that given any $\varepsilon > 0$, there is a $\delta > 0$ such that, for all y in the range of f satisfying $|y - x| < \delta$, we have

$$|f^{-1}(y) - f^{-1}(r)| < \varepsilon.$$

To this end, let $\varepsilon > 0$. By what we have proved already, f is strictly monotonic. We may without loss of generality suppose that f is increasing. Since $s - \varepsilon < s < s + \varepsilon$, we have

$$f(s - \varepsilon) < f(s) < f(s + \varepsilon).$$

Let δ be the smaller of the two numbers $f(s) - f(s - \varepsilon)$ and $f(s + \varepsilon) - f(s)$. Then

$$f(s - \varepsilon) \leq f(s) - \delta < f(s) + \delta \leq f(s + \varepsilon).$$

Hence, for all y satisfying $f(s) - \delta < y < f(s) + \delta$, we have

$$f(s - \varepsilon) < y < f(s + \varepsilon).$$

Since f is increasing, so is f^{-1}. Hence

$$f^{-1}\{f(s - \varepsilon)\} < f^{-1}(y) < f^{-1}\{f(s + \varepsilon)\},$$

that is,

$$s - \varepsilon < f^{-1}(y) < s + \varepsilon.$$

But $s = f^{-1}(r)$ and so

$$|f^{-1}(r) - f^{-1}(y)| < \varepsilon.$$

We therefore have shown that, for all y satisfying $r - \delta < y < r + \delta$,

$$|f^{-1}(r) - f^{-1}(y)| < \varepsilon.$$

Hence f^{-1} is continuous at r. □

Remark Another way of showing that the inverse function f^{-1} in Proposition 35 has to be continuous on the range of f goes as follows.

By Propositions 29 and 31, the range of f is a closed and bounded interval I*. Let $y_0, y_1, y_2, \ldots \in I^*$, $y_n \to y_0$ as $n \to \infty$. The set of points $y_0, y_1, y_2, \ldots$ is closed and bounded; hence, by the remark following Proposition 23, this set of points has a smallest member, say y_m, and a largest member, say y_M. Thus, for all n = 0, 1, 2, ...,

$$y_m \le y_n \le y_M.$$

For j = 0, 1, 2, ..., let $x_j = f^{-1}(y_j)$ Since f^{-1} is strictly monotonic,

$$x_m \le x_n \le x_M \quad \text{or} \quad x_m \ge x_n \ge x_M;$$

in particular, the sequence $(x_n)_{n=0}^{\infty}$ is bounded. Let t be any accumulation point of this sequence. Then $t \in I$ because $x_m, x_M \in I$. There exists a subsequence $(x_{n_j})_{j=0}^{\infty}$ of $(x_n)_{n=0}^{\infty}$ which converges to t. Since f is continuous at t, $y_{n_j} \to f(t)$; thus $f(t) = y_0$. But f is invertible and so $t = x_0$. The sequence $x_n = f^{-1}(y_n)$ is therefore seen to be convergent and having the limit $x_0 = f^{-1}(y_0)$.

Definition Let f be a real-valued function defined on the open interval (a,b). Consider any point x such that $a \le x < b$. We write

$$f(x+) = L$$

if $f(t_n) \to L$ as $n \to \infty$, for all sequences $(t_n)_{n=1}^{\infty}$ in (x,b) such that $t_n \to x$ as $n \to \infty$. To obtain the definition of f(x-), for $a < x \le b$, we restrict ourselves to sequences $(t_n)_{n=1}^{\infty}$ in (a,x).

It is clear that at any point x of (a,b), $\lim_{t \to x} f(t)$ exists if and only if $f(x+) = f(x-) = \lim_{t \to x} f(t)$.

The limits f(x+) and f(x-) are called the *right-hand* and the *left-hand limits of* f *at* x, respectively.

Definition Let f be a real-valued function on an open interval (a,b). If f is discontinuous at a point x of (a,b), and if f(x+) and f(x-) exist, then f is said to have a *discontinuity of the first kind*, or a *simple discontinuity*, *at* x. Otherwise the discontinuity is said to be of the *second kind*.

Illustration Consider the following three functions:

$$f(x) = 1 \quad \text{for} \quad x \geq 0,$$
$$= -1 \quad \text{for} \quad x < 0;$$

$$g(x) = 1 \quad \text{for} \quad x > 0 \text{ and } x < 0,$$
$$= 0 \quad \text{for} \quad x = 0;$$

$$h(x) = \sin\frac{1}{x} \quad \text{for} \quad x > 0 \text{ and } x < 0,$$
$$= 0 \quad \text{for} \quad x = 0.$$

It is clear that $f(0+) = 1$ and $f(0-) = -1$; thus f has a discontinuity of the first kind at $x = 0$. For the function g we have $g(0+) = g(0-) = 1$, but $g(0) = 0$; thus g has a discontinuity of the first kind at $x = 0$. For the function h, neither $h(0+)$ nor $h(0-)$ exists, as we already observed at the beginning of this Section 2; thus h has a discontinuity of the second kind at $x = 0$.

Remark There are two ways in which a function f can have a simple discontinuity at x; either $f(x+) \neq f(x-)$, or $f(x+) = f(x-) \neq f(x)$.

The following proposition shows that a monotonic function f has no discontinuities of the second kind since $f(x+)$ and $f(x-)$ exist.

Proposition 36 Let f be monotonically increasing on (a,b). Then, for any $x \in (a,b)$,

$$\sup_{a<t<x} f(t) = f(x-) \leq f(x) \leq f(x+) = \inf_{x<t<b} f(t). \tag{4}$$

Moreover, if $a < x < y < b$, then

$$f(x+) \leq f(y-).$$

Proof By assumption, the set of numbers $f(t)$, where $a < t < x$, is bounded above by the number $f(x)$, and thus has a supremum which we shall denote by A. Evidently, $A \leq f(x)$. We verify that $A = f(x-)$.

Let $\varepsilon > 0$ be given. It follows from the definition of A as a supremum that there exists $\delta > 0$ such that $a < x - \delta < x$ and

$$A - \varepsilon < f(x - \delta) < A.$$

Since f is monotonic, we have $f(x - \delta) \le f(t) \le A$ for $x - \delta < t < x$. Thus, for $x - \delta < t < x$, we have $|f(t) - A| < \varepsilon$ and so $f(x-) = A$.

The second half of (4) is proved in precisely the same way.

Next, if $a < x < y < b$, we deduce from (4) that

$$f(x+) = \inf_{x<t<b} f(t) = \inf_{x<t<y} f(t).$$

The last inequality in the foregoing is obtained by applying (4) to (a,y) in place of (a,b). Similarly,

$$f(y-) = \sup_{a<t<y} f(t) = \sup_{x<t<y} f(t).$$

Thus $f(x+) \le f(y-)$ is seen to hold for $a < x < y < b$. □

Definition Let f be a monotonically increasing function on a closed interval [a,b] and suppose that $x \in [a,b]$. The numbers

$$f(x) - f(x-) \quad \text{and} \quad f(x+) - f(x)$$

are called the *left-hand* and the *right-hand jumps of* f *at* x, respectively; the number

$$f(x+) - f(x-)$$

is called the *jump of* f *at* x (for the endpoints a and b we consider only one-sided jumps).

Proposition 37 Let f be a monotonically increasing real-valued function on an interval [a,b] and let $x_1, x_2, \ldots, x_n$ be a finite set of points in the interior of [a,b], that is, in (a,b). Then

$$[f(a+) - f(a)] + \sum_{k=1}^{n} [f(x_k+) - f(x_k-)] + [f(b) - f(b-)]$$

is less than or equal to $f(b) - f(a)$.

Proof We assume that $a < x_1 < x_2 < \cdots < x_n < b$. We put $a = x_0$ and $b = x_{n+1}$ and we introduce points $y_0, y_1, \ldots, y_n$ such that for $k = 0, 1, \ldots, n$,

$$x_k < y_k < x_{k+1}.$$

We obtain, for $k = 1, 2, \ldots, n$,

$$f(x_k+) - f(x_k-) \le f(y_k) - f(y_{k-1});$$

also,

$$f(a+) - f(a) \le f(y_1) - f(a) \quad \text{and} \quad f(b) - f(b-) \le f(b) - f(y_n).$$

Adding up all these inequalities, we obtain what we set out to do. □

Corollary Let f be as in Proposition 37. Then f can have at most a finite number of discontinuities at which the jump of f is larger than a given positive number r.

Proof Suppose that $x_1, x_2, \ldots, x_n$ are points of discontinuity situated in the interior of [a,b] at which f has a jump larger than r. Then

$$nr \le f(b) - f(a),$$

implying that n cannot be arbitrarily large. □

Proposition 38 Let f be as in Proposition 37. Then the set of points at which f is discontinuous is at most countably infinite. If $x_1, x_2, \ldots$ are the points of discontinuity of f in (a,b), then

$$[f(a+) - f(a)] + \sum_{k=1}^{\infty} [f(x_k+) - f(x_k-)] + [f(b) - f(b-)]$$

is less than or equal to $f(b) - f(a)$.

Proof Let D denote the set of all discontinuities of f and let D_k denote the set of discontinuities of f at which the jump of f is larger than 1/k with k a positive integer. Clearly, D is the union of the sets D_k with k = 1, 2, ... and therefore D must be countable since each set D_k is finite

Passing to the limit as $n \to \infty$ in the inequality of Proposition 38, we obtain the inequality claimed in the present proposition and the proof is finished. □

Proposition 38 lends meaning to the following definition.

Definition Let f be a real-valued monotonically increasing function on an interval [a,b]. We introduce a function s on [a,b] as follows: we set

$$s(a) = 0 \quad \text{and}$$

$$s(x) = [f(a+) - f(a)] + \sum_{x_k<x} [f(x_k+) - f(x_k-)] + [f(x) - f(x-)]$$

for $a < x \leq b$; the summation is to be understood as follows: sum only over those indices k for which $x_k < x$. The function s is called the *saltus function of* f. Note that s is also increasing.

Remark Note that the series appearing in the definition of the saltus function has positive terms only and converges unconditionally; thus s is welldefined.

Proposition 39 Let f be a real-valued monotonically increasing function on an interval [a,b] and let s be the saltus function of f. Then the function $g(x) = f(x) - s(x)$ is a continuous function on [a,b].

Proof Let $a \leq x < y \leq b$. Instead of the interval [a,b] we use the interval [x,y] and apply to it the inequality occuring in Proposition 38. We get that $s(y) - s(x) \leq f(y) - f(x)$, implying that $g(x) \leq g(y)$; thus g is seen to be monotonically increasing. If in $s(y) - s(x) \leq f(y) - f(x)$ we pass to the limit as y tends to x, we we get that $s(x+) - s(x) \leq f(x+) - f(x)$. On the other hand, from the definition of the saltus function s, we get that

$$f(x+) - f(x) \leq s(y) - s(x) \quad \text{for} \quad x < y;$$

Letting y tend to x, we obtain $f(x+) - f(x) \leq s(x+) - s(x)$. Thus

$$f(x+) - f(x) = s(x+) - s(x),$$

showing that $s(x+) = s(x)$. In a similar way we can verify that $s(x-) = s(x)$. □

Corollary Let f be as in Proposition 39. Then f can be represented as the sum of its saltus function and a continuous function.

Remark If f is a monotonically decreasing function on $[a,b]$, then the function $-f$ is a monotonically increasing function on $[a,b]$. We see therefore that results completely analogous to the statements made concerning monotonically increasing functions also hold for monotonically decreasing functions.

Example Let T be the set of all rational points of an interval (a,b) and let the points T be arranged in a sequence $(x_n)_{n=1}^{\infty}$. Let $(c_n)_{n=1}^{\infty}$ be a sequence of positive real numbers such that the series $\Sigma_{n=1}^{\infty} c_n$ converges. Define, for all x such that $a < x < b$,

$$f(x) = \sum_{x_n < x} c_n.$$

Again, the summation is to be understood as follows: sum over those indices n for which $x_n < x$. If there are no points x_n to the left of x, the sum is of course empty and we define it to be zero. Since the series in the defining equation for the function f is absolutely convergent, the order in which the terms are arranged is immaterial.

It is clear that f is a monotonically increasing function on (a,b); we also see that f is discontinuous at every rational point of (a,b); in fact, $f(x_n+) - f(x_n-) = c_n$. However, f is continuous at every irrational point of (a,b). Note also, if in the definition of f we replace $x_n < x$ by $x_n \le x$, we would have $f(x+) = f(x)$ at every point of (a,b).

Cantor's Function:

Cantor's (ternary) set C was introduced as Application 3 of Proposition 80 in Chapter 1; following Proposition 21 in this chapter we discussed some properties of the set C. We recall that the set C is obtained by removal of the following sequence of open intervals from the closed interval $[0,1]$:

In the first step we remove the interval

$$I\left(\frac{1}{2}\right) = \left(\frac{1}{3},\frac{2}{3}\right);$$

in the second step we remove the two intervals

$$I\left(\frac{1}{4}\right) = \left(\frac{1}{3^2},\frac{2}{3^2}\right), \quad I\left(\frac{3}{4}\right) = \left(\frac{2}{3} + \frac{1}{3^2},\frac{2}{3} + \frac{2}{3^2}\right);$$

in the third step we remove the four intervals

$$I\left(\frac{1}{8}\right) = \left(\frac{1}{3^3},\frac{2}{3^3}\right), \qquad I\left(\frac{3}{8}\right) = \left(\frac{2}{3^2} + \frac{1}{3^3},\frac{2}{3^2} + \frac{2}{3^3}\right),$$

$$I\left(\frac{5}{8}\right) = \left(\frac{2}{3} + \frac{1}{3^3},\frac{2}{3} + \frac{2}{3^3}\right), \quad I\left(\frac{7}{8}\right) = \left(\frac{2}{3} + \frac{2}{3^2} + \frac{1}{3^3},\frac{2}{3} + \frac{2}{3^2} + \frac{2}{3^3}\right);$$

and so forth. The intervals are indexed so that $r > s$ if and only if $I(r)$ lies to the right of $I(s)$.

We define the *Cantor function* c as follows: On $[0,1] - C$ we put

$$c(t) = \frac{m}{2^n} \quad \text{if} \quad t \in I\left(\frac{m}{2^n}\right),$$

and we set

$$c(0) = 0$$

and, for $0 \neq t \in C$, we put

$$c(t) = \sup\left\{c(s): s \in [0,1] - C,\ s < t\right\}.$$

It is easy to see that, for $t \in [0,1]$,

$$0 = c(0) \le c(t) \le c(1) = 1.$$

We next verify that c is monotonically increasing on $[0,1]$. Let $s < t$. If s and t both belong to $[0,1] - C$, then either they are in the same interval $I(r)$, in which case $c(s) = c(t) = r$, or $s \in I(r_1)$, $t \in I(r_2)$ with $r_1 < r_2$, so that

$$c(s) = r_1 < r_2 = c(t).$$

If $s \in [0,1] - C$ and $t \in C$, then

$$c(s) \le \sup\{c(v): v \in [0,1] - C,\ v < t\} = c(t).$$

If $s \in C$, $t \in [0,1] - C$, then $r < s$, $r \in [0,1] - C$ imply $c(r) < c(t)$; hence

$$c(s) = \sup\{c(r): r \in [0,1] - C,\ r < s\} \le c(t).$$

If $s \in C$, $t \in C$, then there is an $r \in [0,1] - C$ such that $s < r < t$, and the foregoing results yield $c(s) \le c(r) \le c(t)$.

Finally, we verify that c is continuous on $[0,1]$. It is of course obvious that c is continuous on $[0,1] - C$ (because it is piecewise constant there). Now let $t_0 \in C$; then

$$c(t_0) = \sup\{c(s): s \in [0,1] - C,\ s < t_0\}.$$

Thus, for $\varepsilon > 0$, there is an s such that $s < t_0$ and $c(t_0) - c(s) < \varepsilon$. The monotonicity of c now yields that $c(t_0-) = c(t_0)$. To show that $c(t_0+) = c(t_0)$, we note that there is a number of the form $m/2^n$ such that

$$c(t_0) < \frac{m}{2^n} < c(t_0) + \varepsilon, \quad \text{with } m \text{ odd.}$$

But for some $s > t_0$ we know that $c(s) = m/2^n$, and we need only apply the monotonicity again to obtain that $c(t_0+) = c(t_0)$. Thus

$$c(t_0+) = c(t_0-) = c(t_0).$$

The proof is complete.

To be able to operate in the extended real number system R_e^1 defined at the end of Section 2 in Chapter 1, we shall now enlarge the scope of earlier definitions.

For any $x \in R^1$, we have already defined a neighborhood of x to be any open interval $(x - \delta, x + \delta)$ with $\delta > 0$.

Definition For any real number a, the set of all real numbers x such that $x > a$ is called a *neighborhood of* $+\infty$ and is written $(a, +\infty)$. Similarly, the set $(-\infty, a)$ is a *neighborhood of* $-\infty$.

Definition Let f be defined on E. We say that f(t) *tends to* L *as* t *tends to* x, and write

$$f(t) \to L \quad \text{as} \quad t \to x,$$

where L and x are in the extended real number system, if for every neighborhood U of L there is a neighborhood V of x such that $V \cap E$ is not empty, and such that $f(t) \in U$ for all $t \in V \cap E$.

Remark It is clear that this definition of limit of a function coincides with the earlier one when L and x are (finite) real numbers.

Proposition 40 Let f and g be defined on $E \subset R_e^1$. Suppose that

$$f(t) \to L, \quad g(t) \to K \quad \text{as } t \to x.$$

Then

(i) $f(t) \to L'$ implies $L' = L$,

(ii) $(f + g)(t) \to L + K$,

(iii) $(fg)(t) \to LK$,

(iv) $(f/g)(t) \to L/K$,

provided that the right-hand members of (ii), (iii), and (iv) are defined.

The proof is left as an exercise.

3. *Functions of Finite Variation*

Let f be a function defined on an interval [a,b] and with range in R^1 and suppose that $x_0, x_1, \ldots, x_n$ is a partition of [a,b], that is, a finite set of points such that

$$a = x_0 < x_1 < \cdots < x_n = b.$$

Definition Consider all partitions of [a,b]. The extended real number

$$V_a^b f = \sup\left\{\sum_{k=0}^{n-1} |f(x_{k+1}) - f(x_k)|: a = x_0 < x_1 < \cdots < x_n = b\right\}$$

(the supremum is taken with respect to all partitions of [a,b]) is called the *total variation of* f *over* [a,b]. If $V_a^b f < \infty$, then f is said to be of *finite* (or *bounded*) *variation over* [a,b].

Proposition 41 A monotonic function is of finite variation.

Proof Suppose that f is increasing on [a,b]. Then for any partition $a = x_0 < x_1 < \cdots < x_n = b$ of [a,b] we have

$$\sum_{k=0}^{n-1} [f(x_{k+1}) - f(x_k)] \le f(b) - f(a).$$ □

Example The function

$$f(x) = x \cos \frac{\pi}{2x} \quad \text{for} \quad 0 < x \le 1,$$

$$= 0 \qquad \text{for} \quad x = 0,$$

is of infinite total variation. Indeed, consider the partition

$$0 < \frac{1}{2n} < \frac{1}{2n - 1} < \cdots < \frac{1}{3} < \frac{1}{2} < 1.$$

Then $\Sigma_{k=1}^n 1/k < V_a^b f$ and so $V_a^b f = +\infty$.

Proposition 42 Any function of finite variation is bounded.

Proof For $a < x' \le b$ we have

$$|f(x') - f(a)| + |f(b) - f(x')| \le V_a^b f$$

and hence

$$|f(x')| \le |f(x') - f(a)| + |f(a)| \le |f(a)| + V_a^b f.$$ □

Proposition 43 The sum, difference, and product of two functions f and g of finite variation over [a,b] are again of finite variation over [a,b].

Proof Let $s = f \pm g$. Then

$$|s(x_{k+1}) - s(x_k)| \le |f(x_{k+1}) - f(x_k)| + |g(x_{k+1}) - g(x_k)|$$

implies that

$$V_a^b s \le V_a^b f + V_a^b g.$$

Let $p = fg$ and (using Proposition 42) let

$A = \sup\{|f(x)|: x \in [a,b]\}$ and $B = \sup\{|g(x)|: x \in [a,b]\}$.

Then

$$\begin{aligned}|p(x_{k+1}) - p(x_k)| &\le |f(x_{k+1})\, g(x_{k+1}) - f(x_k)\, g(x_{k+1})| \\ &\quad + |f(x_k)\, g(x_{k+1}) - f(x_k)\, g(x_k)| \\ &\le B\, |f(x_{k+1}) - f(x_k)| + A\, |g(x_{k+1}) - g(x_k)|,\end{aligned}$$

implying $V_a^b p \le B\, V_a^b f + A\, V_a^b g$. □

Proposition 44 Let f and g be functions of finite variation over [a,b] and suppose that $|g(x)| \ge \lambda > 0$ for all $x \in [a,b]$. Then the quotient f/g is of finite variation over [a,b].

Proof In view of the foregoing proposition, it will be sufficient to show that the function

$$h(x) = \frac{1}{g(x)}$$

is of finite variation over [a,b]. But

$$|h(x_{k+1}) - h(x_k)| = \frac{|g(x_{k+1}) - g(x_k)|}{|g(x_k)|\ |g(x_{k+1})|} \le \frac{1}{\lambda^2} |g(x_{k+1}) - g(x_k)|,$$

and so $V_a^b h \le (1/\lambda^2)\ V_a^b g$. □

Proposition 45 If f is of finite variation over [a,b] and $a < c < b$, then $V_a^b f = V_a^c f + V_c^b f$.

Proof Let

$$a = y_0 < y_1 < \cdots < y_m = c \quad \text{and} \quad c = z_0 < z_1 < \cdots < z_n = b$$

and form the sums

$$v_1 = \sum_{k=0}^{m-1} |f(y_{k+1}) - f(y_k)| \quad \text{and} \quad v_2 = \sum_{k=0}^{n-1} |f(z_{k+1}) - f(z_k)|.$$

Take the union of the above two partitions and let v be the corresponding sum of the variation. Then $v = v_1 + v_2$. It is clear that $v_1 + v_2 \le V_a^b f$ and thus $V_a^c f + V_c^b f \le V_a^b f$.

Now consider the partition $a = x_0 < x_1 < \cdots < x_j = b$ of [a,b] and suppose that the point c is among these, that is, $c = x_m$ for for some m between 0 and j. Then the corresponding sum

$$w = \sum_{k=0}^{m-1} |f(x_{k+1}) - f(x_k)| + \sum_{k=m}^{j-1} |f(x_{k+1}) - f(x_k)|$$

or, more briefly, $w = w_1 + w_2$, where w_1 and w_2 are the sums corresponding to the intervals [a,c] and [c,b]. Then $w \le V_a^c f + V_c^b f$.

The inequality $w \le V_a^c f + V_c^b f$ was only verified for sums w for which the point c is a member of the partition $a = x_0 < x_1 < \cdots < x_j$

= b. But the addition of more points of division will certainly not make the sum w any smaller and so $w \le V_a^c f + V_c^b f$ holds for all sums w. [Note that, if $x_k < x' < x_{k+1}$, then

$$|f(x_{k+1}) - f(x_k)| \le |f(x_{k+1}) - f(x')| + |f(x') - f(x_k)|.]$$

Hence $V_a^b f \le V_a^c f + V_c^b f$ and the proof is finished. □

Corollary 1 If f is of finite variation over [a,b] and $a < c < b$, then f is also of finite variation over [a,c] and [c,b] and conversely.

Corollary 2 Let f be as in Corollary 1 and let

$$g(a) = 0 \quad \text{and} \quad g(x) = V_a^x f \quad \text{for} \quad a < x \le b.$$

Then g is monotonically increasing on [a,b].

Corollary 3 Let f be a real-valued function on [a,b] and suppose that [a,b] can be partitioned into a finite set of subintervals on each of which f is monotonic; then f is of finite variation over [a,b].

Proposition 46 A function f is of finite variation if and only if it can be represented as the difference of two monotonically increasing functions.

Proof Propositions 41 and 43 show that the condition is sufficient. To see that the condition is also necessary, we put

$$g(a) = 0 \quad \text{and} \quad g(x) = V_a^x f \quad \text{for} \quad a < x \le b.$$

By Corollary 2 of Proposition 45, g is an increasing function. We set

$$h(x) = g(x) - f(x)$$

and note that h is an increasing function. In fact, if $a \le x < y \le b$, then, by Proposition 45,

$$h(y) = g(y) - f(y) = g(x) + V_x^y f - f(y)$$

and thus

$$h(y) - h(x) = V_x^y f - [f(y) - f(x)].$$

By the definition of finite variation we get

$$f(y) - f(x) \le V_x^y f,$$

so that $h(y) - h(x) \ge 0$.

To get the representation of f in terms of the difference of two (bounded) monotonically increasing functions, we rewrite $h(x) = g(x) - f(x)$ into $f(x) = g(x) - h(x)$. □

Corollary The set of points of discontinuity of a function of finite variation is at most countably infinite. At every point x_0 of discontinuity of f the one-sided limits $f(x_0+)$ and $f(x_0-)$ exist.

Proposition 47 A function f is of finite variation over [a,b] if and only if there exists a monotonically increasing (bounded) function F on [a,b] such that for any two points x' and x" in [a,b] with x' < x" we have

$$|f(x'') - f(x')| \le F(x'') - F(x').$$

Proof The necessity of the condition follows from the fact that for a function f of finite variation the function

$$g(a) = 0 \quad \text{and} \quad g(x) = V_a^x f \quad \text{for} \quad a < x \le b$$

is monotonically increasing and bounded, thus we may use for F the function g, noting that

$$|f(x'') - f(x')| \le g(x'') - g(x') = V_{x'}^{x''} f.$$

That the condition is also sufficient follows from the inequality

$$\sum_{k=0}^{n-1} |f(x_{k+1}) - f(x_k)| \le \sum_{k=0}^{n-1} [F(x_{k+1}) - F(x_k)] = F(b) - F(a).$$

□

Proposition 48 Let f be a function of finite variation over [a,b]. If f is continuous at $x_0 \in (a,b)$, then the function $g(x) = V_a^x f$ is also continuous at x_0.

Proof We assume that $x_0 < b$ and show that $g(x_0+) = g(x_0)$; the verification of $g(x_0-) = g(x_0)$ for $x_0 > a$ we leave to the reader.

We pick an arbitrary $\varepsilon > 0$ and partition the interval $[x_0,b]$ by the points $x_0 < x_1 < \cdots < x_n = b$ in such a way that

$$v = \sum_{k=0}^{n-1} |f(x_{k+1}) - f(x_k)| > V_{x_0}^b f - \varepsilon$$

holds. Since v can only get larger by the addition of new points, we may assume that

$$|f(x_1) - f(x_0)| < \varepsilon$$

holds. We therefore see that

$$V_{x_0}^b f < \varepsilon + \sum_{k=0}^{n-1} |f(x_{k+1}) - f(x_k)| < 2\varepsilon + \sum_{k=1}^{n-1} |f(x_{k+1}) - f(x_k)|$$

$$< 2\varepsilon + V_{x_1}^b f.$$

Thus the total variation of f over $[x_0,x_1]$ is less than 2ε and so

$$g(x_1) - g(x_0) < 2\varepsilon.$$

Hence $g(x_0+) - g(x_0) < 2\varepsilon$ and since $\varepsilon > 0$ is arbitrary, we arrive at $g(x_0+) = g(x_0)$ □

Corollary A continuous function of finite variation can be represented as the difference of two continuous monotonically increasing functions.

Proposition 49 Let f be continuous on a closed and bounded interval [a,b]. We partition [a,b] into subintervals with the points

$$a = x_0 < x_1 < \cdots < x_n = b$$

and form the sum

$$v = \sum_{k=0}^{n-1} |f(x_{k+1}) - f(x_k)|.$$

Then

$$\lim_{\lambda \to 0} v = V_a^b f,$$

where $\lambda = \max\{x_{k+1} - x_k : 0 \le k \le n - 1\}$, that is, for any $\varepsilon > 0$ there exists a $\delta > 0$ such that for any partition

$$a = x_0 < x_1 < \cdots < x_n = b$$

$$0 \le V_a^b f - v < \varepsilon \text{ if } \max\{x_{k+1} - x_k : k = 0, 1, \ldots, n - 1\} < \delta.$$

Proof As we have noted earlier, v does not get smaller with the addition of a new point to the partition $x_0, x_1, \ldots, x_n$. Moreover, the sum v cannot increase by more than twice the oscillation of f on $[x_k, x_{k+1}]$ in case this new point in the partition of the interval [a,b] is sitated in $[x_k, x_{k+1}]$.

We pick a number A such that $A < V_a^b f$ and look for a sum v* such that

$$v^* > A. \tag{1}$$

Let $a = x_0^* < x_1^* < \cdots < x_m^* = b$ be a partition of [a,b] which corresponds to the sum v*. Then we select $\delta >$ so small that

$$|f(x'') - f(x')| < \frac{v^* - A}{4m}$$

holds provided that $|x'' - x'| < \delta$; this is possible by the uniform continuity of f. We now show that for any partition of [a,b] for which $\lambda < \delta$ we have

$$v > A. \tag{2}$$

From a partition P_1 we obtain a partition P_2 by adjoining all points x_k^*. If v_0 is the sum corresponding to the partition P_2, then

$$v_0 \geq v^*. \tag{3}$$

On the other hand, the partition P_2 is obtained from the partition P_1 by adding at most m times a new point. The sum v increases each time by at most $(v^* - A)/2m$; thus

$$v_0 - v < \frac{v^* - A}{2}.$$

From this we get, using (3) and (1),

$$v > v_0 - \frac{v^* - A}{2} \geq \frac{A + v^*}{2} > A.$$

Thus (2) is satisfied for $\lambda < \delta$. But we always have that $v \leq V_a^b f$ and the proposition is established. □

Corollary Let f and $x_0, x_1, \ldots, x_n$ and λ be as in Proposition 49. Let

$$w = \sum_{k=0}^{n-1} w_k,$$

where w_k denotes the oscillation of f on the subinterval $[x_k, x_{k+1}]$. Then

$$\lim_{\lambda \to 0} w = V_a^b f.$$

Proof Let

$$v = \sum_{k=0}^{n-1} |f(x_{k+1}) - f(x_k)|.$$

It is clear that $w \geq v$.

The sum w corresponds to the partition $x_0, x_1, \ldots, x_n$. When we adjoin to this partition points at which the function f assumes the values

$$m_k = \inf\{f(x): x_k \le x \le x_{k+1}\}$$

and

$$M_k = \sup\{f(x): x_k \le x \le x_{k+1}\},$$

then the sum v' corresponding to the new partition is evidently not smaller than w, thus $w \le V_a^b f$. In other words, we have

$$v \le w \le V_a^b f.$$

Using Proposition 49, we get the desired conclusion by passage to the limit. □

Definition Let $(f_n)_{n=1}^{\infty}$ be a sequence of real-valued functions defined on a set E of real numbers. We say that $(f_n)_{n=1}^{\infty}$ *converges (pointwise) on* E if, for every $x \in E$, the numerical sequence

$$\{f_n(x)\}_{n=1}^{\infty}$$

is convergent. In this case we define a function f on E by taking

$$f(x) = \lim_{n \to \infty} f_n(x)$$

for every $x \in E$. This function f is called the *(pointwise) limit of the sequence* $(f_n)_{n=1}^{\infty}$.

Definition An infinite collection H of real-valued functions h defined on a set E of real numbers is said to be *uniformly bounded* if there exists a (finite) real number K such that

$$\sup\{|h(x)|: x \in E, h \in H\} \le K.$$

In this case, K is called a *uniform bound for* H *on* E.

Proposition 50 Let H be an infinite collection of real-valued functions h defined on an interval [a,b]. If H is uniformly bounded on [a,b], then we may select a sequence $(h_n)_{n=1}^{\infty}$ from H such that this sequence of functions converges (pointwise) on any countable subset D of [a,b].

Proof Let D be any sequence $(x_n)_{n=1}^{\infty}$ of distinct points of [a,b]. Consider the set

$$\{h(x_1): h \in H\}$$

of real numbers. This set is clearly bounded and so has an accumulation point A_1. We may therefore select a sequence $(h_n^{(1)})_{n=1}^{\infty}$ of functions in H such that

$$\lim_{n\to\infty} h_n^{(1)}(x_1) = A_1$$

exists. Next, we consider the sequence $\{h_n^{(1)}(x_2)\}_{n=1}^{\infty}$. There is a subsequence $(h_n^{(2)})_{n=1}^{\infty}$ of $(h_n^{(1)})_{n=1}^{\infty}$ such that

$$\lim_{n\to\infty} h_n^{(2)}(x_2) = A_2$$

exists.

Continuing this process of selection, we get a sequence

$$(h_n^{(m)})_{n=1}^{\infty}$$

such that, for m = 1, 2, ...,

$$\lim_{n\to\infty} h_n^{(m)}(x_m) = A_m$$

exists. We now consider the *diagonal sequence* $(h_n^{(n)})_{n=1}^{\infty}$. For fixed k, $\{h_n^{(n)}(x_k)\}_{n\geq k}$ is a subsequence of $\{h_n^{(k)}(x_k)\}_{n\geq k}$ and hence converges to A_k. Therefore $(h_n^{(n)})_{n=1}^{\infty}$ converges at every point of D. □

Proposition 51 Let F be an infinite collection of monotonically increasing functions on [a,b] and suppose that F is uniformly bounded. Then we may select from F a sequence $(f_n)_{n=1}^{\infty}$ which converges at every point of [a,b] to some monotonically increasing function g.

Proof Using Proposition 50, we let D be the set of all rational points of [a,b] and we adjoin the point a to this collection, in case a is irrational. At every point $x_k \in D$ there exists a finite limit

$$\lim_{n\to\infty} f^{(n)}(x_k)$$

for a sequence $(f^{(n)})_{n=1}^{\infty} = F_0$ selected from F by use of Proposition 50.

We define a function h by putting, for every $x_k \in D$,

$$h(x_k) = \lim_{n\to\infty} f^{(n)}(x_k).$$

This function h is so far only defined on the set D; one can easily see that h is increasing on D.

We now extend the domain of definition of h from D to all of the interval [a,b] by putting, for every $x \in (a,b]$,

$$h(x) = \sup\{h(x_k): x_k < x, x_k \in D\}.$$

It is easy to see that h is monotonically increasing on [a,b]. Hence the set S of discontinuities of h in [a,b] must be countable (by Proposition 38).

At every point x_0 at which h is continuous, we have

$$\lim_{n\to\infty} f^{(n)}(x_0) = h(x_0).$$

Indeed, for any $\varepsilon > 0$, we can find points x_k and x_j in D such that

$$x_k < x_0 < x_j, \quad h(x_j) - h(x_k) < \frac{\varepsilon}{2}$$

holds. Keeping these points fixed, we can find some index n_0 such that for all $n \geq n_0$ we have

$$|f^{(n)}(x_k) - h(x_k)| < \frac{\varepsilon}{2}, \quad |f^{(n)}(x_j) - h(x_j)| < \frac{\varepsilon}{2}.$$

We can see that for all such n

$$h(x_0) - \varepsilon < f^{(n)}(x_k) \leq f^{(n)}(x_j) < h(x_0) + \varepsilon$$

holds. Since

$$f^{(n)}(x_k) \leq f^{(n)}(x_0) \leq f^{(n)}(x_j),$$

we have for all $n \geq n_0$

$$h(x_0) - \varepsilon < f^{(n)}(x_0) < h(x_0) + \varepsilon$$

and therefore

$$\lim_{n \to \infty} f^{(n)}(x_0) = h(x_0).$$

Accordingly, the relation

$$\lim_{n \to \infty} f^{(n)}(x) = h(x)$$

need not be fulfilled for an at most countable set S of points of discontinuity of h.

We apply Proposition 50 once more. By Proposition 50 applied to the set F_0 of functions and to the countable number of points S, we obtain a sequence $(f_n)_{n=1}^{\infty}$ of functions in F_0 which converges at all points of [a,b] [since, whenever the sequence $(f^{(n)})_{n=1}^{\infty}$ converges, its subsequence $(f_n)_{n=1}^{\infty}$ must converge as well]. We put, for $x \in [a,b]$,

$$g(x) = \lim_{n \to \infty} f_n(x).$$

Evidently, g is a monotonically increasing function on [a,b]. □

Proposition 52 (Helly's Selection Principle) Let F be an infinite collection of real-valued functions on [a,b]. Suppose F is uniformly bounded by a constant K and let the total variation of all functions f in F be bounded by the same constant K, that is,

$$|f(x)| \leq K, \quad V_a^b f \leq K \quad \text{with } K < \infty$$

for all $f \in F$ and all $x \in [a,b]$. Then we may select from F a sequence of functions $(f_n)_{n=1}^{\infty}$ such that this sequence converges at every point of $[a,b]$ to some function g of finite variation.

Proof For every $f \in F$, we set

$$v(a) = 0, \quad v(x) = V_a^x f, \quad \text{and} \quad w(x) = v(x) - f(x).$$

Note that for each f we get two increasing functions v and w and

$$|v(x)| \le K \quad \text{and} \quad |w(x)| \le 2K \quad \text{for all} \quad x \in [a,b].$$

We apply to the collection $\{v\}$ of functions Proposition 51 to obtain a convergent sequence $(v_k)_{k=1}^{\infty}$ of functions and put

$$\lim_{n\to\infty} v_k(x) = \alpha(x) \quad \text{for} \quad x \in [a,b].$$

To every function v_k there corresponds a function w_k via the equation by which the function w was introduced at the beginning of the proof. We apply Proposition 51 to the sequence $(w_k)_{k=1}^{\infty}$ and obtain a convergent subsequence $(w_{k_i})_{i=1}^{\infty}$ of functions; we put

$$\lim_{i\to\infty} w_{k_i}(x) = \beta(x) \quad \text{for} \quad x \in [a,b].$$

But then the sequence of functions

$$f_{k_i}(x) = v_{k_i}(x) - w_{k_i}(x)$$

selected from the infinite collection F converges (pointwise) to the function g given by

$$g(x) = \alpha(x) - \beta(x)$$

where $x \in [a,b]$, and the proof is finished. □

Remarks There is of course no loss of generality by assuming in Proposition 52 that the two bounds are the same; if

$$|f(x)| \le K_1 \quad \text{and} \quad V_a^b f \le K_2 \quad \text{for all } f \in F \text{ and all } x \in [a,b],$$

we simply take $K = \max\{K_1, K_2\}$.

Functions of finite variation need not be of a "tame" nature; in Exercise 20 at the end of Chapter 3 we will give an example of a function of finite variation that oscillates wildly as it approaches the origin $x = 0$.

4. *Functional Equations*

Proposition 53 (Cauchy) Any continuous function f mapping R^1 into itself such that

$$f(x + y) = f(x) + f(y) \tag{4}$$

for any $x, y \in R^1$ is of the form

$$f(x) = cx, \tag{4a}$$

where c is a constant.

Proof Indeed, for each integer $n > 0$, we have, by induction on n, $f(nx) = n\,f(x)$; replacing x by x/n, we see that

$$f\left(\frac{x}{n}\right) = \frac{1}{n}\,f(x).$$

On the other hand, $f(0 + x) = f(0) + f(x)$, hence $f(0) = 0$, and

$$f[x + (-x)] = f(x) + f(-x) = f(0) = 0,$$

hence $f(-x) = -f(x)$. Thus, for any pair of integers p, q such that $q > 0$, we have

$$f\left(\frac{px}{q}\right) = \frac{p}{q}\,f(x);$$

in other words, $f(rx) = r\,f(x)$ for any rational number r. If we let $x = 1$ and denote $f(1)$ by c, then we obtain $f(r) = cr$.

Let ρ be any irrational number. We choose for ρ an approximating sequence $(r_n)_{n=1}^{\infty}$ of rational numbers, for example, the sequence of finite decimal fractions approximating ρ. Then, for $n = 1, 2, \ldots,$

$$f(r_n) = cr_n.$$

Since f is assumed continuous on R^1, we obtain, on passing to the limit as $n \to \infty$,

$$f(\rho) = \lim_{n\to\infty} f(r_n) = \lim_{n\to\infty} cr_n = c\rho. \qquad \square$$

Remarks

1. In the foregoing proof we obtained that $f(x) = cx$, where c is a constant, for any rational x without using the hypothesis that f is continuous on R^1. Using Proposition 34, we can readily deduce that this function has a unique continuous extension $f(x) = cx$ with $x \in R^1$.

2. If f is a mapping of R^1 into itself and satisfies the functional equation (4), then the assumption that f is continuous at a single point x_0 implies that f is continuous everywhere on R^1. The verification of this claim is simple and we leave it as an exercise for the reader.

Corollary Any continuous mapping g of R^1 into itself such that

$$g\left(\frac{x + y}{2}\right) = \frac{g(x) + g(y)}{2}$$

for any $x, y \in R^1$ is of the form

$$g(x) = cx + a,$$

where c and a are constants.

Proof Indeed, for $y = 0$, we obtain from the functional equation

$$g\left(\frac{x}{2}\right) = \frac{g(x) + g(0)}{2} = \frac{g(x) + a}{2},$$

where $a = g(0)$; thus

$$\frac{g(x) + g(y)}{2} = g\left(\frac{x + y}{2}\right) = \frac{g(x + y) + a}{2},$$

that is, $g(x + y) = g(x) + g(y) - a$, and, with $f(x) = g(x) - a$,

$$f(x + y) = f(x) + f(y).$$

But g is continuous; hence f is continuous and so, by Proposition 53, f is of the form $f(x) = cx$. Thus g is of the form $g(x) = cx + a$ and the proof is finished. □

It is clear that the function $f(x) = cx$, with c being a constant and $x \in R^1$, satisfies the functional equation (4). If for $a > 0$,

$$f(x) = a^x, \tag{5}$$

then for all $x, y \in R^1$ we have the functional equation

$$f(x + y) = f(x)\,f(y) \tag{5a}$$

corresponding to the familiar law of exponents

$$a^{x+y} = a^x\,a^y.$$

Proposition 54 The only nonzero function continuous on R^1 that satisfies the functional equation (5a) is the exponential function.

Proof Let f be a function satisfying the hypothesis of the proposition. Then there is a point x_0 such that $f(x_0) \neq 0$. In (5a) we replace y by $x_0 - x$ and observe that

$$f(x)\,f(x_0 - x) = f(x_0) \neq 0.$$

Hence, $f(x) \neq 0$ for any x. For $x = y = z/2$, we get from (5a) the relation

$$f(z) = \left[f\left(\frac{z}{2}\right)\right]^2$$

and it follows that f(x) is positive for any $x \in R^1$. We may therefore take the logarithm (say, to the base e):

$$\log_e f(x + y) = \log_e f(x) + \log_e f(y).$$

Putting

$$g(x) = \log_e f(x),$$

we see that g is continuous (by Proposition 28) and that

$$g(x + y) = g(x) + g(y)$$

holds for any $x, y \in R^1$. By Proposition 53, g is of the form

$$g(x) = \log_e f(x) = cx \quad (c \text{ constant})$$

and so

$$f(x) = e^{cx} = (e^c)^x = a^x.$$

The proof is finished. □

If

$$f(x) = \log_a x \quad (a > 0,\ a \neq 1), \tag{6}$$

then for all positive x and y

$$f(xy) = f(x) + f(y). \tag{6a}$$

This corresponds to the familiar law of logarithms

$$\log_a xy = \log_a x + \log_a y.$$

This relation, together with continuity, is enough to characterize the logarithmic function.

Proposition 55 The only nonzero function continuous on $(0,\infty)$ that satisfies the functional equation (6a) is the logarithmic function.

Proof Let f be a function satisfying the stipulations of the proposition. For $x > 0$, we put

$$x = e^t, \quad \text{that is,} \quad t = \log_e x,$$

where t varies over R^1. Moreover, we set

$$g(t) = f(e^t), \quad \text{that is,} \quad f(x) = g(\log_e x).$$

But g is continuous and satisfies

$$g(t + s) = f(e^{t+s}) = f(e^t e^s) = f(e^t) + f(e^s) = g(t) + g(s)$$

and is therefore (by Proposition 53) of the form $g(t) = ct$, that is,

$$f(x) = c \log_e x.$$

Since $f \neq 0$, c cannot be zero; we may therefore set $a = e^{1/c}$ and get

$$f(x) = \log_a x.$$

This completes the proof. □

For

$$f(x) = x^{\alpha} \tag{7}$$

we have the functional equation

$$f(xy) = f(x)\ f(y) \tag{7a}$$

for all positive x and y, because

$$(xy)^{\alpha} = x^{\alpha}y^{\alpha}.$$

This relationship, together with continuity, characterizes the function mapping x to x^{α} with $x > 0$.

Proposition 56 The only nonzero function f continuous on $(0,\infty)$ and and satisfying the functional equation (7a) is the function $f(x) = x^{\alpha}$.

Proof Let f be a continuous function satisfying the assumptions of the proposition. For $x > 0$, we put (as in the proof of Proposition 55)

$$x = e^t, \quad \text{that is,} \quad t = \log_e x.$$

We set

$$g(t) = f(e^t), \quad \text{that is,} \quad f(x) = g(\log_e x).$$

But

$$g(t + s) = f(e^{t+s}) = f(e^t e^s) = f(e^t)\ f(e^s) = g(t)\ g(s);$$

hence g satisfies the functional equation (5a). By Proposition 54,

$$g(t) = a^t \quad (a > 0),$$

and so

$$f(x) = (a)^{\log_e x} = x^{\alpha},$$

where

$$\alpha = \log_e a.$$

This completes the proof. □

Finally, we consider a functional equation for trigonometric and hyperbolic functions.

If, for $a > 0$,

$$f(x) = \cos ax, \quad \text{resp.} \quad f(x) = \cosh ax \tag{8}$$

then for any $x, y \in R^1$

$$f(y + x) + f(y - x) = 2\, f(x)\, f(y). \tag{8a}$$

This follows immediately from the familiar formulas

$$\cos (y \pm x) = \cos x \cos y \mp \sin x \sin y,$$

$$\cosh (y \pm x) = \cosh x \cosh y \pm \sinh x \sinh y.$$

The functional equation (8a), together with continuity, suffices to characterize the two cosine functions.

Proposition 57 The only nonzero continuous functions on R^1 that satisfy the functional equation (8a) are the trigonometric and the hyperbolic cosines (8).

Proof Let some function f satisfy the requirements of the proposition. We let $x = 0$ and choose y such that $f(y) \neq 0$; then

$$f(0) = 1. \tag{9}$$

For $y = 0$ we get

$$f(-x) = f(x), \tag{10}$$

that is, f is an *even function*.

Since the continuous function f is positive for $x = 0$, there exists a positive number c such that $f(x)$ is positive on the entire interval $[0,c]$. The following considerations follow two different lines of reasoning depending on the two cases:

Case 1: $f(c) \le 1$,

Case 2: $f(c) > 1$

We consider Case 1 first. Since $0 < f(c) \le 1$, there is a θ with $0 \le \theta < \pi/2$ such that

$$f(c) = \cos \theta. \tag{11}$$

We now write (8a) in the form

$$f(y + x) = 2 f(x) f(y) - f(y - x)$$

and put, successively,

$$x = c, \quad y = c,$$

$$x = c, \quad y = 2c,$$

$$x = c, \quad y = 3c,$$

etc. Using (9) and (11), we obtain

$$f(2c) = 2 \cos^2 \theta - 1 = \cos 2\theta,$$

$$f(3c) = 2 \cos \theta \cos 2\theta - \cos \theta = \cos 3\theta,$$

$$f(4c) = 2 \cos \theta \cos 3\theta - \cos 2\theta = \cos 4\theta,$$

etc. By induction, we get for any $m \in N$,

$$f(mc) = \cos m\theta. \tag{12}$$

If in (8a) we put $x = y = \frac{1}{2}c$, we get, using (9) and (11),

$$\left[f\left(\frac{1}{2} c\right)\right]^2 = \frac{f(0) + f(c)}{2} = \frac{1 + \cos \theta}{2} = \left(\cos \frac{1}{2}\theta\right)^2;$$

since $f(x)$ is positive between 0 and c and $\cos x$ is positive between 0 and θ, we may take the root and obtain

$$f\left(\frac{1}{2} c\right) = \cos \frac{1}{2}\theta.$$

In the same fashion we get, putting $x = y = (1/2^2)c$ in (8a),

$$f\left(\frac{1}{2^2}c\right) = \cos \frac{1}{2^2}\theta,$$

etc. By induction, for n = 1, 2, 3, ..., we get

$$f\left(\frac{1}{2^n}c\right) = \cos \frac{1}{2^n}\theta. \tag{13}$$

By repeating the process that got us from (11) to (12), we obtain from (13)

$$f\left(\frac{m}{2^n}c\right) = \cos \frac{m}{2^n}\theta.$$

Thus, if x is of the form $m/2^n$, we have

$$f(cx) = \cos \theta x. \tag{14}$$

But every positive real number can be represented as a p-adic fraction with p = 2. Since f and the trigonometric cosine function are continuous, we see that (14) holds for any $x > 0$. That (14) also holds for $x = 0$ and $x < 0$, can be seen from (9) and (10), respectively. If in (14) we replace x by x/c and put $\theta/c = a$, we obtain

$$f(x) = \cos ax.$$

In Case 2 we have $f(c) > 1$. Then there exists a θ such that $f(c) = \cosh \theta$. By repeating the reasoning used in Case 1 and applying the corresponding formulas for the hyperbolic cosine we get, for $a > 0$,

$$f(x) = \cosh ax.$$

For $a = 0$, we get of course in both cases $f(x) = 1$ for all $x \in R^1$.

This completes the proof. □

Remark The foregoing five functional equations were first investigated by A. L. Cauchy who also discovered their continuous solutions.

Exercises

1. Let E be the set of points in [0,1] such that x is in E if and only if the decimal expansion of x does not contain the digit 7. Show that E is a perfect set.

2. Let f be a real-valued function on [a,b] and suppose that the sets

$$\{x \in [a,b]: f(x) \geq \alpha\} \quad \text{and} \quad \{x \in [a,b]: f(x) \leq \alpha\}$$

are closed for any $\alpha \in R^1$. Show that f is continuous.

3. If f is a continuous function on a closed set F in R^1, then there exists a continuous extension f* of f to R^1. Moreover, if $|f(x)| \leq M$ for all $x \in F$, then $|f^*(x)| \leq M$ for all $x \in R^1$.

Proof There exist disjoint open intervals (a_n, b_n) such that $R^1 - F$, being an open set, can be represented uniquely up to the order of summation by the union of these intervals. Define f* on each of these intervals (a_n, b_n) linearly by

$$f^*(x) = f(a_n) + \frac{f(b_n) - f(a_n)}{b_n - a_n}(x - a_n).$$ □

4. Show that $\lim_{x \to 3} (x^2 + 2x) = 15$.

Solution Note that $|(x^2 + 2x) - 15| = |x - 3||x + 5|$. If $|x - 3| < 1$, then $x \in (2,4)$ and $x + 5 \in (7,9)$. Hence, given any $\varepsilon > 0$, there is a $\delta > 0$, (namely $\delta = \min\{1, \varepsilon/9\}$) for which we have $|(x^2 + 2x) - 15| < \varepsilon$ whenever $0 < |x - 3| < \delta$. □

5. Show that $f(x) = x^2$ is not uniformly continuous on R^1.

Solution Suppose that there was a $\delta > 0$ such that

$$|f(x) - f(a)| < 2 \quad \text{whenever} \quad |x - a| < \delta \quad \text{for all } a \in R^1.$$

We would then have for $a > 0$ and $x = a + (\delta/2)$

$$|f(x) - f(a)| = |x - a||x + a| = \frac{\delta}{2}\left|2a + \frac{\delta}{2}\right| < 2,$$

implying that $a\delta < 2$ for all $a > 0$ which is clearly false. □

6. The function f is defined and bounded on a neighborhood of a point c; M(h) and m(h) are respectively its supremum and infimum on $[c - h, c + h]$ for any sufficiently small h. Show that, as $h \to 0+$, M(h) and m(h) tend to limits M and m, say, and that f is continuous if and only if $M = m$.

Solution M(h) decreases with h and is bounded below by $m(h_0)$, if $0 < h \le h_0$. $M = m$ implies $M(h) - m(h) < \varepsilon$ for $0 < h < \delta$ and this in turn implies that $|f(x) - f(c)| < \varepsilon$ for $|x - c| < \delta$. Conversely, $|f(x) - f(c)| < \varepsilon$ for $|x - c| < \delta$ implies that $M(h) - m(h) \le 2\varepsilon$ for $0 < h < \delta$; but this means that $M = m$. ∎

7. Continuation of the foregoing exercise: Is M the largest number which is limit of sequences of the form

$$f(x_1),\ f(x_2),\ f(x_3),\ \ldots,$$

with $x_n \in (c - h, c + h)$, as $x_n \to c$? How can m be characterized?

8. Let f be continuous on [a,b] and let g be defined on [a,b] by $g(a) = f(a)$ and, for $a < x \le b$, $g(x) = \sup\{f(t) : a \le t \le x\}$. Show that g is continuous on [a,b].

9. The function f is defined on [0,1] as follows:

$$f(x) = x \quad \text{if } x \text{ is rational,}$$
$$= 1 - x \quad \text{if } x \text{ is irrational.}$$

Let

$$F(x) = f(x) \quad \text{if} \quad x \ne 1,2,\ 1/3,$$
$$= 1/2 \quad \text{if} \quad x = 1/3,$$
$$= 1/3 \quad \text{if} \quad x = 1/2.$$

Is $f = f^{-1}$ and $F = F^{-1}$ on [0,1]? Is f only continuous at $x = 1/2$ and F nowhere continuous on [0,1]?

10. A real-valued function f defined on (A,B) is said to be *convex* if

$$f[\lambda x + (1 - \lambda)y] \le \lambda f(x) + (1 - \lambda) f(y)$$

whenever $A < x < B$, $A < y < B$, $0 < \lambda < 1$. Geometrically, this means that if P, Q, and R are any three points on the graph of f with Q between P and R, then Q is on or below chord PR. Prove that every convex function is continuous.

Solution Let $[a,b]$ be any closed subinterval of (A,B). We show that the set $\{f(x): x \in [a,b]\}$ is bounded above and below and then establish that there exists a constant K so that for any two points $x, y \in [a,b]$ we have $|f(x) - f(y)| \le K|x - y|$.

Observe that $M = \max\{f(a), f(b)\}$ is an upper bound for f on $[a,b]$, since for any point $z = \lambda a + (1 - \lambda)b$ in $[a,b]$

$$f(z) \le \lambda\, f(a) + (1 - \lambda)\, f(b) \le \lambda M + (1 - \lambda)M = M.$$

But f is also bounded from below because, writing an arbitrary point in the form $(a + b)/2 + t$, we have

$$f\left(\frac{a + b}{2}\right) \le \frac{1}{2}\, f\left(\frac{a + b}{2} + t\right) + \frac{1}{2}\, f\left(\frac{a + b}{2} - t\right)$$

or

$$f\left(\frac{a + b}{2} - t\right) \ge 2\, f\left(\frac{a + b}{2}\right) + f\left(\frac{a + b}{2} - t\right).$$

Using M as the upper bound,

$$-f\left(\frac{a + b}{2} - t\right) \ge -M,$$

so

$$f\left(\frac{a + b}{2} + t\right) \ge 2\, f\left(\frac{a + b}{2}\right) - M = m.$$

This shows that M and m are upper and lower bounds of f on $[a,b]$, respectively.

We next pick $h > 0$ so that $a - h$ and $b + h$ belong to (A,B), and let m and M be the lower and upper bounds for f on $[a - h, b + h]$. If x and y are distinct points of $[a,b]$, set

$$z = y + \frac{h}{|y - x|}\,(y - x), \quad \lambda = \frac{|y - x|}{h + |y - x|}.$$

Then $z \in [a - h, b + h]$, $y = \lambda z + (1 - \lambda)x$, and we have

$$f(y) \le \lambda\, f(z) + (1 - \lambda)\, f(x) = \lambda[f(z) - f(x)] + f(x),$$

$$f(y) - f(x) \le \lambda(M - m) < \frac{|y - x|}{h}(M - m) = K|y - x|,$$

where $K = (M - m)/h$. Since this is true for any $x, y \in [a,b]$, we conclude that $|f(y) - f(x)| \le K|y - x|$ as desired. □

11. Show that f is convex on (A,B) if and only if for all sets of distinct $x_1, x_2, x_3 \in (A,B)$,

$$\frac{(x_3 - x_2) f(x_1) + (x_1 - x_3) f(x_2) + (x_2 - x_1) f(x_3)}{(x_1 - x_2)(x_2 - x_3)(x_3 - x_1)}$$

is nonnegative.

Hint Let $x_1 < x_2 < x_3$ and consider the determinant

$$\begin{vmatrix} 1 & x_1 & f(x_1) \\ 1 & x_2 & f(x_2) \\ 1 & x_3 & f(x_3) \end{vmatrix}.$$

Interpret this determinant geometrically.

12. Let f_1, f_2, and f_3 be continuous real-valued functions on [a,b] and let f(x) denote that one of the three values $f_1(x)$, $f_2(x)$, and $f_3(x)$ that lies between the other two. Show that f is continuous on [a,b].

Hint

$$f(x) = f_1(x) + f_2(x) + f_3(x) - \max\{f_1(x), f_2(x), f_3(x)\}$$
$$- \min\{f_1(x), f_2(x), f_3(x)\}.$$

13. Let f be defined and bounded on [0,1], and $f(ax) = b\, f(x)$ for $0 \le x \le 1/a$, with a and b numbers larger than 1. Show that $f(0+) = f(0)$.

Hint $f(0) = 0$; if M is an upper bound for $|f|$ on [0,1], then, for $0 \le x \le a^{-n}$, $|f(x)| = |b^{-n} f(a^n x)| \le Mb^{-n}$.

14. Let A be a nonempty subset of R^1 and define the *distance from a point* $x \in R^1$ *to* A by

$$d(x,A) = \inf\{|x - p| : p \in A\}.$$

If A and B are two nonempty subsets of R^1, we define the *distance between* A *and* B by

$$d(A,B) = \inf\{|p - q|: p \in A,\ q \in B\}.$$

Prove the following assertions:

(a) The statements $d(x,A) = 0$ and $x \in \bar{A}$ are equivalent.

(b) If A is a fixed subset of R^1, then $d(x,A)$ is uniformly continuous on R^1 because $p \in A$ implies

$$d(x,A) \le |x - p| \le |x - y| + |y - p| \le |x - y| + d(y,A)$$

and so

$$|d(x,A) - d(y,A)| \le |x - y|.$$

(c) Let A be a compact subset of R^1 and $B \subset R^1$. Then there is a point $p \in A$ such that $d(p,A) = d(A,B)$.

(d) If A is compact and B is closed, then there are points $p \in A$ and $q \in B$ such that $|p - q| = d(A,B)$.

(e) Let A and B be two disjoint closed subsets of R^1. Then there exist disjoint open sets G and H such that $A \subset G$ and $B \subset H$. To show this, try the subsets

$$\{x \in R^1: d(x,A) - d(x,B) \lessgtr 0\}$$

for G and H, respectively.

(f) Let A and B be as in part (e). Show that the function

$$f(x) = \frac{d(x,A)}{d(x,A) + d(x,B)}$$

is a continuous function on R^1 whose range lies in $[0,1]$ and $f(x) = 0$ precisely on A and $f(x) = 1$ precisely on B. Setting

$$V = f^{-1}([0,\tfrac{1}{2})), \quad W = f^{-1}((\tfrac{1}{2},1]),$$

we get that V and W are open and disjoint, and that $A \subset V$, $B \subset W$.

15. Let g be a real-valued continuous function on R^1 and let $Z(g)$ be the *zeroset* of g, that is, the set of all $x \in R^1$ at which $g(x) = 0$.

(a) Show that $Z(g)$ is closed.

(b) Prove that every closed set $A \subset R^1$ is the zeroset $Z(f)$ of some continuous real-valued function f on R^1.

Hint For part (b) use part (f) of Exercise 14.

16. Let f be a real-valued function on R^1. Suppose that f is *contracting*, that is, there exists a fixed positive number k less than 1 such that for any x, y $\in R^1$, the inequality

$$|f(x) - f(y)| \le k|x - y|$$

holds. Then the equation x = f(x) possesses a unique solution in R^1 and it is obtained by iteration, or successive approximation, beginning with some $x_0 \in R^1$. By this we mean that no matter how x_0 is chosen in R^1, the sequence $(x_n)_{n=0}^{\infty}$ with $x_1 = f(x_0)$, $x_2 = f(x_1)$, ..., $x_n = f(x_{n-1})$, ... converges to a unique solution of the equation x = f(x). The point x in x = f(x) is called the *fixed point of* f.

Show that a contracting function f on R^1 is uniformly continuous and thus takes Cauchy sequences into Cauchy sequences and that a fixed point of a contracting function must be unique.

Hint See the constructive solution of Kepler's Equation following Proposition 58 in Chapter 1.

17. Let I be an interval and let t be a point of I. Let f be a real-valued function defined for all x $\in$ I, x $\neq$ t. Suppose there is a $\delta > 0$ such that f(x) is bounded above for $0 < |x - t| < \delta$, x $\in$ I, that is, $f(x) \le C$ if $0 < |x - t| < \delta$, x $\in$ I, where C is a constant. We then consider the set G of all points y obtained in the following manner: There is a sequence $(x_n)_{n=1}^{\infty}$ in I, $x_n \neq t$ for all n, such that

$$\lim_{n\to\infty} x_n = t, \quad \lim_{n\to\infty} f(x_n) = y.$$

If the set G is nonempty, then it is bounded above (by the same constant C as above). Denote by A the supremum of the set G. We call A the *limit superior of* f(x) *as* x → t, and we write it in any one of the forms

$$\limsup_{x\to t} f(x), \quad \overline{\lim_{x\to t}}\, f(x).$$

If G is empty, we set

$$\limsup_{x\to t} f(x) = -\infty,$$

and if, for any $\delta > 0$, $f(x)$ is not bounded above on the set given by $0 < |x - t| < \delta$, $x \in I$, we put

$$\limsup_{x \to t} f(x) = \infty.$$

The set of all x such that $0 < |x - t| < \delta$ is called a *deleted neighborhood of* t; it is the open interval $(t - \delta, t + \delta)$ with the midpoint t removed.

A more compact way of defining the limit superior is to put

$$\overline{\lim_{x \to t}}\ f(x) = \inf_{\delta > 0}\ \sup_{0 < |x-t| < \delta} f(x).$$

Prove that A is the limit superior of $f(x)$ as $x \to t$ if and only if it has the following two properties, where ε is an arbitrary positive number:

(i) The inequality $f(x) < A + \varepsilon$ holds for all x in some deleted neighborhood of t: $0 < |x - t| < \delta$.

(ii) The inequality $f(x) > A - \varepsilon$ holds for some x in every deleted neighborhood t: $0 < |x - t| < \delta$.

Hints Let A be a finite real number. Suppose that the assertion (i) is false. Then there is an $\varepsilon_0 > 0$ such that for any $\delta = 1/n$ there is a point x_n satisfying

$$0 < |x_n - t| < \frac{1}{n}, \quad f(x_n) \ge A + \varepsilon_0.$$

Since $f(x)$ is bounded above for $0 < |x - t| < \delta$, $x \in I$ (for some δ larger than zero), it follows that the sequence $\{f(x_n)\}_{n=1}^{\infty}$ is bounded. By the Bolzano-Weierstrass Theorem, there is a convergent subsequence $\{f(x_{n_k})\}_{k=1}^{\infty}$. Denote its limit by y. Since $x_{n_k} \to t$ as $k \to \infty$, the point y belongs to the set G. This implies that $A \ge y$. Thus $y = \lim_{k \to \infty} f(x_{n_k}) \ge A + \varepsilon_0$, a contradiction.

To prove the assertion (ii), we suppose that it is false and derive a contradition. If (ii) is false, then there are positive numbers ε, δ such that

$$f(x) \le A - \varepsilon \quad \text{if} \quad 0 < |x - t| < \delta.$$

But then for any sequence $(x_n)_{n=1}^{\infty}$ in I, $x_n \neq t$ for all n, such that

$$\lim_{n\to\infty} x_n = t, \quad \lim_{n\to\infty} f(x_n) = y,$$

we have that $y \le A - \varepsilon$. It follows that $A - \varepsilon$ is an upper bound of the set G. But this is not possible since A is the least upper bound of G.

18. Let I, t, and f be as in Exercise 17. We define the *limit inferior of* f(x) *as* $x \to t$, denoted by

$$\liminf_{x\to t} f(x) \quad \text{or} \quad \underline{\lim}_{x\to t} f(x),$$

as follows:

$$\liminf_{x\to t} f(x) = \sup_{\delta>0} \inf_{0<|x-t|<\delta} f(x).$$

(a) Formulate and prove the analog of the assertion in Exercise 17.

(b) At the end of Section 4 in Chapter 1 we studied the limit superior and the limit inferior of numerical sequences and discussed some of their properties. Show that the limit superior and the limit inferior of functions satisfy completely analogous properties.

19. A real-valued function f defined on an interval I is said to be *lower semicontinuous at a point* t *of* I if either

$$\liminf_{x\to t} f(x) = \infty \quad \text{or} \quad \liminf_{x\to t} f(x) \ge f(t).$$

It is said to be *upper semicontinuous at a point* t if either

$$\limsup_{x\to t} f(x) = -\infty \quad \text{or} \quad \limsup_{x\to t} f(x) \le f(t).$$

If f is lower (upper) semicontinuous at each point of I, then we say that it is *lower (upper) semicontinuous on* I. It is easy to see that f is continuous if and only if it is both lower semicontinuous and upper semicontinuous.

Show: If f is lower semicontinuous (upper semicontinuous) on a closed bounded interval, then f(x) is bounded below (above) and f assumes its least value (largest value) in the interval.

20. Let I, t, and f be as in Exercise 17. We define the *right-side limit superior of* f(x) *as* $x \to t$ as follows:

$$\overline{\lim_{x \to t+}} f(x) = \inf_{\delta > 0} \sup_{0 < x-t < \delta} f(x).$$

The left-side limit superior and the two one-sided limit inferiors are defined similarly.

Let

$$\begin{aligned} g(x) &= x \sin \frac{1}{x} && \text{for } x > 0 \\ &= 0 && \text{for } x = 0, \\ &= x\left(3 + \sin \frac{1}{x}\right) && \text{for } x < 0. \end{aligned}$$

and define

$$f(x) = \frac{g(x)}{x}$$

for $x \neq 0$. Sketch the graph of the function g and verify that

$$\limsup_{x \to 0+} f(x) = 1, \quad \liminf_{x \to 0+} f(x) = -1,$$

$$\limsup_{x \to 0-} f(x) = 4, \quad \liminf_{x \to 0-} f(x) = 2.$$

21. Let $(g_n)_{n=1}^{\infty}$ be a sequence of functions of finite variation over a closed bounded interval [a,b] with the following property: For every $\varepsilon > 0$, there is an integer n_0 such that $n \geq n_0$, $m \geq n_0$ implies

$$V_a^b(g_n - g_m) < \varepsilon.$$

Assuming that $\{g_n(a)\}_{n=1}^{\infty}$ is a Cauchy sequence, show that there is a function g, of finite variation over [a,b], such that $V_a^b(g_n - g) \to 0$ as $n \to \infty$.

22. Let E be a set of real numbers and $B \subset E$. We say that B is *dense in* E if $\overline{B} = E$. If two continuous real-valued functions coincide

on a dense subset of their common domain of definition, show that they then coincide everywhere.

23. Let (x,y) and (u,v) be ordered pairs. Show that (x,y) = (u,v) if and only if x = u and y = v.

24. Let f be a continuous function mapping the interval [0,1] into itself. Show that for some x in [0,1] we have x = f(x).

25. Let f be continuous on [0,2a] and f(0) = f(2a). Show that for some x in [0,a] we have f(x) = f(x + a).

26. Show that, for $t \neq 0$,

$$\lim_{n \to \infty} \left(\cos \frac{t}{2}\right)\left(\cos \frac{t}{2^2}\right)\left(\cos \frac{t}{2^3}\right) \cdots \left(\cos \frac{t}{2^n}\right) = \frac{\sin t}{t}.$$

Solution Clearly,

$$\sin t = 2\left(\cos \frac{t}{2}\right)\left(\sin \frac{t}{2}\right) = 2^2\left(\cos \frac{t}{2}\right)\left(\cos \frac{t}{2^2}\right)\left(\sin \frac{t}{2^2}\right) = \cdots$$

$$= 2^n\left(\cos \frac{t}{2}\right)\left(\cos \frac{t}{2^2}\right) \cdots \left(\cos \frac{t}{2^n}\right)\left(\sin \frac{t}{2^n}\right)$$

and so

$$\frac{\sin t}{2^n \sin(t/2^n)} = \frac{\sin t}{t} \frac{1/2^n}{\sin(t/2^n)}$$

holds. But $x_n = t/2^n \to 0$ as $n \to \infty$ and $(\sin x)/x \to 1$ as $x \to 0$. □

27. An *algebraic number* is one that satisfies some equation of the form

$$x^n + a_1x^{n-1} + \cdots + a_n = 0$$

with rational coefficients $a_1, \ldots, a_n$.

Show:

(a) The set of all algebraic numbers is countable.

(b) If α and β are algebraic numbers, then so are $\alpha\beta$ and $\alpha + \beta$.

28. Let g be a real-valued continuous function defined on an interval [a,b], and let E be the set of points x interior to this interval such that there exists a point t lying to the right of x and satisfying g(t) > g(x). Then the set E is either empty or it

decomposes into countably many disjoint open intervals (a_k, b_k), and

$$g(a_k) \leq g(b_k) \tag{E.1}$$

for each of these intervals.

Proof Note that the set E is open, since if $t > x_0$ and $g(t) > g(x_0)$, then by virtue of the continuity, the relations $t > x$, $g(t) > g(x)$ remain valid when x varies in the neighborhood of the point x_0. Thus E, if nonempty, decomposes into countably many disjoint open intervals (a_k, b_k), by Proposition 3; the points a_k, b_k do not belong to the set E. Let x_0 be a point between a_k and b_k; we shall verify that

$$g(x_0) \leq g(b_k); \tag{E.2}$$

(E.1) will follow by letting x_0 tend to a_k. To prove (E.2), let x_1 be a point at which the function g assumes its largest value on the interval $[x_0, b]$. The point x_1 cannot belong to E, for there exists no t such that $x_1 < t \leq b$ and $g(t) > g(x_1)$. Since the part of $[x_0, b]$ to the left of b_k belongs entirely to E, we necessarily have $b_k \leq x_1 \leq b$. We cannot have $g(b_k) < g(x_1)$ since b_k does not belong to E. Thus $g(b_k) = g(x_1)$ and so $g(x_0) \leq g(b_k)$.

Incidentally, it is even true that $g(a_k) = g(b_k)$ except possibly when $a_k = a$. □

29. In Chapter 1 we defined the number e as

$$e = \lim_{n \to \infty} \left(1 + \frac{1}{n}\right)^n.$$

Show that

$$\text{(i)} \quad \lim_{x \to \infty} \left(1 + \frac{1}{x}\right)^x = e$$

and

$$\text{(ii)} \quad \lim_{x \to -\infty} \left(1 + \frac{1}{x}\right)^x = e.$$

It therefore follows that

$$\text{(iii)} \quad \lim_{\alpha \to 0} (1 + \alpha)^{1/\alpha}.$$

Solution To see the validity of (i), we observe that

$$\left(1 + \frac{1}{n_k}\right)^{n_k} \to e$$

when $(n_k)_{k=1}^{\infty}$ is any sequence of natural numbers which increases indefinitely with the index k. Let $(x_k)_{k=1}^{\infty}$ be any sequence of real numbers tending to ∞ and suppose that $x_k > 1$ for all k. We put $n_k = [x_k]$ so that

$$n_k \le x_k < n_k + 1 \quad \text{and} \quad n_k \to +\infty.$$

Since

$$\frac{1}{n_k + 1} < \frac{1}{x_k} \le \frac{1}{n_k},$$

we have

$$\left(1 + \frac{1}{n_k + 1}\right)^{n_k} < \left(1 + \frac{1}{x_k}\right)^{x_k} < \left(1 + \frac{1}{n_k}\right)^{n_k+1}.$$

But

$$\left(1 + \frac{1}{n_k + 1}\right)^{n_k} = \frac{\left(1 + \frac{1}{n_k + 1}\right)^{n_k+1}}{1 + \frac{1}{n_k + 1}}$$

and

$$\left(1 + \frac{1}{n_k}\right)^{n_k+1} = \left(1 + \frac{1}{n_k}\right)^{n_k}\left(1 + \frac{1}{n_k}\right),$$

where

$$\left(1 + \frac{1}{n_k}\right)^{n_k} \to e, \qquad \left(1 + \frac{1}{n_k + 1}\right)^{n_k+1} \to e,$$

$$1 + \frac{1}{n_k} \to 1, \quad \text{and} \quad 1 + \frac{1}{n_k + 1} \to 1.$$

This shows that part (i) holds.

To see the validity of (ii), we suppose that $x_k \to -\infty$ and $x_k < -1$ for all k. By putting $x_k = -y_k$, we get that $y_k \to \infty$ and $y_k > 1$ for all k. Clearly,

$$\left(1 + \frac{1}{x_k}\right)^{x_k} = \left(1 - \frac{1}{y_k}\right)^{-y_k} = \left(\frac{y_k}{y_k - 1}\right)^{y_k}$$

$$= \left(1 + \frac{1}{y_k - 1}\right)^{y_k - 1}\left(1 + \frac{1}{y_k - 1}\right).$$

However, by (i),

$$\left(1 + \frac{1}{y_k - 1}\right)^{y_k - 1} \to e;$$

on the other hand

$$1 + \frac{1}{y_k - 1} \to 1.$$

To obtain (iii) we merely have to replace x by $1/\alpha$ in (i) and (ii). □

30. Let $\log_a t$ denote the logarithm of t to the base a. Using the fact that the exponential function a^x with basis $a > 0$ is continuous (see the remark following Proposition 34), show:

(a) $$\lim_{\alpha \to 0} \frac{\log_a (1 + \alpha)}{\alpha} = \log_a e,$$

(b) $$\lim_{\alpha \to 0} \frac{a^\alpha - 1}{\alpha} = \log_e a,$$

(c) $$\lim_{\alpha \to 0} \frac{(1 + \alpha)^\lambda - 1}{\alpha} = \lambda.$$

N.B.: If in (b) we set $\alpha = 1/n$ with n = 1, 2, 3, ..., we get

$$\lim_{n \to \infty} n(\sqrt[n]{a} - 1) = \log_e a.$$

31. Let $a > 0$ and $b > 0$. Show that

$$\lim_{n\to\infty}\left(\frac{\sqrt[n]{a}+\sqrt[n]{b}}{2}\right)^n = \sqrt{ab}.$$

<u>32</u>. Let $x_n \to x$ with x being a finite real number. Show that

$$\lim_{n\to\infty}\left(1+\frac{x_n}{n}\right)^n = e^x.$$

<u>33</u>. Let f be a function which maps the closed bounded interval [a,b] into itself and satisfies $|f(x) - f(y)| \le |x - y|$. Show that the sequence obtained by choosing x_1 in [a,b] and defining

$$x_{n+1} = \frac{1}{2}[x_n + f(x_n)]$$

converges to some z in [a,b] and $f(z) = z$.

<u>34</u>. Let u_n denote the "ramp" function

$$\begin{aligned} u_n(x) &= -n \quad \text{for} \quad x \le -n, \\ &= x \quad \text{for} \quad -n < x \le n, \\ &= n \quad \text{for} \quad x > n, \end{aligned}$$

and f denote a real-valued function of a real variable. Show that f is continuous if and only if g_n, where

$$g_n(x) = u_n[f(x)],$$

is continuous for all n.

<u>*Solution*</u> Evidently u_n is continuous. Hence, if f is continuous, then g_n is the composition of continuous functions and thus is continuous. Conversely, suppose that g_n is continuous for all n. To show that f is continuous, it is enough to show that the inverse image of (a,b) under f, that is, $f^{-1}[(a,b)]$, is open for every bounded open interval (a,b). We let $n > \max\{|a|,|b|\}$. Then $u_n^{-1}[(a,b)] = (a,b)$ and so

$$f^{-1}[(a,b)] = f^{-1}\{u_n^{-1}[(a,b)]\} = g_n^{-1}[(a,b)],$$

which is an open set by the continuity of g_n. □

35. If f is a real-valued function defined on (0,1), then the formula $f(x) = o(x)$ is an abbreviation for the statement that

$$\frac{f(x)}{x} \to 0 \quad \text{as} \quad x \to 0.$$

Keeping this in mind, show that if

$$\lim_{x\to 0} f(x) = 0 \quad \text{and} \quad f(x) - f\left(\frac{x}{2}\right) = o(x),$$

then $f(x) = o(x)$.

Solution Given $\lambda > 0$ there is a $\sigma > 0$ such that

$$\left|f(x) - f\left(\frac{x}{2}\right)\right| < \lambda|x| \quad \text{for} \quad 0 < |x| < \sigma.$$

Then for $0 < |x| < \sigma$ we have

$$\left|f(x) - f\left(\frac{x}{2^n}\right)\right| \le \sum_{j=0}^{n-1} \left|f\left(\frac{x}{2^j}\right) - f\left(\frac{x}{2^{j+1}}\right)\right| < \sum_{j=0}^{n-1} \frac{\lambda|x|}{2^{j+1}} < \lambda|x|$$

for every positive integer n. Since n is arbitrary we have

$$|f(x)| \le \lambda|x| \quad \text{for} \quad 0 < |x| < \sigma.$$ □

36. Let $a > 0$ and put

$$S_1 = \log_e a \quad \text{and} \quad S_n = \sum_{i=1}^{n-1} \log_e(a - S_i) \quad \text{for} \quad n = 2, 3, \ldots .$$

Show that

$$\lim_{n\to\infty} S_n = a - 1.$$

Solution On the range $x \le a - 1$ the function $x + \log_e(a - x)$ is increasing and its values lie between x and $a - 1$. Since for $a > 0$, $\log_e a \le a - 1$, it follows that the sequence $(S_n)_{n=1}^{\infty}$ is monotonically increasing and bounded above by $a - 1$, and hence convergent. The relation $S_{n+1} = S_n + \log_e(a - S_n)$ implies, by continuity, that the limit is $a - 1$. □

37. Let $a_0 = 0$, $a_1 = 1 + \sin(-1)$, ..., $a_n = 1 + \sin(a_{n-1} - 1)$, ... Evaluate $\lim_{n\to\infty} (1/n) \sum_{k=1}^{n} a_k$.

Solution Let $b_n = a_n - 1$. Then $b_0 = -1$ and $b_n = \sin b_{n-1}$. By induction it follows that every b_n is negative, and that the b_n form an increasing sequence. Hence they have a limit L which must satisfy $L = \sin L$. Thus $L = 0$, $b_n \to 0$, and $a_n \to 1$ as $n \to \infty$. However, if the sequence $(a_n)_{n=1}^{\infty}$ has a limit, then the sequence of averages

$$\frac{1}{n}\sum_{k=1}^{n} a_k \qquad (n = 1, 2, \ldots)$$

has the same limit. □

38. Let f and g be real-valued functions of a real variable. Show that there are numbers x and y such that $0 \le x \le 1$, $0 \le y \le 1$, and $|xy - f(x) - g(y)| \ge 1/4$.

Solution If $|xy - f(x) - g(y)| < 1/4$ for all x and y in the unit interval, then, in particular,

$$|f(0) + g(1)| < 1/4, \quad |f(0) + g(0)| < 1/4, \text{ and}$$
$$|f(1) + g(0)| < 1/4.$$

But then by the triangle inequality

$$\begin{aligned}|1 - f(1) - g(1)| &\ge 1 - |f(1) + g(1)| \\ &\ge 1 - |f(1) + g(0)| - |-g(0) - f(0)| \\ &\quad - |g(1) + f(0)| \\ &> \frac{1}{4}.\end{aligned}$$

□

39. Given an infinite number of points in a plane, prove that if all distances determined between them are integers then the points are all in a straight line.

Solution Assume that P_1, P_2, P_3 are not collinear and are in the set. Any point P_4 in the set lies on the line P_iP_j or on one of the hyperbolas

$$|d(P,P_i) - d(P,P_j)| = 1, 2, \ldots, d(P_i,P_j) - 1,$$

where $i \ne j$ and $i, j = 1, 2, 3$; $d(P,P_i)$ denotes distance between P and P_i. Since any two such loci have at most four intersections, the possible positions for P_4 are finite in number. □

40. Find all solutions of $n^m = m^n$ in integers n and m with $n \neq m$. In other words, show that the only solutions are $2^4 = 4^2$ and $(-2)^{-4} = (-4)^{-2}$.

41. Show that there does not exist a sequence $(K_n)_{n=1}^{\infty}$ of compact sets of rational numbers such that each compact set of rationals is contained in at least one K_n.

Solution Let $(K_n)_{n=1}^{\infty}$ be any sequence of compact sets of rational numbers. For each n, there is a rational $r_n \notin K_n$, with $0 \le r_n < 1/n$. Otherwise, it would be that K_n contained all rationals in the interval $[0,1/n]$, and hence some irrationals (since K_n is closed). The set $\{0, r_1, r_2, \ldots\}$ is compact and not included in any K_n. □

42. Let

$$f(x) = a_1 \sin x + a_2 \sin 2x + \cdots + a_n \sin nx,$$

where $a_1, a_2, \ldots, a_n$ are real numbers and n is a positive integer. Given that

$$|f(x)| \le |\sin x| \quad \text{for all real } x,$$

prove that $|a_1 + 2a_2 + \cdots + na_n| \le 1$.

43. Consider polynomial forms $ax^2 - bx + c$ with integer coefficients which have two distinct zeros in the open interval $0 < x < 1$. Exhibit with a proof the least positive integer value of a for which such a polynomial exists.

Solution Let $f(x) = ax^2 - bx + c = a(x - r)(x - s)$. Then $f(0)\ f(1) = a^2 r(r - 1)s(s - 1)$. The graph of $r(r - 1)$ shows that $0 < r < 1$ implies $0 < r(r - 1) \le 1/4$, with equality if and only if $r = 1/2$. Similarly, $0 < s(s - 1) \le 1/4$. Since $r \neq s$,

$$r(r - 1)s(s - 1) < \frac{1}{16} \quad \text{and} \quad 0 < f(0)\ f(1) < \frac{a^2}{16}.$$

The coefficients a, b, c are integers, and thus $1 \le f(0)\ f(1)$. Consequently $a^2 > 16$, that is, $a \ge 5$. The discriminant $b^2 - 4ac$ shows that the minimum possible value for b is 5. Furthermore, $5x^2 - 5x + 1$ has two distinct roots between 0 and 1. □

44. Show that after deleting the square numbers from the list of positive integers the number we find in the n-th position is equal to $n + \{\sqrt{n}\}$, where $\{\sqrt{n}\}$ denotes the integer closest to $\sqrt{n}$.

Solution To prove the formula by induction, it suffices to show that the difference $D = n + \{\sqrt{n}\} - (n - 1 + \{\sqrt{n - 1}\}) = 1$ or 2, with the value 2 occuring if and only if the number $n + \{\sqrt{n - 1}\}$ is a square number. For convenience, let $\{\sqrt{n - 1}\} = q$. Then of course $q - 1/2 < \sqrt{n - 1} < q + 1/2$ or better

$$q^2 - q + \frac{1}{4} < n - 1 < q^2 + q + \frac{1}{4}$$

This gives

$$q^2 + \frac{5}{4} < n + \{\sqrt{n - 1}\} < (q + 1)^2 + \frac{1}{4}.$$

Therefore the number $n + \{\sqrt{n - 1}\}$ is a square number if and only if $n = (q + 1)^2 - q$. However, then and only then $\sqrt{n} > q + 1/2 > \sqrt{n - 1}$. In other words, then and only then $\{\sqrt{n}\} - \{\sqrt{n - 1}\} = 1$, because this difference is never greater than 1. □

45. In Proposition 3 we proved that every open set of real numbers is the union of a countable collection of disjoint open intervals. We consider another proof of this result here.

If A is an open set of real numbers, we say that the real numbers x and y are "equivalent" and write $x \sim y$ if and only if there is an open interval (a,b) with the set $\{x, y\} \subset (a,b) \subset A$. The relation $\sim$ is clearly an equivalence relation on A and the resulting equivalence classes are disjoint open intervals whose union is A. The fact that there can be only countably many follows since each must contain a distinct rational number.

N.B.: A *relation* $\sim$ *on* A is any subset of $A \times A$. Let $x, y \in A$; we write $x \sim y$ if the ordered pair (x,y) belongs to $\sim$. The relation $\sim$ is an *equivalence relation on* A if

(i) $x \sim x$ for each $x \in A$,

(ii) $x \sim y$ implies $y \sim x$ for all $x, y \in A$, and

(iii) $x \sim y$ and $y \sim z$ implies $x \sim z$ for all $x, y, z \in A$.

The *equivalence class* $x \in A$ is the set $[x] = \{x' \in A: x' \sim x\}$. If $x, y \in A$, note that either $[x] = [y]$ (and this happens precisely when $x \sim y$) or else $[x] \cap [y] = \emptyset$. Since $x \in [x]$ for each $x \in A$, the sets $[x]$, for $x \in A$, are disjoint sets whose union is A.

<u>46</u>. Prove that any uncountably infinite set of real numbers includes a sequence $(x_n)_{n=1}^{\infty}$ of distinct numbers such that

$$\sum_{n=1}^{\infty} x_n = \pm\infty.$$

<u>47</u>. Show that

$$\sin x - \sin \sin x + \sin \sin \sin x - \cdots$$

converges for any $x \in R^1$.

<u>48</u>. Let $x \neq 0$, $\sin x \neq 0$. Show that

$$v_0^a + v_1^a + v_2^a + \cdots,$$

where $v_0 = x$, $v_1 = \sin v_0$, $v_2 = \sin v_1$, ..., converges for $a > 2$ and diverges for $a \leq 2$.

Chapter 3

DIFFERENTIATION AND INTEGRATION

1. *Differentiability*

Definition Let f be a real-valued function defined on an open interval (a,b) containing the point x. We form the quotient

$$\frac{f(t) - f(x)}{t - x}$$

with a < t < b and $t \neq x$. If the limit

$$\lim_{t \to x} \frac{f(t) - f(x)}{t - x}$$

exists (as a finite real number), its value is denoted by f'(x); in this case f'(x) is called the *derivative of* f *at* x and f is said to be *differentiable at* x.

Many functions fail to have derivatives at certain points and the following considerations are therefore of interest.

Let f, (a,b), and x be as in the foregoing definition. For every sufficiently small positive real number h, let m(x;h) be the infimum and M(x;h) the supremum of the set of values

$$\frac{f(t) - f(x)}{t - x}$$

when $0 < |t - x| < h$; then, as h decreases,

m(x;h) increases (or remains constant)

and

M(x;h) decreases (or remains constant);

hence we may define

$$\underline{D}f(x) = \lim_{h \to 0} m(x;h) \quad \text{and} \quad \bar{D}f(x) = \lim_{h \to 0} M(x;h).$$

$\underline{D}f(x)$ is called the *lower derivative of* f *at* x, and $\bar{D}f(x)$ is called the *upper derivative of* f *at* x; clearly,

$$\underline{D}f(x) \le \bar{D}f(x) \quad \text{and} \quad \underline{D}f(x) = -\bar{D}\{-f(x)\}.$$

We can also see that f is differentiable at x if and only if the upper and the lower derivatives of f at x are finite and coincide.

Recalling the definitions of the limit superior and the limit inferior (given in Exercise 17 at the end of Chapter 2), we see that

$$\bar{D}f(x) = \overline{\lim_{t \to x}} \frac{f(t) - f(x)}{t - x} \quad \text{and} \quad \underline{D}f(x) = \underline{\lim_{t \to x}} \frac{f(t) - f(x)}{t - x}.$$

The limit superior and the limit inferior of

$$\frac{f(x + h) - f(x)}{h}$$

as $h \to 0$ are denoted by $D^+f(x)$ and $D_+f(x)$, respectively, and the limit superior and the limit inferior of

$$\frac{f(x - h) - f(x)}{-h}$$

as $h \to 0$ by $D^-f(x)$ and $D_-f(x)$, respectively. The numbers $D^+f(x)$, $D_+f(x)$, $D^-f(x)$, and $D_-f(x)$ are called the *upper right, lower right, upper left,* and *lower left Dini derivatives of* f *at* x, respectively.

It is clear that a function is differentiable at a point if and only if all four Dini derivatives at that point are finite and equal.

Examples:

1. Let $f(x) = x \sin(1/x)$ if $x \neq 0$ and $f(0) = 0$. Then $x \neq 0$ implies

$$\frac{f(x) - f(0)}{x} = \sin \frac{1}{x}$$

and so $\underline{D}f(0) = -1$ and $\bar{D}f(0) = 1$.

2. Let f be as in Exercise 20 at the end of Chapter 2. Then $D^+f(0) = 4$, $D_+f(0) = 2$, $D^-f(0) = 1$, and $D_-f(0) = -1$.

3. Let $f(x) = \sqrt{|x|} \sin(1/x)$ if $x \neq 0$ and $f(0) = 0$. Then $D^+f(x) = +\infty$, $D_+f(0) = -\infty$, $D^-f(0) = +\infty$, $D_-f(0) = -\infty$.

4. Let $f(x) = |x|$. Then $D^+f(0) = D_+f(0) = 1$ and $D^-f(0) = D_-f(0) = -1$.

Example 4 can easily be modified to give an example of a function that has no derivative at an infinite number of points; we indicate the construction by the diagram in Fig. 3 - 1.

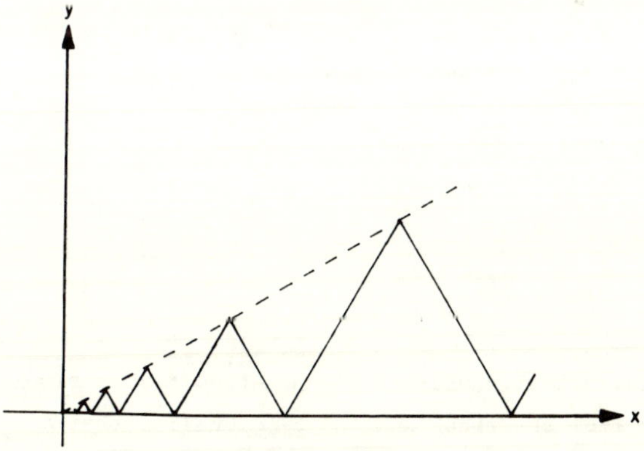

Figure 3 - 1

In Fig. 3 - 2 we give a sketch of the graph of $y = x \sin(1/x)$; the dashed lines in the diagram indicate the lines $y = \pm x$.

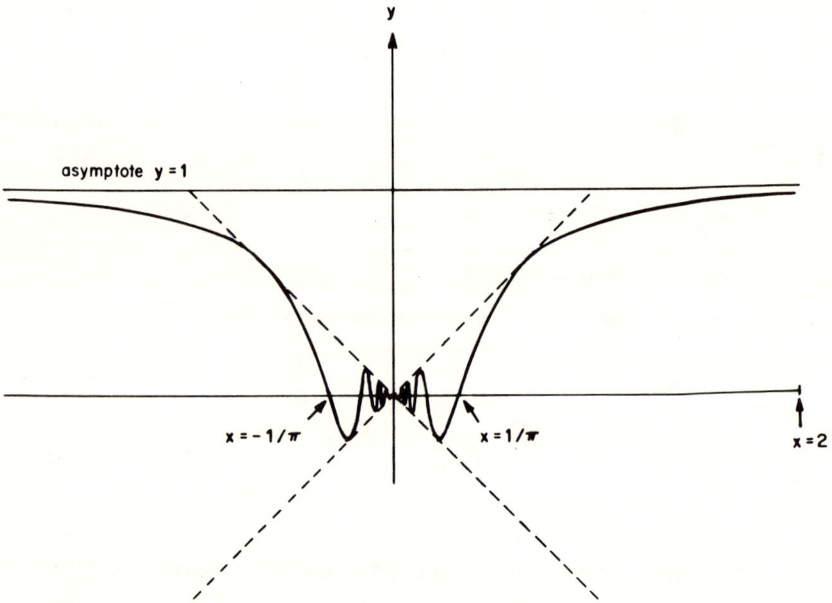

Figure 3 - 2

Definition Let f be a real-valued function defined on a closed interval [a,b]. Then f is said to be *differentiable at the endpoints* a *and* b if the right-hand and the left-hand derivatives

$$f'(a) = \lim_{t \to a+} \frac{f(t) - f(a)}{t - a}$$

and

$$f'(b) = \lim_{t \to b-} \frac{f(t) - f(b)}{t - b}$$

exist (as finite real numbers); the notation $t \to a+$ signifies that t approaches a from the right and $t \to b-$ signifies that t approaches b from the left. If f is differentiable at every point of a set S which is a subset of [a,b], we say that f is *differentiable on* S.

Remark While a function f is automatically continuous at an isolated point of its domain of definition, f cannot be differentiable at such a point since the derivative cannot be formed at such a point.

Proposition 1 Let f be defined on [a,b]. If f is differentiable at a point $x \in [a,b]$, then f is continuous at x.

Proof As $t \to x$, we have (by Corollary 2 of Proposition 2 in Chapter 2),

$$f(t) - f(x) = \frac{f(t) - f(x)}{t - x}(t - x) \to f'(x) \cdot 0 = 0. \qquad \square$$

Remark The converse of the foregoing proposition is not true. We have already seen in connection with the Dini derivatives of a function examples of functions which are continuous everywhere on their domains of definition, but are not differentiable at certain points. In the next chapter we shall discuss a function which is continuous on the whole line R^1 without being differentiable at any point.

An example of a function which is differentiable everywhere but whose derivative fails to be continuous at some point is provided by the function

$$f(x) = x^2 \sin\frac{1}{x} \quad \text{for} \quad x \neq 0,$$

$$= 0 \quad \text{for} \quad x = 0.$$

Here

$$f'(x) = 2x \sin\frac{1}{x} - \cos\frac{1}{x} \quad \text{for } x \neq 0,$$

$$= 0 \quad \text{for } x = 0.$$

We give a sketch of the graph of $y = x^2 \sin(1/x)$ in Fig. 3 - 3; the dashed curves in the diagram represent the parabolas $y = \pm x^2$. Since

$\lim_{x \to 0} \cos(1/x)$ does not exist, f' is not continuous at x = 0.

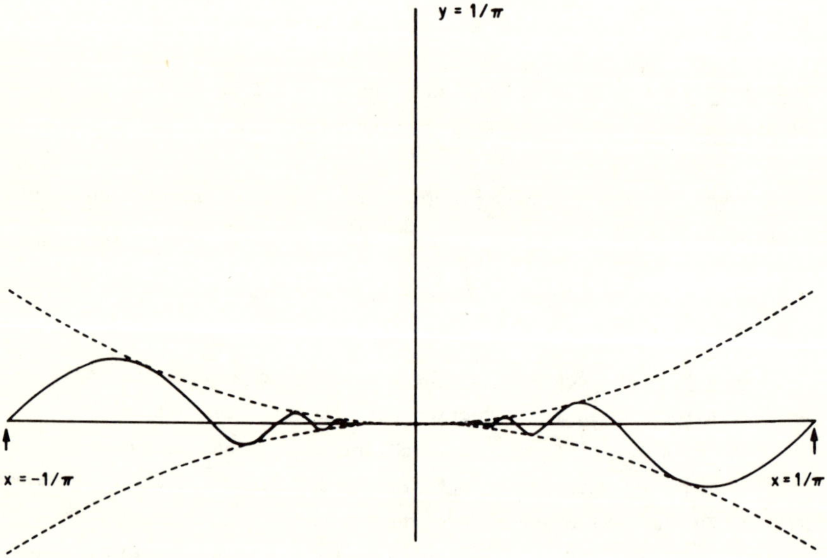

Figure 3 - 3

Proposition 2 Let f and g be functions defined on [a,b] and be differentiable at a point $x \in [a,b]$. Then f + g, fg, and f/g (where $g(x) \neq 0$) are differentiable at x, and

(i) $(f + g)'(x) = f'(x) + g'(x)$;

(ii) $(fg)'(x) = f'(x)\, g(x) + f(x)\, g'(x)$;

(iii) $\left(\frac{f}{g}\right)'(x) = \frac{g(x)\, f'(x) - g'(x)\, f(x)}{g^2(x)}$.

Proof Part (i) is trivial. To verify part (ii), we let h be equal to fg and note that

$$h(t) - h(x) = f(t)\left[g(t) - g(x)\right] + g(x)\left[f(t) - f(x)\right].$$

Dividing by t - x and observing that f and g are continuous at x (by Proposition 1), we let $t \to x$ and see that part (ii) follows.

Finally, let $h = f/g$. Then

$$\frac{h(t) - h(x)}{t - x} = \frac{1}{g(t)\, g(x)} \left[g(x) \frac{f(t) - f(x)}{t - x} - f(x) \frac{g(t) - g(x)}{t - x} \right].$$

Letting $t \to x$, we obtain part (iii). □

Proposition 3 Let f be continuous on [a,b], $f'(x)$ exist at some point $x \in [a,b]$; let g be defined on an interval I which contains the range of f, and let g be differentiable at the point $f(x)$. If, for $t \in [a,b]$,

$$h(t) = g[f(t)],$$

then h is differentiable at x and

$$h'(x) = g'[f(x)]\, f'(x).$$

Proof Let $y = f(x)$. By the definition of the derivative, we have

$$f(t) - f(x) = (t - x)[f'(x) + u(t)],$$

$$g(s) - g(y) = (s - y)[g'(y) + v(s)],$$

where $t \in [a,b]$, $s \in I$, and $u(t) \to 0$ as $t \to x$, $v(s) \to 0$ as $s \to y$. We let $s = f(t)$. Then

$$\begin{aligned} h(t) - h(x) &= g[f(t)] - g[f(x)] = [f(t) - f(x)][g'(y) + v(s)] \\ &= (t - x)[f'(x) + u(t)][g'(y) + v(s)], \end{aligned}$$

or, if $t \neq x$,

$$\frac{h(t) - h(x)}{t - x} = [g'(y) + v(s)][f'(x) + u(t)].$$

Letting $t \to x$, we see that $s \to y$, by the continuity of f at x; thus, as $t \to x$,

$$\frac{h(t) - h(x)}{t - x} \to g'(y)\, f'(x).$$ □

Definition A point x_0, which is the midpoint of an open interval contained in the domain of definition of a real-valued function f, is said to be a *local maximum of* f if there is some $\delta > 0$ such that whenever $|x - x_0| < \delta$ we have $f(x) \le f(x_0)$, but x_0 is said to be a *local mimimum of* f if there is some $\delta > 0$ such that whenever $|x - x_0| < \delta$ we have $f(x) \ge f(x_0)$.

Proposition 4 (Fermat's Theorem) Let f be defined on [a,b]; if f has a local maximum at a point $x \in (a,b)$, and if $f'(x)$ exists, then $f'(x) = 0$. (The analogous statement for local minima is of course also true.)

Proof Let δ be chosen in accordance with the definition of a local maximum so that $a < x - \delta < x < x + \delta < b$. If $x - \delta < t < x$, then

$$\frac{f(t) - f(x)}{t - x} \ge 0.$$

Letting $t \to x$, we see that $f'(x) \ge 0$.

If $x < t < x + \delta$, then

$$\frac{f(t) - f(x)}{t - x} \le 0,$$

which shows that $f'(x) \le 0$. Hence $f'(x) = 0$. □

Proposition 5 (Rolle's Theorem) If f is a continuous real-valued function on a closed interval [a,b] and is differentiable on the open interval (a,b), and if f(a) = f(b), then there is a point $x \in (a,b)$ such that $f'(x) = 0$.

Proof First, if f is constant then $f'(x) = 0$ for every x in [a,b]. Suppose, then, that f is not constant. By Proposition 29 of Chapter 2, f assumes its largest and its smallest value in [a,b]. We therefore see that the largest or smallest value of f is attained at some point $x \in (a,b)$. This is so since there must be some point $x \in (a,b)$ at which the value of f is either greater than the value at the endpoints or is less than the value at the endpoints of the interval. Suppose that f(x) is greater than the value at the endpoints. Then the largest value of f in [a,b] is attained at some x in (a,b). If f(x) is less than the value at the endpoints, the smallest value of f in [a,b] is attained at some $x \in (a,b)$. But this is then either a local maximum or a local minimum. By Proposition 4 it follows that $f'(x) = 0$. □

Proposition 6 (Mean-Value Theorem) If f is continuous on a closed interval [a,b] and is differentiable on the open interval (a,b), then there is a point $x \in (a,b)$ such that

$$f'(x) = \frac{f(b) - f(a)}{b - a}.$$

Proof The proof is obtained by applying Proposition 5 to a properly chosen auxiliary function F. To this end, define

$$F(t) = f(t) - f(a) - \frac{f(b) - f(a)}{b - a}(t - a).$$

Now, F is continuous on [a,b] and differentiable on (a,b) and $F(a) = F(b) = 0$. So F satisfies the conditions of Proposition 5. There is accordingly an $x \in (a,b)$ for which $F'(x) = 0$. But this implies

$$f'(x) - \frac{f(b) - f(a)}{b - a} = 0.$$ □

Proposition 7 (Generalized Mean-Value Theorem) If f and g are continuous on a closed interval [a,b] and are differentiable on the open interval (a,b), and if $g'(t) \neq 0$ for every $t \in (a,b)$, there is an $x \in (a,b)$ such that

$$\frac{f'(x)}{g'(x)} = \frac{f(b) - f(a)}{g(b) - g(a)}.$$

Proof To prove the proposition we apply Proposition 5 to the auxiliary function F given by

$$F(t) = [f(b) - f(a)]g(t) - [g(b) - g(a)]f(t).$$

Then F is continuous on [a,b] and differentiable on (a,b) and $F(a) = F(b) = f(b)\,g(a) - f(a)\,g(b)$. Thus, by Proposition 5, there is an

$x \in (a,b)$ for which $F'(x) = 0$. It then follows that

$$\frac{f'(x)}{g'(x)} = \frac{f(b) - f(a)}{g(b) - g(a)}. \qquad \square$$

Remarks Letting $g(x) = x$, we get as a special case of Proposition 7 the Proposition 6. Using determinants, there is an interesting way of looking at the proof of Proposition 6 which easily leads to more general results; we refer to Fig. 3 - 4. The area of

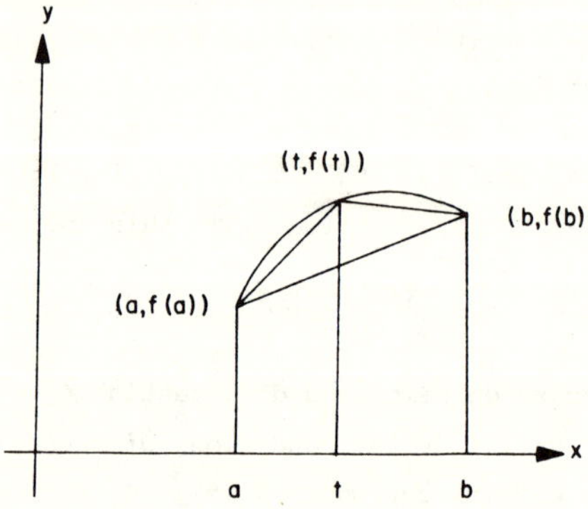

Figure 3 - 4

the triangle with vertices at the points $(a,f(a))$, $(b,f(b))$, and $(t,f(t))$ is one half of the absolute value of the determinant

$$F(t) = \begin{vmatrix} f(t) & t & 1 \\ f(a) & a & 1 \\ f(b) & b & 1 \end{vmatrix}.$$

Now, if f is continuous on $[a,b]$ and differentiable on (a,b), the same is true for the function F. Clearly, $F(a) = F(b) = 0$, so that by Proposition 5 there is a point x in (a,b) such that $F'(x) = 0$. The derivative F' is obtained by differentiating the top row in the determinant. If we do this at $t = x$ and set the resulting determinant equal to zero, we get Proposition 6.

The foregoing procedure can be generalized in the following manner. Suppose that f, g, and h are continuous on [a,b] and differentiable on (a,b). Put

$$F(t) = \begin{vmatrix} f(t) & g(t) & h(t) \\ f(a) & g(a) & h(a) \\ f(b) & g(b) & h(b) \end{vmatrix}.$$

Then F is continuous on [a,b] and differentiable on (a,b), and $F(a) = F(b) = 0$. By Proposition 5, there is a point x in (a,b) such that $F'(x) = 0$. But

$$F'(x) = \begin{vmatrix} f'(x) & g'(x) & h'(x) \\ f(a) & g(a) & h(a) \\ f(b) & g(b) & h(b) \end{vmatrix}.$$

Note that if h is the constant function 1, this gives Proposition 7.

Proposition 8 Let f be differentiable in (a,b).

(i) If $f'(x) \geq 0$ for all $x \in (a,b)$, then f is monotonically increasing.

(ii) If $f'(x) = 0$ for all $x \in (a,b)$, then f is constant.

(iii) If $f'(x) \leq 0$ for all $x \in (a,b)$, then f is monotonically decreasing.

Proof All conclusions can be obtained directly from the equation

$$f(x_2) - f(x_1) = (x_2 - x_1)\, f'(x),$$

which is valid for any pair of numbers $x_1, x_2 \in (a,b)$ and for some x between x_1 and x_2 (by Proposition 6). □

Following Proposition 1 we considered an example of a differentiable function whose derivative is not continuous everywhere. While a derivative need not be continuous, it cannot have a simple discontinuity, as we shall see next.

Proposition 9 (Darboux's Theorem) Let f be a real-valued differentiable function on [a,b] and suppose that $f'(a) < y_0 < f'(b)$. Then there is a point $x_0 \in (a,b)$ such that $f'(x_0) = y_0$.

Proof We first consider the special case where $f'(a) < 0$, $f'(b) > 0$ and show that there is an x in (a,b) such that $f'(x) = 0$.

We note that since f is differentiable it must be continuous. It accordingly attains its smallest value on [a,b]. Since $f'(a) < 0$ there are points $x_1 \in (a,b)$ with $f(x_1) < f(a)$. Since $f'(b) > 0$ there are points $x_2 \in (a,b)$ with $f(x_2) < f(b)$. Thus the least value of f in [a,b] is attained at an $x \in (a,b)$. But then $f'(x) = 0$.

Now suppose f is differentiable and only that $f'(a) < y_0 < f'(b)$. We show that there is an $x_0 \in (a,b)$ such that $f'(x_0) = y_0$.

Consider the auxiliary function $g(t) = f(t) - y_0 t$. Then

$$g'(a) = f'(a) - y_0 < 0 \quad \text{and} \quad g'(b) = f'(b) - y_0 > 0.$$

Since g satisfies the conditions of the special case, there is an $x_0 \in (a,b)$ for which $g'(x_0) = 0$. Now $f'(x_0) = g'(x_0) + y_0 = y_0$, establishing the desired result. □

Proposition 10 Suppose that, for some $\delta_0 > 0$, the real-valued functions f and g are continuous on $[c,c + \delta_0]$ and differentiable on $(c,c + \delta_0)$, and that $f(c) = g(c) = 0$. Then

$$\lim_{t \to c+} \frac{f(t)}{g(t)} = \lim_{t \to c+} \frac{f'(t)}{g'(t)},$$

whenever the second limit exists (as a finite real number).

Proof Suppose that the second limit exists and is L. Its existence implies the existence of f'/g' and so the nonvanishing of g' on some interval $(c,c + \delta')$, where $\delta' \le \delta_0$. Then, for any $\varepsilon > 0$, there exists a positive $\delta = \delta(\varepsilon) < \delta'$ such that

$$\left|\frac{f'(t)}{g'(t)} - L\right| < \varepsilon \quad \text{for all} \quad t \in (c,c + \delta).$$

Now pick any positive $h < \delta$. Then f and g satisfy the conditions of Proposition 7 with $a = c$, $b = c + h$. Hence there is a point x in $(c,c + h)$ such that

$$\frac{f(c + h)}{g(c + h)} = \frac{f(c + h) - f(c)}{g(c + h) - g(c)} = \frac{f'(x)}{g'(x)}.$$

But $x \in (c,c + \delta)$ such that

$$\left|\frac{f(c + h)}{g(c + h)} - L\right| = \left|\frac{f'(x)}{g'(x)} - L\right| < \varepsilon.$$

This holds for every positive $h < \delta$, and so

$$\lim_{t \to c+} \frac{f(t)}{g(t)} = \lim_{h \to 0+} \frac{f(c + h)}{g(c + h)} = L. \qquad \square$$

In an entirely similar way we can prove the following proposition.

Proposition 11 Suppose that, for some $\delta_0 > 0$, the real-valued functions f and g are continuous on $[c - \delta_0,c]$ and differentiable on $(c - \delta_0,c)$, and that $f(c) = g(c) = 0$. Then

$$\lim_{t \to c-} \frac{f(t)}{g(t)} = \lim_{t \to c-} \frac{f'(t)}{g'(t)},$$

provided the second limit exists (as a finite real number).

From Propositions 10 and 11 we obtain immediately the next proposition.

Proposition 12 (L'Hospital's Rule) Suppose that the functions f and g are differentiable on some neighborhood of a point c and that $f(c) = g(c) = 0$. Then

$$\lim_{t \to c} \frac{f(t)}{g(t)} = \lim_{t \to c} \frac{f'(t)}{g'(t)},$$

whenever the second limit exists (as a finite real number).

N.B.: In connection with the foregoing three propositions, especially Proposition 10, the reader may want to look up Exercise 7 at the end of this chapter.

Definition If f has a derivative f' on an interval, and if f' is itself differentiable, we denote the derivative of f' by f'' and call f'' the *second derivative of* f. Continuing in this manner, we obtain functions

$$f,\ f',\ f'',\ f^{(3)},\ \ldots,\ f^{(n)},$$

each of which is the derivative of the preceding one. The function $f^{(n)}$ is called the *n-th derivative of* f.

In order for $f^{(n)}(x)$ to exist at a point x, $f^{(n-1)}(t)$ must exist in a neighborhood of x (or in a one-sided neighborhood, if x is an endpoint of the interval on which f is defined), and $f^{(n-1)}$ must be differentiable at x. Since $f^{(n-1)}$ must exist in a neighborhood of x, $f^{(n-2)}$ has to be differentiable in that neighborhood.

Proposition 13 (Taylor's Theorem) Suppose that f is a real-valued function on [a,b], n is a positive integer, $f^{(n-1)}$ is continuous on [a,b], $f^{(n)}(t)$ exists for every $t \in (a,b)$. Let α, β be distinct points of [a,b], and let

$$P(t) = \sum_{k=0}^{n-1} \frac{f^{(k)}(\alpha)}{k!} (t - \alpha)^k. \tag{1}$$

Then there exists a point x between α and β such that

$$f(\beta) = P(\beta) + \frac{f^{(n)}(x)}{n!} (\beta - \alpha)^n. \tag{2}$$

Proof Let M be the number defined by

$$f(\beta) = P(\beta) + M(\beta - \alpha)^n$$

and put, for $t \in [a,b]$,

$$g(t) = f(t) - P(t) - M(t - \alpha)^n. \tag{3}$$

We have to verify that $n!M = f^{(n)}(x)$ for some x between α and β. By (1) and (3),

$$g^{(n)}(t) = f^{(n)}(t) - n!M,$$

where $t \in (a,b)$. Hence the proof will be accomplished if we can show that $g^{(n)}(x) = 0$ for some x between α and β.

Since $P^{(k)}(\alpha) = f^{(k)}(\alpha)$ for $k = 0, \ldots, n - 1$, we have

$$g(\alpha) = g'(\alpha) = \cdots = g^{(n-1)}(\alpha) = 0.$$

Our choice of M ensures that $g(\beta) = 0$, so that $g'(x_1) = 0$ for some x_1 between α and β, by Proposition 6. Since $g'(\alpha) = 0$, we conclude similarly that $g''(x_2) = 0$ for some x_2 between α and x_1. After n steps we arrive at the conclusion that $g^{(n)}(x_n) = 0$ for some x_n between α and x_{n-1}, that is, between α and β. □

Remarks For $n = 1$, Proposition 13 reduces to Proposition 6. In general, Proposition 13 shows that f can be approximated by a polynomial of degree $n - 1$; and (2) allows us to estimate the error of the approximation, if we know bounds on $|f^{(n)}(x)|$.

Proposition 14 (Leibniz's Theorem) Let n be a positive integer, and suppose that the functions f and g have derivatives of order n at the point c. Then so does the function $F = fg$, and

$$F^{(n)}(c) = \sum_{j=0}^{n} \binom{n}{j} f^{(j)}(c)\, g^{(n-j)}(c).$$

Proof We prove the proposition by induction. The proposition is true for $n = 1$, by Proposition 2. We assume that it is true for n, and show that if $f^{(n+1)}(c)$ and $g^{(n+1)}(c)$ exist, so does $F^{(n+1)}(c)$, and that

$$F^{(n+1)}(c) = \sum_{j=0}^{n+1} \binom{n+1}{j} f^{(j)}(c)\, g^{(n+1-j)}(c).$$

Since $f^{(n+1)}$ and $g^{(n+1)}$ exist at c, $f^{(j)}$ and $g^{(j)}$ exist on some neighborhood $(c - \delta, c + \delta)$, with $\delta > 0$, for $0 \le j \le n$, and so, by our inductive assumption, for $x \in (c - \delta, c + \delta)$,

$$F^{(n)}(x) = \sum_{j=0}^{n} \binom{n}{j} f^{(j)}(x)\, g^{(n-j)}(x).$$

We now apply Proposition 2 to each of the n + 1 terms on the right-hand side, and get

$$F^{(n+1)}(c) = \sum_{j=0}^{n} \binom{n}{j} \left(f^{(j+1)}(c)\, g^{(n-j)}(c) + f^{(j)}(c)\, g^{(n+1-j)}(c) \right)$$

$$= \sum_{j=0}^{n+1} a_j\, f^{(j)}(c)\, g^{(n+1-j)}(c),$$

where

$$a_0 = \binom{n}{0} = \binom{n+1}{0}, \quad a_{n+1} = \binom{n}{n} = \binom{n+1}{n+1},$$

and, for $0 < j < n + 1$,

$$a_j = \binom{n}{j} + \binom{n}{j-1} = \binom{n+1}{j}.$$

Thus the result holds for n + 1, and so, by induction, the proposition is established. □

Remark Various applications of the foregoing propositions on differential calculus will be found in the exercise section at the end of this chapter.

2. *The Riemann-Stieltjes Integral*

Let $[a,b]$ be a given interval. We recall that a *partition* P *of* $[a,b]$ is any finite set of points $x_0, x_1, \ldots, x_n$ such that

$$a = x_0 < x_1 < \cdots < x_n = b.$$

We write, for $1 \le k \le n$,

$$\Delta x_k = x_k - x_{k-1}$$

and denote by

$$|P| = \max\{\Delta x_k : 1 \le k \le n\};$$

$|P|$ is called the *mesh of the partition* P.

Definition A partition P^* of $[a,b]$ is called a *refinement of* a partition P of $[a,b]$ (in symbols, $P^* \supset P$) if every point of P is a point of P^*. Given two partitions P_1 and P_2 of $[a,b]$, we call $P^* = P_1 \cup P_2$ their *common refinement*. P^* is *finer than* P if $P^* \supset P$.

It is clear that the common refinement of two partitions of an interval is actually a refinement of each.

Definition Let f be a bounded real-valued function on a bounded closed interval $[a,b]$. Corresponding to each partition P of $[a,b]$ we put

$$M_k = \sup\{f(x) : x_{k-1} \le x \le x_k\},$$

$$m_k = \inf\{f(x) : x_{k-1} \le x \le x_k\},$$

and define the *upper* and the *lower Darboux sums of* f *relative to* P, respectively, by

$$U(P,f) = \sum_{k=1}^{n} M_k \,\Delta x_k \quad \text{and} \quad L(P,f) = \sum_{k=1}^{n} m_k \,\Delta x_k;$$

finally we put

$$\overline{\int_a^b} f(x)\,dx = \inf U(P,f), \tag{4}$$

$$\underline{\int_a^b} f(x)\,dx = \sup L(P,f), \tag{5}$$

where the infimum and the supremum are taken over all partitions P of [a,b]. The left-hand members of (4) and (5) are called the *upper* and the *lower Riemann integrals of* f *over* [a,b], respectively. If the upper and lower Riemann integrals are equal, we say that f is *Riemann integrable on* [a,b] and we denote the common value of (4) and (5) by the symbol

$$\int_a^b f(x)\,dx \quad \text{or} \quad \int_a^b f(x)\,dx \tag{6}$$

and call it the *Riemann integral of* f *over* [a,b]; in this case we write $f \in \mathcal{R}$ (where $\mathcal{R}$ denotes the set of Riemann integrable functions).

Remark To see that the upper and lower Riemann integrals exist for every bounded function f on a compact interval [a,b], we observe that the numbers L(P,f) and U(P,f) form a bounded set. Indeed, since f is bounded, there are two numbers, m and M, such that

$$m \le f(x) \le M$$

for $a \le x \le b$; hence, for any partition P, we have

$$m(b - a) \le L(P,f) \le U(P,f) \le M(b - a).$$

Definition Let α be a real-valued monotonically increasing function on [a,b] (since $\alpha(a)$ and $\alpha(b)$ are finite, we have that α is bounded on [a,b]). Corresponding to each partition of the interval [a,b] we write

$$\Delta\alpha_k = \alpha(x_k) - \alpha(x_{k-1})$$

and note that $\Delta\alpha_k \ge 0$. For any real-valued function f which is bounded on the compact interval $[a,b]$ we define the *upper* and the *lower Darboux sums of* f *with respect to* α *relative to* P, respectively, by

$$U(P,f,\alpha) = \sum_{k=1}^{n} M_k \, \Delta\alpha_k \quad \text{and} \quad L(P,f,\alpha) = \sum_{k=1}^{n} m_k \, \Delta\alpha_k,$$

where M_k, m_k have the same meaning as in the foregoing definition, and we put

$$\overline{\int_a^b} f(x) \, d\alpha(x) = \inf U(P,f,\alpha), \tag{7}$$

$$\underline{\int_a^b} f(x) \, d\alpha(x) = \sup L(P,f,\alpha), \tag{8}$$

where the infimum and the supremum are again taken over all partitions P of $[a,b]$. If the left-hand members of (7) and (8) are equal, we denote their common value by

$$\int_a^b f(x) \, d\alpha(x) \quad \text{or} \quad \int_a^b f(x) \, d\alpha(x)$$

and call it the *Riemann-Stieltjes integral of* f *with respect to* α, *over* $[a,b]$; in this case we say that f is *integrable with respect to* α, *in the Riemann sense*, and we write $f \in \mathcal{R}(\alpha)$.

Remark If $\alpha(x) = x$, the Riemann integral is seen to be a special case of the Riemann-Stieltjes integral.

Proposition 15 If P* is a refinement of P, then

$$L(P,f,\alpha) \le L(P^*,f,\alpha) \quad \text{and} \quad U(P^*,f,\alpha) \le U(P,f,\alpha).$$

Proof To prove the first inequality, we suppose first that P* contains only one more point than P. Let x^* be this extra point, and assume that $x_{k-1} < x^* < x_k$, where x_{k-1} and x_k are two consecutive points of P. We put

$$w_1 = \inf\{f(x): x_{k-1} \le x \le x^*\},$$

$$w_2 = \inf\{f(x): x^* \le x \le x_k\}.$$

Letting, as before,

$$m_k = \inf\{f(x): x_{k-1} \le x \le x_k\}$$

we see that $w_1 \ge m_k$ and $w_2 \ge m_k$. Thus

$$\begin{aligned} &L(P^*,f,\alpha) - L(P,f,\alpha) \\ &\quad = w_1[\alpha(x^*) - \alpha(x_{k-1})] + w_2[\alpha(x_k) - \alpha(x^*)] - m_k[\alpha(x_k) - \alpha(x_{k-1})] \\ &\quad = (w_1 - m_k)[\alpha(x^*) - \alpha(x_{k-1})] + (w_2 - m_k)[\alpha(x_k) - \alpha(x^*)] \\ &\quad \ge 0. \end{aligned}$$

If P* contains j points more than P, we repeat the above reasoning j times, and arrive at the first inequality of the proposition. The proof of the second inequality in the proposition is completely analogous to the proof of the first inequality. □

Corollary We have

$$\underline{\int_a^b} f(x)\ d\alpha(x) \le \overline{\int_a^b} f(x)\ d\alpha(x).$$

Proof Let P* be the common refinement of two partitions P_1 and P_2 of the closed bounded interval [a,b]. By Proposition 15,

$$L(P_1,f,\alpha) \le L(P^*,f,\alpha) \le U(P^*,f,\alpha) \le U(P_2,f,\alpha)$$

and so

$$L(P_1,f,\alpha) \le U(P_2,f,\alpha).$$

By keeping P_2 fixed, and taking the supremum over all P_1, the foregoing inequality gives

$$\underline{\int_a^b} f(x)\ d\alpha(x) \le U(P_2,f,\alpha).$$

The claim of the corollary now follows by taking the infimum over all P_2 in the foregoing inequality. □

Proposition 16 We have $f \in \mathcal{R}(\alpha)$ on $[a,b]$ if and only if for any $\varepsilon > 0$ there is a partition P such that

$$U(P,f,\alpha) - L(P,f,\alpha) < \varepsilon. \tag{9}$$

Proof For any P we have

$$L(P,f,\alpha) \le \underline{\int_a^b} f(x)\ d\alpha(x) \le \overline{\int_a^b} f(x)\ d\alpha(x) \le U(P,f,\alpha).$$

Thus (9) implies

$$0 \le \overline{\int_a^b} f(x)\ d\alpha(x) - \underline{\int_a^b} f(x)\ d\alpha(x) < \varepsilon.$$

Thus, if (9) holds for any $\varepsilon > 0$, then

$$\overline{\int_a^b} f(x)\ d\alpha(x) = \underline{\int_a^b} f(x)\ d\alpha(x),$$

implying that $f \in \mathcal{R}(\alpha)$.

Conversely, suppose that $f \in \mathcal{R}(\alpha)$, and let $\varepsilon > 0$ be given. Then there are partitions P_1 and P_2 of $[a,b]$ such that

$$U(P_2,f,\alpha) - \int_a^b f(x)\ d\alpha(x) < \frac{\varepsilon}{2} \tag{10}$$

and

$$\int_a^b f(x)\ d\alpha(x) - L(P_1,f,\alpha) < \frac{\varepsilon}{2} \tag{11}$$

Let P be the common refinement of P_1 and P_2. Then, by Proposition 15, together with (10) and (11), we get

$$U(P,f,\alpha) \le U(P_2,f,\alpha) < \int_a^b f(x)\ d\alpha(x) + \frac{\varepsilon}{2}$$

$$< L(P_1,f,\alpha) + \varepsilon \le L(P,f,\alpha) + \varepsilon;$$

thus (9) holds for the partition $P = P_1 \cup P_2$. □

Remark In Proposition 35 we shall again take up the question of characterizing the class of functions $\mathcal{R}(\alpha)$.

Proposition 17 If f is continuous on [a,b], then $f \in \mathcal{R}(\alpha)$ on [a,b]. Moreover, For any $\varepsilon > 0$ there is a $\delta > 0$ such that

$$\left| \sum_{k=1}^{n} f(t_k)\ \Delta\alpha_k - \int_a^b f(x)\ d\alpha(x) \right| < \varepsilon \tag{12}$$

for any partition $P = \{x_0, x_1, \ldots, x_n\}$ of [a,b] with $|P| < \delta$, and for any choice of points $t_k \in [x_{k-1}, x_k]$.

Proof Let $\varepsilon > 0$ be given, and pick $\gamma > 0$ such that

$$[\alpha(b) - \alpha(a)]\gamma < \varepsilon.$$

Since f is uniformly continuous on [a,b] (Proposition 32 of Chapter 2), there is a $\delta > 0$ such that

$$|f(x) - f(t)| < \gamma \tag{13}$$

if $|x - t| < \delta$ and $x \in [a,b]$, $t \in [a,b]$.

We choose P such that $|P| < \delta$. Then (13) implies $M_k - m_k \leq \gamma$ for $1 \leq k \leq n$. Hence

$$U(P,f,\alpha) - L(P,f,\alpha) = \sum_{k=1}^{n} (M_k - m_k)\ \Delta\alpha_k \leq \gamma \sum_{k=1}^{n} \Delta\alpha_k$$

$$= \gamma[\alpha(b) - \alpha(a)] < \varepsilon.$$

By Proposition 16, $f \in \mathcal{R}(\alpha)$.

Since the two numbers

$$\sum_{k=1}^{n} f(t_k)\ \Delta\alpha_k \quad \text{and} \quad \int_a^b f(x)\ d\alpha(x)$$

are between $U(P,f,\alpha)$ and $L(P,f,\alpha)$, (12) is also proved. □

Proposition 18 Let f be a monotonic real-valued function on a closed bounded interval [a,b] and let α be a continuous real-valued monotonically increasing function on [a,b]. Then $f \in \mathcal{R}(\alpha)$ on [a,b].

Proof Let $\varepsilon > 0$ be given. For any positive integer n, we select a partition P of [a,b] such that, for $1 \le k \le n$,

$$\Delta\alpha_k = \frac{\alpha(b) - \alpha(a)}{n}.$$

As α is continuous, this can be done by Proposition 31 of Chapter 2.

Assume that f is increasing (the proof is similar in the other case). Then, for $1 \le k \le n$, $M_k = f(x_k)$ and $m_k = f(x_{k-1})$ and so

$$U(P,f,\alpha) - L(P,f,\alpha) = \frac{\alpha(b) - \alpha(a)}{n} \sum_{k=1}^{n} [f(x_k) - f(x_{k-1})]$$

$$= \frac{\alpha(b) - \alpha(a)}{n} [f(b) - f(a)] < \varepsilon,$$

provided that n is taken large enough. By Proposition 16, $f \in R(\alpha)$.

□

Proposition 19 We have the following:

(i) If $f_1, f_2 \in R(\alpha)$ on [a,b], then $f_1 + f_2 \in R(\alpha)$ on [a,b] and

$$\int_a^b [f_1(x) + f_2(x)]\, d\alpha(x) = \int_a^b f_1(x)\, d\alpha(x) + \int_a^b f_2(x)\, d\alpha(x).$$

(ii) If $f \in R(\alpha)$ on [a,b] and if c is a fixed real number, then $cf \in R(\alpha)$ on [a,b] and

$$\int_a^b c\, f(x)\, d\alpha(x) = c \int_a^b f(x)\, d\alpha(x).$$

(iii) If $f_1, f_2 \in R(\alpha)$ on [a,b] and $f_1(x) \le f_2(x)$ on [a,b], then

$$\int_a^b f_1(x)\, d\alpha(x) \le \int_a^b f_2(x)\, d\alpha(x).$$

(iv) If $f \in R(\alpha)$ on [a,b] and if $a < c < b$, then $f \in R(\alpha)$ on [a,c] and on [c,b], and

$$\int_a^c f(x)\ d\alpha(x) + \int_c^b f(x)\ d\alpha(x) = \int_a^b f(x)\ d\alpha(x).$$

(v) If $f \in \mathcal{R}(\alpha)$ on $[a,b]$ and if $|f(x)| \le M$ on $[a,b]$, then

$$\left|\int_a^b f(x)\ d\alpha(x)\right| \le M[\alpha(b) - \alpha(a)].$$

(vi) If $f \in \mathcal{R}(\alpha_1)$ and $f \in \mathcal{R}(\alpha_2)$ on $[a,b]$, then $f \in \mathcal{R}(\alpha_1 + \alpha_2)$ on $[a,b]$ and

$$\int_a^b f(x)\ d[\alpha_1(x) + \alpha_2(x)] = \int_a^b f(x)\ d\alpha_1(x) + \int_a^b f(x)\ d\alpha_2(x).$$

(vii) If $f \in \mathcal{R}(\alpha)$ on $[a,b]$ and c is a fixed positive real number, then $f \in \mathcal{R}(c\alpha)$ on $[a,b]$ and

$$\int_a^b f(x)\ d[c\alpha(x)] = c\int_a^b f(x)\ d\alpha(x).$$

Proof The claims under (ii), (iii), (v), (vi) and (vii) are rather simple to verify; we shall therefore only prove the claims under (i) and (iv).

Verification of part (i): Let $\varepsilon > 0$ be given. Choose P_j such that

$$U(P_j,f_j,\alpha) - L(P_j,f_j,\alpha) < \frac{\varepsilon}{2}$$

for $j = 1, 2$, and let $P = P_1 \cup P_2$. Using Proposition 15 and the simple inequalities

$$\inf\{f_1 + f_2\} \le \inf f_1 + \inf f_2$$

and

$$\sup\{f_1 + f_2\} \le \sup f_1 + \sup f_2$$

(where the infima and the suprema are evaluated for the ranges of the respective functions, that is, $\inf f_1$, for example, stands for $\inf\{f_1(x): x \in [a,b]\}$), we get

$$\begin{aligned} L(P_1,f_1,\alpha) + L(P_2,f_2,\alpha) &\le L(P,f_1,\alpha) + L(P,f_2,\alpha) \\ &\le L(P,f_1 + f_2,\alpha) \le U(P,f_1 + f_2,\alpha) \\ &\le U(P,f_1,\alpha) + U(P,f_2,\alpha) \\ &\le U(P_1,f_1,\alpha) + U(P_2,f_2,\alpha) \\ &< L(P_1,f_1,\alpha) + L(P_2,f_2,\alpha) + \varepsilon. \end{aligned}$$

It follows that

$$U(P,f_1 + f_2,\alpha) - L(P,f_1 + f_2,\alpha) < \varepsilon$$

and hence $f_1 + f_2 \in \mathcal{R}(\alpha)$.

Next, let $\int_a^b f_j(x)\, d\alpha(x) = A_j$ and D_j be such that

$$|L(D_j,f_j,\alpha) - A_j| < \frac{\varepsilon}{6} \tag{14}$$

and

$$|U(D_j,f_j,\alpha) - A_j| < \frac{\varepsilon}{6}. \tag{15}$$

It follows that $0 \le U(D_j,f_j,\alpha) - L(D_j,f_j,\alpha) < \varepsilon/3$ for $j = 1, 2$. Putting $D = D_1 \cup D_2$, we have

$$\begin{aligned} L(D_1,f_1,\alpha) + L(D_2,f_2,\alpha) &\le L(D,f_1,\alpha) + L(D,f_2,\alpha) \\ &\le L(D,f_1 + f_2,\alpha) \le U(D,f_1 + f_2,\alpha) \\ &\le U(D,f_1,\alpha) + U(D,f_2,\alpha) \\ &\le U(D_1,f_1,\alpha) + U(D_2,f_2,\alpha) \\ &< L(D_1,f_1,\alpha) + L(D_2,f_2,\alpha) + \frac{2\varepsilon}{3} \end{aligned}$$

from which we get that

$$L(D,f_1 + f_2,\alpha) - L(D_1,f_1,\alpha) - L(D_2,f_2,\alpha) < \frac{2\varepsilon}{3} \tag{16}$$

and

$$U(D_1,f_1,\alpha) + U(D_2,f_2,\alpha) - U(D,f_1 + f_2,\alpha) < \frac{2\varepsilon}{3}. \tag{17}$$

From (14) and (16) we infer that

$$|L(D,f_1 + f_2,\alpha) - (A_1 + A_2)| < \varepsilon,$$

and from (15) and (17) that

$$|U(D,f_1 + f_2,\alpha) - (A_1 + A_2)| < \varepsilon.$$

Verification of part (iv): Let $\varepsilon > 0$ be given. Choose a partition P of $[a,b]$ such that

$$U(P,f,\alpha) - L(P,f,\alpha) < \varepsilon.$$

In view of Proposition 15 we may (and do) suppose that $c \in P$; say

$$P = \{a = x_0 < \cdots < x_m = c < x_{m+1} < \cdots < x_n = b\}.$$

Let

$$P_1 = \{a = x_0 < \cdots < x_m = c\} \quad \text{and} \quad P_2 = \{c = x_m < \cdots < x_n = b\}.$$

Then

$$[U(P_1,f,\alpha) - L(P_1,f,\alpha)] + [U(P_2,f,\alpha) - L(P_2,f,\alpha)]$$

$$= U(P,f,\alpha) - L(P,f,\alpha) < \varepsilon.$$

It follows that $f \in \mathcal{R}(\alpha)$ on both $[a,c]$ and $[c,b]$. Let

$$\int_a^c f(x)\, d\alpha(x) = K_1 \quad \text{and} \quad \int_c^b f(x)\, d\alpha(x) = K_2.$$

Clearly

$$0 \le U(P_1,f,\alpha) - K_1 < \varepsilon \quad \text{and} \quad 0 \le U(P_2,f,\alpha) - K_2 < \varepsilon.$$

Adding these two inequalities, we obtain

$$0 \le U(P,f,\alpha) - (K_1 + K_2) < 2\varepsilon.$$

Since a similar statement is true for $L(P,f,\alpha)$, we conclude that $\int_a^b f(x)\, d\alpha(x) = K_1 + K_2$. □

Proposition 20 Suppose $f \in \mathcal{R}(\alpha)$ on $[a,b]$, $m \le f(x) \le M$ for all x in $[a,b]$, g is continuous on $[m,M]$, and $h(x) = g[f(x)]$ on $[a,b]$. Then $h \in \mathcal{R}(\alpha)$ on $[a,b]$.

Proof Let $\varepsilon > 0$ be given. By the uniform continuity of g on $[m,M]$ we can find some $\delta_1 > 0$ such that

$$|g(s) - g(t)| < \varepsilon$$

if $|s - t| < \delta_1$ and $s, t \in [m,M]$. Let $\delta = \min\{\delta_1, \varepsilon\}$. Corresponding to δ^2, choose a partition $P = \{x_0, x_1, \ldots, x_n\}$ of $[a,b]$ such that

$$U(P,f,\alpha) - L(P,f,\alpha) < \delta^2, \tag{18}$$

which is possible by Proposition 16.

As usual, let

$$m_k = \inf\{f(x): x \in [x_{k-1},x_k]\} \quad \text{and} \quad M_k = \sup\{f(x): x \in [x_{k-1},x_k]\}$$

and let m_k^* and M_k^* be the analogous numbers for the function h. Divide the numbers 1, 2, ..., n into two classes: $k \in A$ if $M_k - m_k < \delta$ and $k \in B$ if $M_k - m_k \ge \delta$.

For $k \in A$, our choice of δ shows that $M_k^* - m_k^* \le \varepsilon$.

For $k \in B$, $M_k^* - m_k^* \le 2K$, where $K = \sup\{|g(t)|: m \le t \le M\}$.

By (18) we have

$$\delta \sum_{k \in B} \Delta\alpha_k \le \sum_{k \in B} (M_k - m_k)\, \Delta\alpha_k < \delta^2$$

so that $\Sigma_{k \in B}\, \Delta\alpha_k < \delta$. It follows that

$$\begin{aligned} U(P,h,\alpha) - L(P,h,\alpha) &= \sum_{k \in A} (M_k^* - m_k^*)\, \Delta\alpha_k + \sum_{k \in B} (M_k^* - m_k^*)\, \Delta\alpha_k \\ &\le \varepsilon[\alpha(b) - \alpha(a)] + 2K\delta \\ &< \varepsilon[\alpha(b) - \alpha(a) + 2K]. \end{aligned}$$

Since $\varepsilon > 0$ was arbitrary, Proposition 16 implies that $h \in \mathcal{R}(\alpha)$ on $[a,b]$. □

Proposition 21 If $u \in \mathcal{R}(\alpha)$ and $v \in \mathcal{R}(\alpha)$ on $[a,b]$, then

(i) $uv \in \mathcal{R}(\alpha)$,

(ii) $|u| \in \mathcal{R}(\alpha)$ and $\left|\int_a^b u(x)\ d\alpha(x)\right| \le \int_a^b |u(x)|\ d\alpha(x)$.

Proof If we choose $g(t) = t^2$, Proposition 20 shows that u^2 belongs to $\mathcal{R}(\alpha)$. Applying the identity

$$4uv = (u + v)^2 - (u - v)^2$$

we get claim (i).

If we take $g(t) = |t|$, Proposition 20 shows that $|u| \in \mathcal{R}(\alpha)$. We choose c as either -1 or 1 to make

$$c \int_a^b u(x)\ d\alpha(x) \ge 0.$$

Then

$$\left|\int_a^b u(x)\ d\alpha(x)\right| = c \int_a^b u(x)\ d\alpha(x) = \int_a^b cu(x)\ d\alpha(x)$$

$$\le \int_a^b |u(x)|\ d\alpha(x),$$

because $c\,u(x) \le |u(x)|$ for all $x \in [a,b]$. □

Definition Let [a,b] be a bounded closed interval, f a bounded real-valued function on [a,b], and α a real-valued monotonically increasing function on [a,b]. Take a partition

$$P = \{a = x_0 < x_1 < \cdots < x_n = b\}$$

of [a,b]. With t_k selected such that $x_{k-1} \le t_k \le x_k$ for $k = 1, \ldots, n$, form the sum

$$S(P,f,\alpha) = \sum_{k=1}^{n} f(t_k)\ \Delta\alpha_k$$

where

$$\Delta\alpha_k = \alpha(x_k) - \alpha(x_{k-1});$$

the number $S(P,f,\alpha)$ is called a *Riemann-Stieltjes sum for* f *with respect to* α *relative to* P. We use the notation

$$\lim_{|P| \to 0} S(P,f,\alpha) = A,$$

where $|P|$ denotes the mesh of P, if, for each given $\varepsilon > 0$, there is a $\delta > 0$ such that $|P| < \delta$ implies

$$|S(P,f,\alpha) - A| < \varepsilon.$$

Remark Note that $S(P,f,\alpha)$ depends also on the choice of the points t_k, subject only to the conditions $x_{k-1} \le t_k \le x_k$; however

$$\lim_{|P| \to 0} S(P,f,\alpha) = A$$

means that $|S(P,f,\alpha) - A| < \varepsilon$ is true for all P and all admissible choices of t_k, provided that $|P| < \delta$.

Proposition 22 If f is continuous on [a,b], then for any partition P of [a,b] the set $S(P,f,\alpha)$ is a closed interval.

Proof With $P = \{x_0, x_1, \ldots, x_n\}$, let $\underline{t}_k$ and $\bar{t}_k$ be in the closed interval $[x_{k-1},x_k]$ which furnish the smallest and the largest values of $f(t)\,\Delta\alpha_k$ with $t \in [x_{k-1},x_k]$, for $k = 1, \ldots, n$, respectively; since f is continuous, this is possible by Proposition 29 of Chapter 2. Hence, for any choice of $t_k \in [x_{k-1},x_k]$,

$$f(\underline{t}_k)\,\Delta\alpha_k \le f(t_k)\,\Delta\alpha_k \le f(\bar{t}_k)\,\Delta\alpha_k.$$

By summing from $k = 1$ to n and letting c and d be the values of the left-hand and the right-hand sums, then

$$c,\ d \in S(P,f,\alpha) \subset [c,d].$$

Define F for $0 \le \lambda \le 1$ by

$$F(\lambda) = \sum_{k=1}^{n} f[(1 - \lambda)\underline{t}_k + \lambda\bar{t}_k]\,\Delta\alpha_k.$$

For each such λ, $x_{k-1} \leq (1 - \lambda)\underline{t}_k + \lambda\bar{t}_k \leq x_k$ and hence $F(\lambda)$ belongs to $S(P,f,\alpha)$. Moreover, F is continuous, has smallest value $F(0) = c$, largest value $F(1) = d$, and assumes all intermediate values between c and d (by Proposition 31 of Chapter 2). Hence $S(P,f,\alpha) = [c,d]$. □

Proposition 23 If f and α have a common point of discontinuity $c \in [a,b]$, then

$$\lim_{|P| \to 0} S(P,f,\alpha)$$

does not exist.

Proof Let $a < c < b$ and suppose that α is continuous from the left at c, that is, $\lim_{x \to c-} \alpha(x) = \alpha(c)$. We suppose moreover that f is discontinuous from the left at c and that α is discontinuous from the right at c and let $\varepsilon > 0$ be such that corresponding to any $\delta > 0$ there are points t and x such that

$$c - \delta < t < c \quad \text{and} \quad |f(t) - f(c)| \geq \varepsilon,$$

$$c < x < c + \delta \quad \text{and} \quad |\alpha(x) - \alpha(c)| \geq \varepsilon.$$

Since α is continuous from the left at c, corresponding to $\varepsilon/2$, let $\delta_1 > 0$ be such that

$$c - \delta_1 < x < c \quad \text{implies} \quad |\alpha(x) - \alpha(c)| < \varepsilon/2.$$

Now, suppose that the limit of $S(P,f,\alpha)$ as $|P| \to 0$ exists and is equal to A. Let $\delta_2 > 0$ be such that $|P| < \delta_2$ implies

$$S(P,f,\alpha) \subset \left(A - \frac{1}{4}\varepsilon^2, A + \frac{1}{4}\varepsilon^2\right).$$

We put $\delta = \min\{\delta_1,\delta_2\}$ and select x' such that $c - \delta/2 < x' < c$ and x" such that $c < x'' < c + \delta/2$ and $|\alpha(x'') - \alpha(c)| \geq \varepsilon$. Then $|\alpha(x') - \alpha(c)| < \varepsilon/2$, and hence

$$0 < x'' - x' < \delta \quad \text{and} \quad |\alpha(x'') - \alpha(x')| \geq \frac{\varepsilon}{2}.$$

Next we select a partition P of [a,b] with $|P| < \delta$ and such that x' and x" are consecutive points of P:

$$P = \{x_0, x_1, \ldots, x_{j-1} = x', x_j = x'', \ldots, x_n\}.$$

Take $t_j' = c$ and t_j'' such that

$$x_{j-1} < t_j'' < c \quad \text{and} \quad |f(t_j'') - f(c)| \geq \varepsilon.$$

Also choose $t_k' = t_k'' \in [x_{k-1}, x_k]$, $1 \leq k \leq n$, $k \neq j$. Put

$$\Sigma' = \sum_{k=1}^{n} f(t_k')\,\Delta\alpha_k \quad \text{and} \quad \Sigma'' = \sum_{k=1}^{n} f(t_k'')\,\Delta\alpha_k.$$

Then

$$\begin{aligned} |\Sigma' - \Sigma''| &= |f(t_j') - f(t_j'')|\,|\alpha(x_j) - \alpha(x_{j-1})| \\ &= |f(c) - f(t_j'')|\,|\alpha(x'') - \alpha(x')| \\ &\geq \frac{\varepsilon^2}{4}. \end{aligned}$$

But Σ' and Σ'' are both points of $S(P,f,\alpha)$ and since $|P| < \delta$,

$$|\Sigma' - \Sigma''| \leq |\Sigma' - A| + |\Sigma'' - A| < \frac{\varepsilon^2}{4}.$$

This contradiction proves the proposition in the case considered. □

Example Let

$$\begin{aligned} f(x) &= 1 \quad \text{for} \quad 0 \leq x \leq \tfrac{1}{2} \\ &= 2 \quad \text{for} \quad \tfrac{1}{2} < x \leq 1 \end{aligned}$$

and

$$\begin{aligned} \alpha(x) &= 3 \quad \text{for} \quad 0 \leq x < \tfrac{1}{2} \\ &= 4 \quad \text{for} \quad \tfrac{1}{2} < x \leq 1 \end{aligned}$$

and

$$P_1 = \left\{0, \frac{1}{3}, \frac{2}{3}, 1\right\}.$$

Then $\Delta\alpha_1 = 0$, $\Delta\alpha_2 = 1$, and $\Delta\alpha_3 = 0$. Thus

$$\sum_{k=1}^{3} f(t_k)\ \Delta\alpha_k = 1 \quad \text{if} \quad \frac{1}{3} \leq t_2 \leq \frac{1}{2}$$

$$= 2 \quad \text{if} \quad \frac{1}{2} < t_2 \leq \frac{2}{3},$$

and so $S(P_1,f,\alpha) = \{1, 2\}$ has two points. In general, if P_2 and P_3 are partitions of $[0,1]$ such that $\frac{1}{2} \notin P_2$ but $\frac{1}{2} \in P_3$, then

$$S(P_2,f,\alpha) = \{1, 2\} \quad \text{and} \quad S(P_3,f,\alpha) = \{1\}.$$

It is easy to see that for these functions f and α

$$\lim_{|P| \to 0} S(P,f,\alpha)$$

does not exist; however, f is readily seen to belong to $\mathcal{R}(\alpha)$ on $[0,1]$.

Remark In connection with Proposition 23 and the foregoing example see Exercises 34 and 35 at the end of this chapter.

Proposition 24 We have the following:

(i) If $\lim S(P,f,\alpha)$ exists, as $|P| \to 0$, then $f \in \mathcal{R}(\alpha)$ and

$$\lim_{|P| \to 0} S(P,f,\alpha) = \int_a^b f(x)\ d\alpha(x). \tag{19}$$

(ii) If (a) f is continuous, or if (b) $f \in \mathcal{R}(\alpha)$ and α is continuous (monotonic) on $[a,b]$, then (19) holds.

Proof Suppose that the limit on the left-hand side of (19) exists and is equal to A. Let $\varepsilon > 0$ be given.

There exists $\delta > 0$ such that $|P| < \delta$ implies

$$A - \frac{\varepsilon}{2} < S(P,f,\alpha) < A + \frac{\varepsilon}{2}. \tag{20}$$

We choose one such partition P. If we let the points t_k range over the intervals $[x_{k-1}, x_k]$ and take the supremum and the infimum of the numbers $S(P,f,\alpha)$ so obtained, (20) yields

$$A - \frac{\varepsilon}{2} \le L(P,f,\alpha) \le U(P,f,\alpha) \le A + \frac{\varepsilon}{2}.$$

By Proposition 16, $f \in \mathcal{R}(\alpha)$, and the proof of (i) is complete, since

$$L(P,f,\alpha) \le \int_a^b f(x)\, d\alpha(x) \le U(P,f,\alpha).$$

Part (a) of (ii) is contained in Proposition 17. To verify part (b), suppose $f \in \mathcal{R}(\alpha)$, α is continuous (monotonically increasing), and $\varepsilon > 0$. There is a partition P^* such that

$$U(P^*,f,\alpha) < \int_a^b f(x)\, d\alpha(x) + \frac{\varepsilon}{4}. \tag{21}$$

Let $M = \sup\{|f(x)| : x \in [a,b]\}$. Since α is uniformly continuous on the bounded closed interval $[a,b]$, there is a $\delta_1 > 0$ with the following property: If P is any partition of $[a,b]$ with $|P| < \delta_1$, then

$$\Delta\alpha_k < \frac{\varepsilon}{4Mn}$$

for all k, where n is the number of subintervals into which $[a,b]$ is divided by P^*. Let P be any partition such that $|P| < \delta_1$, and consider the sum which defines $U(P,f,\alpha)$. Those intervals of P which contain a point of P^* in their interior contribute no more than

$$(n - 1) \max_k\{\Delta\alpha_k\}\, M < \frac{(n - 1)\varepsilon M}{4Mn} < \frac{\varepsilon}{4}.$$

This, together with (21), gives

$$U(P,f,\alpha) < \int_a^b f(x)\, d\alpha(x) + \frac{\varepsilon}{2} \tag{22}$$

for all P with $|P| < \delta_1$.

In the same manner we can show that there exists a $\delta_2 > 0$ such that

$$L(P,f,\alpha) > \int_a^b f(x)\, d\alpha(x) - \frac{\varepsilon}{2} \tag{23}$$

for all P with $|P| < \delta_2$.

Taking $\delta = \min\{\delta_1, \delta_2\}$, we see that (22) and (23) hold for every P such that $|P| < \delta$.

But

$$L(P,f,\alpha) \le S(P,f,\alpha) \le U(P,f,\alpha)$$

holds for any partition P and so (22) and (23) give

$$S(P,f,\alpha) < \int_a^b f(x)\, d\alpha(x) + \frac{\varepsilon}{2} \quad \text{and} \quad S(P,f,\alpha) > \int_a^b f(x)\, d\alpha(x) - \frac{\varepsilon}{2},$$

respectively; thus

$$\left| S(P,f,\alpha) - \int_a^b f(x)\, d\alpha(x) \right| < \varepsilon$$

for all P such that $|P| < \delta$. □

Remark Note that in case $\alpha(x) = x$ the foregoing proposition gives a necessary and sufficient condition for $f \in \mathcal{R}$ on $[a,b]$ and a way to compute $\int_a^b f(x)\, dx$. In Proposition 35 we shall give an analogous result for $f \in \mathcal{R}(\alpha)$ on $[a,b]$. In Exercise 49 at the end of this chapter we shall encounter another necessary and sufficient condition for f to be in $\mathcal{R}$ on $[a,b]$.

Definition If α is a real-valued function of finite variation on a closed bounded interval $[a,b]$ and if (see Proposition 46 of Chapter 2) $\alpha = \beta - \gamma$, where β and γ are increasing functions, we define

$$\int_a^b f(x)\, d\alpha(x) = \int_a^b f(x)\, d\beta(x) - \int_a^b f(x)\, d\gamma(x), \tag{24}$$

provided the two integrals on the right-hand side exist in the sense in which we have defined them earlier.

If $\alpha = \beta_1 - \gamma_1$ is another decomposition of α into a difference of increasing functions, then $\beta + \gamma_1 = \beta_1 + \gamma$ and so, by part (vi) of Proposition 19,

$$\int_a^b f(x)\ d\beta(x) + \int_a^b f(x)\ d\gamma_1(x) = \int_a^b f(x)\ d\beta_1(x) + \int_a^b f(x)\ d\gamma(x).$$

This shows that $\int_a^b f(x)\ d\alpha(x)$, as defined in (24), is independent of the particular decomposition of α that we choose, provided that f belongs to $R(\beta)$, $R(\beta_1)$, $R(\gamma)$, and $R(\gamma_1)$.

Remark In the following two cases the existence of the integrals in question is assured:

(i) If f is continuous and α is of finite variation.

(ii) If f and α are of finite variation, and α is continuous.

In case (i), we appeal to Proposition 17. In case (ii), we agree to decompose α only into a difference of continuous monotonic functions (which can be done, by the corollary to Proposition 48 of Chapter 2), we decompose f into a difference of monotonic functions, and we appeal to Proposition 18. We further observe that the full force of Proposition 24 is at our disposal in cases (i) and (ii).

Proposition 25 Suppose that f and α are real-valued functions on a closed bounded interval [a,b] which satisfy (i) or (ii) of the foregoing remark. Let v be the *total variation function of* α *on* the interval [a,b], that is,

$$v(a) = 0 \quad \text{and} \quad v(x) = V_a^x \alpha \quad \text{for} \quad a < x \le b.$$

Then

$$\left| \int_a^b f(x)\ d\alpha(x) \right| \le \int_a^b |f(x)|\ dv(x).$$

Proof If (i) holds, then |f| is continuous. If (ii) holds, then v is continuous by Proposition 48 of Chapter 2 and since f is of finite variation, so is $|f|$ (because, for any t', t" in [a,b], $||f(t')| - |f(t")|| \le |f(t') - f(t")|$ by a consequence of the triangle inequality). Hence $\int_a^b |f(x)|\, dv(x)$ exists in any case.

For any partition $P = \{x_0, \ldots, x_n\}$ of [a,b] we have

$$\left|\sum_{k=1}^{n} f(t_k)\, \Delta\alpha_k\right| \le \sum_{k=1}^{n} |f(t_k)||\Delta\alpha_k| \le \sum_{k=1}^{n} |f(t_k)|\, \Delta v_k, \tag{25}$$

if $x_{k-1} \le t_k \le x_k$ for $1 \le k \le n$. As $|P| \to 0$, the first sum in (25) tends to $\int_a^b |f(x)\, d\alpha(x)|$ and the third tends to $\int_a^b |f(x)|\, dv(x)$, by Proposition 24. □

Corollary We have

$$\left|\int_a^b f(x)\, d\alpha(x)\right| \le \left(V_a^b\alpha\right)\left(\sup\{|f(x)|: x \in [a,b]\}\right).$$

Lemma Given are two sequences $(a_n)_{n=0}^{\infty}$ and $(b_n)_{n=0}^{\infty}$. We put

$$A_{-1} = 0 \quad \text{and} \quad A_n = \sum_{k=0}^{n} a_k \quad \text{if} \quad n \ge 0.$$

Then, if $0 \le p \le q$, we have

$$\sum_{n=p}^{q} a_n b_n = \sum_{n=p}^{q-1} A_n(b_n - b_{n+1}) + A_q b_q - A_{p-1} b_p.$$

Proof An easy calculation shows that

$$\sum_{n=p}^{q} a_n b_n = \sum_{n=p}^{q} (A_n - A_{n-1}) b_n = \sum_{n=p}^{q} A_n b_n - \sum_{n=p-1}^{q-1} A_n b_{n+1}$$
$$= \sum_{n=p}^{q-1} A_n(b_n - b_{n+1}) + A_q b_q - A_{p-1} b_p.$$
□

Proposition 26 (Integration-by-Parts Formula) Let f and α be of finite variation over [a,b], and f also be continuous. Then

$$\int_a^b f(x)\, d\alpha(x) = f(b)\,\alpha(b) - f(a)\,\alpha(a) - \int_a^b \alpha(x)\, df(x).$$

Proof Choose a partition $P = \{x_0, \ldots, x_n\}$ of $[a,b]$. Pick $t_1, \ldots, t_n$ such that $x_{k-1} \le t_k \le x_k$, put $t_0 = a$, $t_{n+1} = b$, and let Q be the partition $\{t_0, t_1, \ldots, t_{n+1}\}$ of $[a,b]$. Then, by the foregoing lemma,

$$S(P,f,\alpha) = \sum_{k=1}^{n} f(t_k)[\alpha(x_k) - \alpha(x_{k-1})]$$

$$= f(b)\,\alpha(b) - f(a)\,\alpha(a) - \sum_{k=1}^{n+1} \alpha(x_{k-1})[f(t_k) - f(t_{k-1})]$$

$$= f(b)\,\alpha(b) - f(a)\,\alpha(a) - S(Q,\alpha,f),$$

since $t_{k-1} \le x_{k-1} \le t_k$. If $|P| \to 0$, then $|Q| \to 0$, and Proposition 24 shows that

$$S(P,f,\alpha) \to \int_a^b f(x)\, d\alpha(x) \quad \text{and} \quad S(Q,\alpha,f) \to \int_a^b \alpha(x)\, df(x). \qquad \square$$

Remark If f is continuous and α is of finite variation, then $\int_a^b f(x)\, d\alpha(x)$ exists and Proposition 26 may be used to define the integral $\int_a^b \alpha(x)\, df(x)$ even if f is not of finite variation.

Proposition 27 (First Mean-Value Theorem for Integrals) If f is continuous and α is monotonically increasing on $[a,b]$, then there is a point x such that $a \le x \le b$, and

$$\int_a^b f(t)\, d\alpha(t) = f(x)[\alpha(b) - \alpha(a)].$$

Proof Let

$$M = \sup\{f(t) : t \in [a,b]\} \quad \text{and} \quad m = \inf\{f(t) : t \in [a,b]\}.$$

Then

$$m[\alpha(b) - \alpha(a)] \le \int_a^b f(t)\, d\alpha(t) \le M[\alpha(b) - \alpha(a)].$$

Hence there is a number s, $m \le s \le M$, such that

$$\int_a^b f(t)\, d\alpha(t) = s[\alpha(b) - \alpha(a)].$$

By Proposition 31 of Chapter 2, there exists a point $x \in [a,b]$ for which $f(x) = s$, and the proposition follows. □

Remark It may not be possible to choose x in (a,b); if $\alpha(a) = 0$ and $\alpha(t) = 1$ for $a < t \le b$, and f continuous, we have

$$\int_a^b f(t)\, d\alpha(t) = f(a) = f(a)[\alpha(b) - \alpha(a)].$$

However, if α is strictly increasing, it is possible to choose x in (a,b); see Exercise 38 at the end of this chapter.

Proposition 28 (Second Mean-Value Theorem for Integrals) Let f be monotonic and α be continuous and of finite variation on [a,b]. Then there is a point $x \in [a,b]$ such that

$$\int_a^b f(t)\, d\alpha(t) = f(a)[\alpha(x) - \alpha(a)] + f(b)[\alpha(b) - \alpha(x)].$$

Proof By Propositions 26 and 27,

$$\int_a^b f(t)\, d\alpha(t) = f(b)\,\alpha(b) - f(a)\,\alpha(a) - \int_a^b \alpha(t)\, df(t)$$

$$= f(b)\,\alpha(b) - f(a)\,\alpha(a) - \alpha(x)[f(b) - f(a)]$$

for some $x \in [a,b]$. □

Remark In connection with Proposition 28, see Exercises 44 and 63 at the end of this chapter.

Proposition 29 (Change-of-Variable Formula) Let f and g be continuous on [a,b], g be strictly increasing, and h be the inverse function of g, that is, $h = g^{-1}$ (see Proposition 35 of Chapter 2). Then

$$\int_a^b f(x)\ dx = \int_c^d f[h(y)]\ dh(y), \tag{26}$$

where c = g(a) and d = g(b).

Proof For any partition P,

$$P = \{a = x_0 < x_1 < \cdots < x_{n-1} < x_n = b\},$$

put $y_k = g(x_k)$ for $0 \le k \le n$, and consider the partition Q,

$$Q = \{g(a) = y_0 < y_1 < \cdots < y_{n-1} < y_n = g(b)\}.$$

Putting s(y) = f[h(y)], we have

$$\sum_{k=1}^{n} f(x_k)\ (x_k - x_{k-1}) = \sum_{k=1}^{n} s(y_k)[h(y_k) - h(y_{k-1})]. \tag{27}$$

By the uniform continuity of g on [a,b] (Proposition 32 of Chapter 2), $|P| \to 0$ implies $|Q| \to 0$. Thus, if we let $|P| \to 0$ in (27), we get (26) since s is continuous by Propositions 35 and 28 of Chapter 2.

□

Remark In connection with Proposition 29 see Exercise 46 at the end of this chapter.

Proposition 30 Let $f \in R$ on [a,b]. For $a \le x \le b$, put

$$F(x) = \int_a^x f(t)\ dt.$$

Then F is uniformly continuous on [a,b]; moreover, if f is continuous at a point x_0 of [a,b], then F is differentiable at x_0 and $F'(x_0) = f(x_0)$.

Proof Since $f \in \mathcal{R}$, f is bounded. Suppose $|f(t)| \le M$ on $[a,b]$. If $a \le x < y \le b$, then

$$|F(y) - F(x)| = \left|\int_x^y f(t)\, dt\right| \le M(y - x),$$

by Proposition 19, parts (iv) and (v). Given $\varepsilon > 0$, we get

$$|F(y) - F(x)| < \varepsilon$$

provided that $|y - x| < \varepsilon/M$, showing the uniform continuity of F on $[a,b]$.

Suppose now that f is continuous at x_0. Given $\varepsilon > 0$, pick $\delta > 0$ such that

$$|f(t) - f(x_0)| < \varepsilon$$

if $|t - x_0| < \delta$, and $t \in [a,b]$. Hence, if

$$x_0 - \delta < s \le x_0 \le t < x_0 + \delta \quad \text{and} \quad a \le s < t \le b,$$

we have by part (v) of Proposition 19

$$\left|\frac{F(t) - F(s)}{t - s} - f(x_0)\right| = \left|\frac{1}{t - s}\int_s^t [f(u) - f(x_0)]\, du\right| < \varepsilon.$$

It follows that $f(x_0) = F'(x_0)$. □

Remark For a further property of the function F defined in Proposition 30 consult part (a) of Exercise 61 at the end of this chapter.

Proposition 31 (Fundamental Theorem of Calculus) If $f \in \mathcal{R}$ on $[a,b]$ and if there exists a differentiable function F on $[a,b]$ such that $F' = f$, then

$$\int_a^b f(x)\, dx = F(b) - F(a).$$

Proof Given any partition $P = \{x_0, x_1, \ldots, x_n\}$ of $[a,b]$, pick t_k, $1 \le k \le n$, such that

$$F(x_k) - F(x_{k-1}) = f(t_k)\,(x_k - x_{k-1});$$

this is possible by Proposition 6. Then

$$F(b) - F(a) = \sum_{k=1}^{n} [F(x_k) - F(x_{k-1})] = \sum_{k=1}^{n} f(t_k)\,\Delta x_k,$$

and the last sum tends to $\int_a^b f(x)\,dx$ as $|P| \to 0$ [by Proposition 24 with $\alpha(x) = x$]. □

Proposition 32 If $f \in R$ and $\alpha' \in R$ on $[a,b]$, then $f \in R(\alpha)$ on $[a,b]$ and

$$\int_a^b f(x)\,d\alpha(x) = \int_a^b f(x)\,\alpha'(x)\,dx.$$

Proof Let $\varepsilon > 0$ be given and pick M so that $|f(x)| \le M$ on $[a,b]$. Since $f\alpha' \in R$ [by part (i) of Proposition 21] and $\alpha' \in R$ on $[a,b]$, part (ii) of Proposition 24 shows that there exist $\delta_1 > 0$ and $\delta_2 > 0$ such that

$$\left|\sum_{k=1}^{n} f(t_k)\,\alpha'(t_k)\,\Delta x_k - \int_a^b f(x)\,\alpha'(x)\,dx\right| < \varepsilon \qquad (28)$$

if $|P| < \delta_1$ and $x_{k-1} \le t_k \le x_k$, and

$$\left|\sum_{k=1}^{n} \alpha'(t_k)\,\Delta x_k - \int_a^b \alpha'(x)\,dx\right| < \varepsilon \qquad (29)$$

if $|P| < \delta_2$ and $x_{k-1} \le t_k \le x_k$. Letting t_k vary in (29) we see that

$$\sum_{k=1}^{n} |\alpha'(t_k) - \alpha'(s_k)|\,\Delta x_k < 2\varepsilon \qquad (30)$$

if $|P| < \delta_2$ and $x_{k-1} \le t_k \le x_k$, $x_{k-1} \le s_k \le x_k$.

Now choose P so that $|P| < \delta = \min\{\delta_1, \delta_2\}$, and choose t_k in $[x_{k-1}, x_k]$ such that

$$\Delta\alpha_k = \alpha'(s_k)\,\Delta x_k.$$

Then

$$\sum_{k=1}^{n} f(t_k)\,\Delta\alpha_k = \sum_{k=1}^{n} f(t_k)\,\alpha'(t_k)\,\Delta x_k + \sum_{k=1}^{n} f(t_k)[\alpha'(s_k) - \alpha'(s_{k-1})]\,\Delta x_k. \tag{31}$$

By (28) and (30), the left-hand side of (31) differs from the integral $\int_a^b f(x)\,\alpha'(x)\,dx$ by less than $(2M + 1)\varepsilon$. This shows that

$$\lim_{|P|\to 0} S(P,f,\alpha) = \int_a^b f(x)\,\alpha'(x)\,dx$$

and completes the proof, by part (i) of Proposition 24. □

We leave the proof of the next proposition as an exercise for the reader.

Proposition 33 Let f be continuous and α be of finite variation on [a,b]. Moreover, let v be the total variation function of α on [a,b] introduced in Proposition 25 and define

$$F(a) = 0 \quad \text{and} \quad F(x) = \int_a^x f(t)\,d\alpha(t) \quad \text{for} \quad a < x \le b.$$

Then F is of finite variation on [a,b] and

$$V_a^x F \le \int_a^x |f(t)|\,dv(t).$$

Moreover, for $a \le x < b$,

$$F(x+) - F(x) = f(x)[\alpha(x+) - \alpha(x)]$$

and, for $a < x \le b$,

$$F(x) - F(x-) = f(x)[\alpha(x) - \alpha(x-)].$$

Proposition 34 Let f, α, v, and F be as in Proposition 33. Then

$$V_a^b F = \int_a^b |f(t)|\, dv(t).$$

Proof By Proposition 33, F is of finite variation over [a,b] and the total variation of F over [a,b] does not exceed

$$\int_a^b |f(t)|\, dv(t).$$

We put

$$J = \int_a^b |f(t)|\, dv(t).$$

For a given partition $P = \{x_0, x_1, \ldots, x_n\}$ with mesh $|P| = \delta$, put

$$S_\delta = \sum_{k=1}^{n} \left| \int_{x_{k-1}}^{x_k} f(t)\, d\alpha(t) \right|.$$

Then $0 \le S_\delta \le J$. We wish to show that by the choice of P the sum S_δ may be brought arbitrarily near to J.

Let $\varepsilon > 0$ be given. Pick δ_0 and a partition P of mesh $|P| = \delta$ less than δ_0 such that

$$0 \le v(b) - \sum_{k=1}^{n} |\alpha(x_k) - \alpha(x_{k-1})| < \varepsilon \tag{32}$$

and

$$|f(t') - f(t'')| < \varepsilon$$

with $|t' - t''| < \delta_0$, $a \le t' < t'' \le b$. This is possible by the definition of v(b) and the uniform continuity of f on [a,b].

By Proposition 27, for $x_{k-1} \le s_k \le x_k$,

$$J = \sum_{k=1}^{n} |f(s_k)| \int_{x_{k-1}}^{x_k} dv(t).$$

Hence

$$J - S_\delta = \sum_{k=1}^{n} |f(s_k)| \left\{ \int_{x_{k-1}}^{x_k} dv(t) - \left| \int_{x_{k-1}}^{x_k} d\alpha(t) \right| \right\}$$

$$+ \sum_{k=1}^{n} \left\{ \left| \int_{x_{k-1}}^{x_k} f(s_k)\, d\alpha(t) \right| + \left| \int_{x_{k-1}}^{x_k} d\alpha(t) \right| \right\}.$$

The first of these two sums does not exceed

$$\sup_{a \le t \le b} |f(t)| \left\{ v(b) - \sum_{k=1}^{n} |\alpha(x_k) - \alpha(x_{k-1})| \right\} \le \sup_{a \le t \le b} |f(t)|\, \varepsilon$$

by (32); the second is not greater than

$$\sum_{k=1}^{n} \left| \int_{x_{k-1}}^{x_k} \left\{ f(s_k) - f(t) \right\} d\alpha(t) \right| \le \varepsilon\, v(b)$$

by (33) and the corollary of Proposition 25. □

Suppose that $[a,b]$ is a closed bounded interval, f and α real-valued functions on $[a,b]$, f being bounded and α being monotonically increasing on $[a,b]$. Let

$$P = \{a = x_0 < x_1 < \cdots < x_n = b\}$$

be a partition of $[a,b]$ and let t_k be a point in the subinterval $[x_{k-1}, x_k]$. We defined a Riemann-Stieltjes sum of f with respect to α relative to P by

$$S(P,f,\alpha) = \sum_{k=1}^{n} f(t_k)\, \Delta\alpha_k.$$

We recall that

$$\lim_{|P| \to 0} S(P,f,\alpha) = A$$

means: There exists a real number A having the property that for any

$\varepsilon > 0$ there exists a $\delta > 0$ such that for every partition P of $[a,b]$ with mesh $|P| < \delta$ and for every choice t_k in $[x_{k-1},x_k]$, we have

$$|S(P,f,\alpha) - A| < \varepsilon.$$

We know from Proposition 24 and the example preceding it that $f \in \mathcal{R}(\alpha)$ does not imply in general the existence of $\lim S(P,f,\alpha)$, as $|P| \to 0$; thus the following proposition is of special interest.

Proposition 35 Suppose that $[a,b]$ is a closed bounded interval, f and α real-valued functions on $[a,b]$, f being bounded and α being monotonically increasing on $[a,b]$. Then the following statements are equivalent:

(i) $f \in \mathcal{R}(\alpha)$ on $[a,b]$.

(ii) There exists a real number A [namely, $\int_a^b f(x)\, d\alpha(x)$] having the following property: For any $\varepsilon > 0$, there is a partition P_ε of $[a,b]$ such that for every refinement P of P_ε and for any choice of the points t_k in $[x_{k-1},x_k]$ we have $|S(P,f,\alpha) - A| < \varepsilon$.

(iii) For any $\varepsilon > 0$ there exists a partition P_ε such that P being a refinement of P_ε implies
$$0 \le U(P,f,\alpha) - L(P,f,\alpha) < \varepsilon.$$

Proof We will show that part (i) implies (ii), part (ii) implies (iii), and part (iii) implies (i).

Assume that (i) holds, that is,

$$\overline{\int_a^b} f(x)\, d\alpha(x) = \underline{\int_a^b} f(x)\, d\alpha(x)$$

and let A denote their common value. Given $\varepsilon > 0$, choose P'_ε so that $U(P,f,\alpha) < \overline{\int_a^b} f(x)\, d\alpha(x) + \varepsilon$ for all P finer than P'_ε. Also choose P''_ε such that $L(P,f,\alpha) > \underline{\int_a^b} f(x)\, d\alpha(x) - \varepsilon$ for all P finer than P''_ε. If P_ε is the common refinement of P'_ε and P''_ε, we get

$$\underline{\int_a^b} f(x)\ d\alpha(x) - \varepsilon < L(P,f,\alpha) \le S(P,f,\alpha) \le U(P,f,\alpha)$$
$$< \overline{\int_a^b} f(x)\ d\alpha(x) + \varepsilon$$

for every P finer than P_ε. But since

$$\overline{\int_a^b} f(x)\ d\alpha(x) = \underline{\int_a^b} f(x)\ d\alpha(x) = A,$$

this means that $|S(P,f,\alpha) - A| < \varepsilon$ whenever P is a refinement of P_ε. This shows that (ii) is satisfied and $A = \int_a^b f(x)\ d\alpha(x)$.

Assume that (ii) holds. If $\alpha(b) = \alpha(a)$, then (iii) holds trivially, so we can suppose that $\alpha(a) < \alpha(b)$. Given $\varepsilon > 0$, pick P_ε such that for any finer P and all choices of t_k and t_k' in $[x_{k-1},x_k]$ we have

$$\left|\sum_{k=1}^{n} f(t_k)\ \Delta\alpha_k - A\right| < \frac{\varepsilon}{3} \quad\text{and}\quad \left|\sum_{k=1}^{n} f(t_k')\ \Delta\alpha_k - A\right| < \frac{\varepsilon}{3}.$$

Combining these inequalities, we get

$$\left|\sum_{k=1}^{n} [f(t_k) - f(t_k')]\ \Delta\alpha_k\right| < \frac{2}{3}\ \varepsilon.$$

Since $M_k - m_k = \sup\{f(x) - f(x') : x,\ x' \in [x_{k-1},x_k]\}$, it follows that for every $h > 0$ we can select t_k and t_k' such that

$$f(t_k) - f(t_k') > M_k - m_k - h.$$

Making a choice corresponding to

$$h = \frac{\varepsilon}{3[\alpha(b) - \alpha(a)]},$$

we get

$$U(P,f,\alpha) - L(P,f,\alpha) = \sum_{k=1}^{n} (M_k - m_k)\ \Delta\alpha_k$$
$$< \sum_{k=1}^{n} [f(t_k) - f(t_k')]\ \Delta\alpha_k + h\sum_{k=1}^{n} \Delta\alpha_k$$
$$< \varepsilon.$$

Hence (ii) implies (iii).

Finally, assume that (iii) holds. If $\varepsilon > 0$ is given, there is a partition P_ε such that P finer than P_ε implies

$$U(P,f,\alpha) < L(P,f,\alpha) + \varepsilon.$$

Hence, for such P we have

$$\overline{\int_a^b} f(x)\, d\alpha(x) \le U(P,f,\alpha) < L(P,f,\alpha) + \varepsilon \le \underline{\int_a^b} f(x)\, d\alpha(x) + \varepsilon.$$

Since $\varepsilon > 0$ is arbitrary, we have that $\overline{\int_a^b} f(x)\, d\alpha(x) \le \underline{\int_a^b} f(x)\, d\alpha(x)$. But, by the corollary of Proposition 15, we also have the opposite inequality. Hence (iii) implies (i). □

We conclude this section with an example which is of particular interest to Riemann integration.

Example: Every rational number x can be written in the form $x = p/q$, where $q > 0$, and p and q are integers without any common divisors. When $x = 0$, we take $q = 1$. Consider the function f on the interval $[0,1]$ defined by

$$f(x) = 0 \quad \text{for } x \text{ irrational},$$
$$= \frac{1}{q} \quad \text{for } x = \frac{p}{q}.$$

Then f is continuous at every irrational point, and f has a simple discontinuity at every rational point.

Indeed, let x_0 be any point of $[0,1]$. Given $\varepsilon > 0$, there is only a finite number of positive integers q that are not larger than $1/\varepsilon$ and this means that in $[0,1]$ there are only finitely many rational points p/q for which $f(p/q) = 1/q \ge \varepsilon$. Thus one may construct around the point x_0 a neighborhood $(x_0 - \delta, x_0 + \delta)$ with $\delta > 0$ such that in this neighborhood there is no point x for which $f(x) \ge \varepsilon$ (except possibly the point x_0 itself). Thus, if $0 < |x - x_0| < \delta$, then for both rational and irrational x we have $|f(x)| < \varepsilon$. Hence

$$f(x_0+) = f(x_0-) = 0$$

for every point x_0. If x_0 is irrational, then $f(x_0) = 0$, that is, f is continuous at x_0; if x_0 is irrational, then $f(x_0) \neq 0$, that is, f is discontinuous at x_0.

We claim that f is Riemann integrable on [0,1].

Indeed, let [0,1] be partitioned into subintervals of length $\Delta x_k \leq \lambda$ by a partition P with $|P| = \lambda$. We pick an arbitrary positive integer I. The subintervals fall into two classes:

1. To the first class belong those intervals that contain the points p/q with q at most equal to I; since there are only finitely many such points p/q, say $k = k_I$, there are only at most $2k_I$ such intervals; their total length does not exceed $2k_I\lambda$.

2. To the second class belong those intervals that do not contain the points mentioned in 1; over these intervals the oscillation $M_k - m_k$ of the function f is clearly smaller than 1/I.

We now observe that

$$U(P,f) - L(P,f) = \sum_{k=1}^{n} (M_k - m_k)\, \Delta x_k < 2k_I\lambda + \frac{1}{I}.$$

Taking $I > 2/\varepsilon$ and then $\lambda < \varepsilon/(4k_I)$, we get

$$\sum_{k=1}^{n} (M_k - m_k)\, \Delta x_k < \varepsilon.$$

Thus $f \in \mathcal{R}$ on [0,1].

We also note that if

$$g(y) = 1 \quad \text{for } 0 < y \leq 1,$$
$$= 0 \quad \text{for } y = 0,$$

then the function

$$h(x) = g[f(x)]$$

for $x \in [0,1]$ is not Riemann integrable over [0,1], because $h(x) = 1$ for rational x and $h(x) = 0$ for irrational x; however,

$$\overline{\int_0^1} h(x)\,dx = 1 \quad \text{and} \quad \underline{\int_0^1} h(x)\,dx = 0.$$

Remark In Exercise 32 at the end of Chapter 5 we shall take up H. Lebesgue's elegant characterization of the Riemann integral.

3. The Number e Is Transcendental

Lemma For $c > 0$,

$$\lim_{n \to \infty} \frac{c^n}{n!} = 0.$$

Proof Let

$$x_n = \frac{c^n}{n!}.$$

It is clear that $x_n \to 0$, as $n \to \infty$, when $0 < c \leq 1$. Since

$$x_{n+1} = \frac{c}{n+1} x_n, \tag{34}$$

we see that (x_n) is a monotonically decreasing sequence for $n > c - 1$; moreover, the sequence is bounded below by zero. Thus (x_n) converges to some real number a, by Proposition 54 of Chapter 1. Passing to the limit in (34), as $n \to \infty$, we see that $a = 0 \cdot a$, or $a = 0$. □

The following form of the Integration-by-Parts Formula gets much use in elementary calculus:

If u' and v' are continuous functions on a closed and bounded interval [a,b], then

$$\int_a^b u(x)\, v'(x)\, dx = u(b)\, v(b) - u(a)\, v(a) - \int_a^b u'(x)\, v(x)\, dx.$$

A simpler notation for the foregoing formula is:

$$\int_a^b uv'\,dx = uv\Big|_a^b - \int_a^b u'v\,dx.$$

It is easy to see that the Integration-by-Parts Formula generalizes to the following statement:

If $u^{(n+1)}$ and $v^{(n+1)}$ are continuous functions on a closed and bounded interval $[a,b]$, then

$$\int_a^b uv^{(n+1)}\,dx = \left(uv^{(n)} - u'v^{(n-1)} + \cdots + (-1)^n u^{(n)}v\right)\Big|_a^b + (-1)^{n+1}\int_a^b u^{(n+1)}v\,dx. \tag{35}$$

We shall use this generalized version of the Integration-by-Parts Formula in the proof of the next proposition.

Proposition 36 (Hermite's Theorem) The number e, where

$$e = \lim_{n\to\infty}\left(1 + \frac{1}{n}\right)^n,$$

satisfies no relation of the form

$$c_0 + c_1e + c_2e^2 + \cdots + c_me^m = 0 \tag{36}$$

having integral coefficients not all zero. (Stated otherwise, e is a *transcendental number*.)

Proof Suppose that e is a root of the polynomial

$$c_0 + c_1x + c_2x^2 + \cdots + c_mx^m$$

with integral coefficients $c_0, c_1, \ldots, c_m$ which are not all zero; we shall show that this assumption leads to a contradiction.

In (35), let $u = f$ be an arbitrary polynomial of degree n; moreover, let $v(x) = (-1)^{n+1} e^{-x}$. Setting a = 0, (35) gives

$$\int_0^b f(x)\, e^{-x}\, dx = -e^{-x}\left(f(x) + f'(x) + \cdots + f^{(n)}(x)\right)\Big|_0^b,$$

because $f^{(n+1)} = 0$. We put

$$F(x) = f(x) + f'(x) + \cdots + f^{(n)}(x).$$

Then

$$e^b F(0) = F(b) + e^b \int_0^b f(x)\, e^{-x}\, dx. \tag{37}$$

In (37) we let b = 0, 1, 2, ..., m, successively, and multiply the resulting equations by $c_0, c_1, c_2, \ldots, c_m$, respectively; finally, we add up the resulting relations and obtain, by (36),

$$\begin{aligned} 0 &= c_0 F(0) + c_1 F(1) + \cdots + c_m F(m) \\ &\quad + \sum_{j=1}^{m} c_j e^j \int_0^j f(x)\, e^{-x}\, dx, \end{aligned} \tag{38}$$

which, according to our assumption, must hold for <u>every</u> polynomial f (with integral coefficients). We shall show that there is in fact an f for which (38) is false; this will of course prove the claim of the proposition. To this end we put

$$f(x) = \frac{1}{(p-1)!} x^{p-1} (x-1)^p (x-2)^p \cdots (x-m)^p,$$

where p is an odd prime larger than both m and $|c_0|$. The derivatives of order p and larger of this polynomial have integral coefficients which are divisible by p; this is an immediate consequence of the fact that the product of p consecutive natural numbers is divisible by p!. Since the polynomial f and its first p - 1 derivatives vanish for x = 1, 2, ..., m, we have F(1), F(2), ..., F(m) are integral multiples of p. But the situation is different for F(0). For x = 0, besides f(x) only the first p - 2 derivatives vanish and so

$$F(0) = f^{(p-1)}(0) + f^{(p)}(0) + \cdots$$

holds. All summands beginning with the second summand are integral multiples of p; but, since

$$f^{(p-1)}(0) = [(-1)^m m!]^p,$$

F(0) is not divisible by p. Since p is a prime number both larger tham m and $|c_0|$, it follows in particular that c_0 is not divisible by p. We therefore see that the first sum in (38), namely,

$$c_0 F(0) + c_1 F(1) + \cdots + c_m F(m),$$

is not divisible by p and thus cannot be equal to zero. We now turn our attention to the second sum in (38). On the interval [0,m] we obviously have

$$|f(x)| < \frac{1}{(p-1)!} m^{p-1} m^p m^p \cdots = \frac{m^{mp+p-1}}{(p-1)!}.$$

Thus

$$\left|\int_0^j f(x) e^{-x} dx\right| < \frac{m^{mp+p-1}}{(p-1)!} \int_0^j e^{-x} dx < \frac{m^{mp+p-1}}{(p-1)!}$$

and, on setting

$$C = |c_0| + |c_1| + \cdots + |c_m|,$$

we obtain

$$\left|\sum_{j=1}^{m} c_j e^j \int_0^j f(x) e^{-x} dx\right| < C e^m \frac{m^{mp+p-1}}{(p-1)!} = C e^m m^m \frac{(m^{m+1})^{p-1}}{(p-1)!}.$$

By the foregoing lemma we know that the last factor tends to zero as $p \to \infty$. Thus, by taking p large enough, the absolute value of the second sum in (38) can be made as small as we please. In other words, by choosing p suitably large, the total sum on the right-hand side of (38) cannot be equal to zero. We have thus arrived at a contradiction. □

Formula (35) may also be used as a starting point in obtaining a Taylor-type formula with remainder (see Proposition 13). Let

$$v(x) = (b - x)^n.$$

Then

$$v'(x) = -n(b - x)^{n-1},$$

$$v''(x) = n(n - 1)(b - x)^{n-2},$$

$$\vdots$$

$$v^{(n)}(x) = (-1)^n\, n(n - 1) \cdots 1,$$

$$v^{(n+1)}(x) = 0;$$

moreover, $v(b) = v'(b) = \cdots = v^{(n-1)}(b) = 0$. Letting $u = f$, formula (35) assumes the form

$$0 = (-1)^n\Big[n!\, f(b) - n!\, f(a) - n!\, f'(a)\,(b - a) - \frac{n!}{2!} f''(a)\,(b - a)^2 - \cdots - f^{(n)}(a)\,(b - a)^n\Big] + (-1)^{n+1} \int_a^b f^{(n+1)}(x)\,(b - x)^n\, dx.$$

(We are assuming of course that $f^{(n+1)}$ is continuous on $[a,b]$.) From this in turn we obtain

$$f(b) = f(a) + \frac{f'(a)}{1!}(b - a) + \frac{f''(a)}{2!}(b - a)^2 + \cdots + \frac{f^{(n)}(a)}{n!}(b - a)^n + \frac{1}{n!}\int_a^b f^{(n+1)}(x)\,(b - x)^n\, dx.$$

Replacing a by α and b by β, we get

$$f(\beta) = f(\alpha) + \frac{f'(\alpha)}{1!}(\beta - \alpha) + \frac{f''(\alpha)}{2!}(\beta - \alpha)^2 + \cdots + \frac{f^{(n)}(\alpha)}{n!}(\beta - \alpha)^n + \frac{1}{n!}\int_\alpha^\beta f^{(n+1)}(t)\,(\beta - t)^n\, dt.$$

However, $(\beta - t)^n$ does not change sign on the interval $[\alpha,\beta]$ and we may therefore apply Proposition 27 (and we do). We obtain

$$\frac{1}{n!}\int_{\alpha}^{\beta} f^{(n+1)}(t)\,(\beta - t)^{n}\,dt = \frac{1}{n!} f^{(n+1)}(x) \int_{\alpha}^{\beta} (\beta - t)^{n}\,dt$$

$$= \frac{f^{(n+1)}(x)}{(n+1)!} (\beta - \alpha)^{n+1}.$$

where x is a point of the interval $[\alpha,\beta]$.

Remark In the proof of Proposition 36 we used the fact that the product of p consecutive natural numbers is divisible by $p!$ To see that this is so, observe that

$$n(n - 1) \cdots (n - p + 1) = \binom{n}{p} p!\,.$$

Exercises

1. Let f be differentiable on the right at a point c, that is, let

$$\lim_{t \to c+} \frac{f(t) - f(c)}{t - c} = f'_R(c)$$

exist (as a finite real number), and suppose that $f'_R(c)$ is positive (negative). Show that there is a positive δ such that $f(t) - f(c)$ is positive (negative) for all $t \in (c, c + \delta)$. Similarly, let f be differentiable on the left at a point c, that is, let

$$\lim_{t \to c-} \frac{f(t) - f(c)}{t - c} = f'_L(c)$$

exist (as a finite real number), and suppose that $f'_L(c)$ is positive (negative). Verify that there exists a positive δ such that the difference $f(t) - f(c)$ is negative (positive) for all $t \in (c - \delta, c)$.

2. Let f be differentiable on $[a,b]$, $f'_R(a) = f'_L(b) = 0$, and $f(a) < f(b)$. If

$$g(a) = 0 \quad\text{and}\quad g(t) = \frac{f(t) - f(a)}{t - a} \quad \text{for } a < t \le b,$$

show that g is continuous on [a,b] and differentiable on (a,b) and that $g_L'(b) < 0$. Show also that g attains its supremum at a point of (a,b) and deduce that there exists a point c in (a,b) for which

$$f'(c) = \frac{f(c) - f(a)}{c - a}.$$

3. Let f be defined on [a,b] and suppose that f assumes its largest (smallest) value at a point $c \in (a,b)$. Show that if $f'(c)$ exists (as a finite real number), then necessarily $f'(c) = 0$.

4. If f is defined on [a,b] and

$$|f(x) - f(y)| < A|x - y|^2$$

for all points x, y of [a,b], where A is a constant, prove that f is constant on [a,b].

5. If f is a polynomial of degree n, show that

$$f(x) = f(a) + (x - a)f'(a) + \frac{(x - a)^2}{2!} f''(a) + \cdots$$

$$+ \frac{(x - a)^n}{n!} f^{(n)}(a).$$

6. Let f be defined on [a,b] and satisfy the property: for every pair of points c_1, c_2 in [a,b] with $f(c_1) > f(c_2)$ and for every γ satisfying $f(c_1) > \gamma > f(c_2)$ there exists a point c lying between c_1 and c_2 such that $f(c) = \gamma$. If, moreover, f is differentiable on (a,b) and has a bounded derivative on (a,b), show that f is then continuous on [a,b].

Hint To show that f is continuous at a, suppose $|f'(x)| < K$ for $a < x < b$ and take $a < x < a + \varepsilon/(2K)$. If $f(x) \neq f(a)$, choose γ [$= f(t)$] between f(x) and f(a) to make $|\gamma - f(a)| < \varepsilon/2$. Then $|f(x) - f(a)| \leq |f(a) - f(t)| + |\gamma - f(a)| < |x - t|K + \varepsilon/2 < \varepsilon$.

7. The two functions f and g are differentiable on $(c,c + \delta)$ with $\delta > 0$, and $f'(x)/g'(x) \to L$ as $x \to c+$. If $g(x) \to \infty$ as $x \to c+$, show that $f(x)/g(x) \to L$ as $x \to c+$.

Hint For any $\varepsilon > 0$ take $\delta \leq \delta_0$ such that $|f'(x)/g'(x) - L| < \varepsilon$ for $c < x < c + \delta$. Take a fixed $x_0 \in (c,c + \delta)$ and any x with

$c < x < x_0 < c + \delta$; then $g(x) > g(x_0)$ as otherwise g' would vanish somewhere in $(c, c + \delta)$. By Proposition 7 there exists a point t in $(c, c + \delta)$ such that

$$\left| \frac{f(x) - f(x_0)}{g(x) - g(x_0)} - L \right| = \left| \frac{f'(t)}{g'(t)} - L \right| < \varepsilon.$$

Thus $(L - \varepsilon)[g(x) - g(x_0)] + f(x_0) < f(x) < (L + \varepsilon)[g(x) - g(x_0)] + f(x_0)$.

8. Let f be a continuous function on [a,b] and be differentiable on (a,b). If $a \geq 0$, show that there exist three numbers x_1, x_2, and x_3 in (a,b) such that

$$f'(x_1) = (b + a) \frac{f'(x_2)}{2x_2} = (b^2 + 2ba + a^2) \frac{f'(x_3)}{3x_3^2}.$$

Verify also that the three numbers x_1, x_2, and x_3 are all equal only if they coincide with a point at which f' vanishes.

Hint Use Proposition 7 with $g(t) = t$, t^2, and t^3.

9. If f'' exists and is continuous on a neighborhood of a, show that

$$\frac{f(a + 2h) - 2f(a + h) + f(a)}{h^2} \to f''(a) \quad \text{as} \quad h \to 0.$$

Hint Apply Proposition 12 twice.

10. Let f be such that $f''(x)$ exists, $|f(x)| \leq A$ and $|f''(x)| \leq B$ for all $x > 0$, where A and B are positive constants. Show that, for all $x > 0$, $|f'(x)| \leq 2\sqrt{AB}$.

Hint The claim is trivial for $B = 0$. For $B > 0$, by Proposition 13 with $x > 0$, $\beta - \alpha = 2\sqrt{A/B}$.

11. Let f be such that $f''(x)$ exists in (a,b) and $f'(a) = f'(b) = 0$. Show that there is a point c in (a,b) such that

$$f''(c) \geq \frac{4}{(b - a)^2} [f(b) - f(a)].$$

Solution Since $f'(a) = f'(b) = 0$, we get, by Proposition 13,

$$f\left(\frac{a+b}{2}\right) = f(a) + \frac{f''(x_1)}{2!}\left(\frac{b-a}{2}\right)^2$$

and

$$f\left(\frac{a+b}{2}\right) = f(b) + \frac{f''(x_2)}{2!}\left(\frac{b-a}{2}\right)^2,$$

where x_1 is between a and $(a + b)/2$ and x_2 is between $(a + b)/2$ and b. But

$$|f(b) - f(a)| \le \left|f(b) - f\left(\frac{a+b}{2}\right)\right| + \left|f\left(\frac{a+b}{2}\right) - f(a)\right|$$

$$\le \left(\frac{b-a}{2}\right)^2 \left(\frac{1}{2!}\right) [|f''(x_1)| + |f''(x_2)|].$$

Let $|f''(c)|$ be the larger of the two numbers $|f''(x_1)|$ and $|f''(x_2)|$; then $\frac{1}{2}[|f''(x_1)| + |f''(x_2)|] \le |f''(c)|$ and the claim follows immediately. □

12. A sequence of functions $f_0, f_1, f_2, \ldots$ is defined on R^1 by the relations

$$f_0(x) = 1, \quad f_{k+1}(x) = x\, f_k(x) - f_k'(x)$$

for $k \ge 0$, $x \in R^1$. Show that f_n is a polynomial of degree n with n distinct real roots arranged symmetrically about 0.

Hint First verify that f_n is a polynomial of degree n, with leading term x^n, and $f_n(-x) = f_n(x)$ if n is even and $f_n(-x) = -f_n(x)$ if n is odd, for x any real number.

Let P_n be the property that f_n has distinct zeros symmetrically situated about the origin at which f_n' is alternately positive and negative, being positive at the greatest zero a_n of f_n. Then P_1 is true. Assuming P_n, show that f_{n+1} vanishes between every pair of consecutive zeros of f_n and also at $\pm a_{n+1}$, where $a_{n+1} > a_n$, since $f_{n+1}(x) \to \infty$ as $x \to \infty$. This proves part of P_{n+1} and the rest follows easily.

13. If $f(x) = |x|^3$, show that $f^{(3)}(0)$ does not exist.

14. If $c_0, c_1, \ldots, c_n$ are real numbers such that

$$c_0 + \frac{c_1}{2} + \cdots + \frac{c_{n-1}}{n} + \frac{c_n}{n+1} = 0,$$

show that the equation

$$c_0 + c_1x + \cdots + c_{n-1}x^{n-1} + c_nx^n = 0$$

has at least one real root between 0 and 1.

15. Given that $0 < a < b$, show that for $n = 2, 3, \ldots$

$$n(b - a)a^{n-1} < b^n - a^n < n(b - a)b^{n-1}$$

holds.

16. Let $x > 0$. Show that

$$\log(x + 1) - \log x = \frac{1}{t}$$

for some $t \in (x,x + 1)$.

17. Let f be differentiable on $[a,b]$. Show that f' is the pointwise limit of a sequence of continuous functions.

Solution Since f is only defined on $[a,b]$, we extend the domain of definition of f by setting, for $x \in (b,b + 1]$, $f(x) = f(b)$. Then $f(t + 1/n)$ exists for $n = 1, 2, \ldots$ and all $t \in [a,b]$. The sequence

$$f_n(t) = n\left[f\left(t + \frac{1}{n}\right) - f(t)\right]$$

has the desired properties. □

18. For $n = 1, 2, 3, \ldots$, let $g_n(x) = (x^2 - 1)^n$. Show that

$$f_n(x) = g_n^{(n)}(x)$$

has exactly n distinct real roots between -1 and 1.

Solution Note that

$$(x^2 - 1)^n = (x - 1)^n (x + 1)^n$$

and that the $n - 1$ consecutive derivatives of g_n vanish at $x = \pm 1$. By Proposition 5, g_n' has one real root between -1 and 1; again by Proposition 5, g_n'' has two distinct real roots between $- 1$ and 1, etc.; $g_n^{(n-1)}$ has n - 1 distinct real roots between -1 and 1. Thus $g_n^{(n)} = f_n$ has n distinct real roots between -1 and 1, by Proposition 5. □

19. Let f_n be as in Exercise 18. Using Proposition 14, show that $f_n(-1) = f_n(1) = 2^n\, n!$; setting

$$X_n(x) = \frac{1}{2^n\, n!} f_n(x),$$

verify that X_n satisfies

$$(x^2 - 1)\, X_n''(x) + 2x\, X_n'(x) - n(n + 1)\, X_n(x) = 0.$$

The functions X_n are called *Legendre polynomials.*

20. Show that the function $f(x) = x^2 \sin(\pi/x)$ for $x \neq 0$ and $f(0) = 0$ has finite total variation over any interval of finite length.

Solution Note that $f'(x) = 2x \sin(\pi/x) - \pi \cos(\pi/x)$ for $x \neq 0$ and $f'(0) = 0$; f' is therefore bounded. Hence, by Proposition 6,

$$|f(b) - f(a)| = |f'(x)\ (b - a)| \leq K(b - a)$$

for some $x \in (a,b)$ and K denoting the supremum of $|f'(x)|$ as x ranges over (a,b). Thus the total variation $V_a^b f \leq K(b - a)$. □

N.B.: The function $g(x) = x^2 \sin(\pi/x^2)$ for $x \neq 0$ and $g(0) = 0$ has infinite total variation over [0,2]. Indeed,

$$\sum_{k=1}^{n} \left| g\left(\sqrt{\frac{2}{2k - 1}}\right) - g\left(\frac{1}{\sqrt{k}}\right)\right| \geq \sum_{k=1}^{n} \frac{1}{k}.$$

21. Let $f \in R$ on [a,b]. Show that, for any $\varepsilon > 0$, there exists a continuous function g_ε on [a,b] such that $\int_a^b |f(x) - g_\varepsilon(x)|\, dx < \varepsilon$.

22. If $|f| \in R$ on [a,b], does it follow in general that $f \in R$?

Hint Let f(x) be 1 for rational x and -1 for irrational x.

23. Verify that

$$\int_0^3 x\, d([x] - x) = \frac{3}{2}.$$

24. Let X_n be the Legendre polynomials on [-1,1] (see Exercise 19) for n = 1, 2, ... and define $X_0(x) = 1$. Show that

$$\int_{-1}^{1} X_n(x)\, X_m(x)\, dx = 0 \qquad \text{for} \quad n \neq m,$$

$$= \frac{2}{2n + 1} \qquad \text{for} \quad n = m.$$

25. Suppose that $f(x) \geq 0$ on [a,b] and f is continuous on [a,b]. Show that $\int_a^b f(x)\, dx = 0$ implies that $f = 0$.

Solution If $f(x_1) = \alpha > 0$ for some $x_1 \in (a,b)$, then there is a $\delta > 0$ such that $f(x) > \alpha/2$ for all $x \in [x - \varepsilon, x + \varepsilon]$ with $\varepsilon > 0$. Thus $\int_a^b f(x)\, dx \geq (\alpha/2)(2\delta) = \alpha\delta > 0$. If $f = 0$, there is of course nothing to prove. □

26. (a) Show by a suitable example that the value of a Riemann-Stieltjes integral can be altered by changing the value of f at a single point. In fact, the existence of the integral can also be affected by such a change; consider

$\alpha(x) = 0$ if $x \neq 0$, $\alpha(0) = 1$,

$f(x) = 1$ if $-1 \leq x \leq 1$,

then redefine $f(0) = 2$ and $f(x) = 1$ if $x \neq 0$.

(b) Show that in a Riemann integral the values of f can be changed at a finite number of points without affecting either the existence or the value of the integral.

27. Consider the function f defined on [0,1] by

$$f(x) = 0 \quad \text{if } x \text{ is irrational}$$
$$= \frac{1}{q} \quad \text{if } x = \frac{p}{q}.$$

We have seen at the end of Section 2 that $f \in \mathcal{R}$ on [0,1]. Noting that there is only a finite number of points at which the functional value of f exceeds an assigned positive real number, show that $\int_0^1 f(x)\, dx = 0$.

28. Let $E_1, E_2, \ldots, E_n$ be n intervals which are situated in

the unit interval [0,1]. If each point of [0,1] belongs to at least q of these intervals E_j with $j = 1, \ldots, n$, show that at least one of these intervals must have length $\geq q/n$.

Solution For $x \in [0,1]$, define $f_j(x) = 1$ if $x \in E_j$ and $f_j(x) = 0$ if $x \notin E_j$; then let

$$f(x) = \sum_{j=1}^{n} f_j(x).$$

Evidently, $f(x) \geq q$ for every x in the interval [0,1] and so

$$q \leq \int_0^1 f(x)\,dx = \int_0^1 \sum_{j=1}^{n} f_j(x)\,dx = \sum_{j=1}^{n} \int_0^1 f_j(x)\,dx$$

$$= \sum_{j=1}^{n} |E_j|,$$

where $|E_j|$ denotes the length of the interval E_j. It is clear that not every summand in the last sum can be less than q/n, for if it were then we would have $q < n(q/n)$. □

29. Hölder's, Minkowski's, and Jensen's Inequalities also hold for integrals.

(a) If $f(x) \geq 0$, $g(x) \geq 0$, $k > 1$, $1/k + 1/k' = 1$, and f^k and $g^{k'}$ are integrable on [a,b], the fg is integrable on [a,b] and

$$\int_a^b f(x)\,g(x)\,dx \leq \left(\int_a^b f^k(x)\,dx\right)^{1/k} \left(\int_a^b g^{k'}(x)\,dx\right)^{1/k'}.$$

(b) If $f(x) \geq 0$, $g(x) \geq 0$, $k > 1$, and f^k and g^k are integrable on [a,b], then $(f + g)^k$ is integrable on [a,b] and

$$\left(\int_a^b \left[f(x) + g(x)\right]^k dx\right)^{1/k} \leq \left(\int_a^b f^k(x)\,dx\right)^{1/k} + \left(\int_a^b g^k(x)\,dx\right)^{1/k}.$$

(c) Let f be convex on (c,d) and g be integrable on [a,b] with

$c < g(x) < d$. If p is positive on [a,b] and gp is integrable on [a,b], then

$$f\left(\frac{\int_a^b p(x)\ g(x)\ dx}{\int_a^b p(x)\ dx}\right) \le \frac{\int_a^b p(x)\ f\big(g(x)\big)\ dx}{\int_a^b p(x)\ dx}.$$

Verify the foregoing inequalities by imitating the corresponding proofs in Section 3 of Chapter 1.

30. Let $f \in \mathcal{R}$ on [a,b]. Show that

$$\lim_{p\to\infty} \int_a^b f(x) \begin{matrix}\sin px \\ \cos px\end{matrix}\ dx = 0.$$

Solution For any bounded interval $[\alpha,\beta]$ we have

$$\left|\int_\alpha^\beta \sin px\ dx\right| = \left|\frac{\cos p\alpha - \cos p\beta}{p}\right| \le \frac{2}{p}.$$

Let $|f(x)| \le A$ on [a,b] and ε be the usual arbitrary positive number. There is a partition of [a,b], say

$$a = x_0 < x_1 < \cdots < x_n = b,$$

such that $U(P,f) - L(P,f) < \varepsilon/2$. Thus

$$\begin{aligned}\left|\int_a^b f(x)\ \sin px\ dx\right| &= \left|\sum_{k=1}^n \int_a^b [f(x_k) + f(x) - f(x_k)]\ \sin px\ dx\right| \\ &\le \sum_{k=1}^n \left\{|f(x_k)|\left|\int_{x_{k-1}}^{x_k} \sin px\ dx\right| \right. \\ &\quad \left. + \int_{x_{k-1}}^{x_k} |f(x) - f(x_k)||\sin px|\ dx\right\} \\ &< \frac{2nA}{p} + [U(P,f) - L(P,f)] < \frac{2nA}{p} + \frac{\varepsilon}{2} < \varepsilon\end{aligned}$$

when $p \ge (4nA)/\varepsilon$. Hence $\int_a^b f(x)\ \sin px\ dx$ tends to 0 as $p \to \infty$. In the same way, $\int_a^b f(x) \cos px\, dx$ tends to 0 as $p \to \infty$. □

31. Let $a < c < b$ and α be monotonically increasing on $[a,b]$. Show that if two of the integrals below exist, then the third also exists and

$$\int_a^c f(x)\, d\alpha(x) + \int_c^b f(x)\, d\alpha(x) = \int_a^b f(x)\, d\alpha(x).$$

32. Let $f \in R(\alpha)$ on $[a,b]$. Show that $f \in R(\alpha)$ on every subinterval $[c,d] \subset [a,b]$.

33. Assume that α is of finite variation on $[a,b]$. Let $v(x)$ denote the total variation of α on $[a,x]$ if $a < x \le b$, and let $v(a) = 0$. Verify that if $f \in R(\alpha)$ on $[a,b]$, then $f \in R(v)$ on $[a,b]$.

34. Let $[a,b]$ be a closed bounded interval, f and α real-valued functions on $[a,b]$ with f being bounded and α being monotonically increasing on $[a,b]$. Let c be such that $a < c < b$. Assume further that f and α are discontinuous from the right at $x = c$; that is, assume that there exists an $\varepsilon > 0$ such that for every $\delta > 0$ there are values of x and y in the interval $(c, c + \delta)$ for which

$$|f(x) - f(c)| \ge \varepsilon \quad \text{and} \quad |\alpha(y) - \alpha(c)| \ge \varepsilon.$$

Show that the integral $\int_a^b f(x)\, d\alpha(x)$ does not exist, and that the integral also fails to exist if f and α are discontinuous from the left at $x = c$.

35. Given $a < c < b$, let α be defined on $[a,b]$ as follows: The values $\alpha(a)$, $\alpha(c)$, $\alpha(b)$ are arbitrary;

$$\alpha(x) = \alpha(a) \quad \text{if} \quad a \le x < c,$$
$$= \alpha(b) \quad \text{if} \quad c < x \le b.$$

Let f be bounded on $[a,b]$ and be such that at least one of the functions f or α is continuous from the left at $x = c$ and the other is continuous from the right at $x = c$. Assuming that $[a,b]$ is bounded, show that $f \in R(\alpha)$ on $[a,b]$ and

$$\int_a^b f(x)\, d\alpha(x) = f(c)[\alpha(c+) - \alpha(c-)].$$

36. Let α be of finite variation on $[a,b]$ and suppose that $f \in \mathcal{R}(\alpha)$ on $[a,b]$. Let F be defined by

$$F(x) = \int_a^x f(t)\, d\alpha(t)$$

for $a < x \le b$ and $F(a) = 0$. Show:

(a) F is of finite variation over $[a,b]$.

(b) Every point of continuity of α is also a point of continuity of F.

(c) If α is increasing on $[a,b]$, the derivative $F'(x)$ exists at each point x in (a,b) where $\alpha'(x)$ exists and where f is continuous. For such x, we have $F'(x) = f(x)\,\alpha'(x)$.

37. Prove Proposition 33. Note in particular that $F(a+)$ need not be zero (take for example $f = 1$) even though we have assumed that $F(a) = 0$.

38. If α is a strictly increasing function in Proposition 27, does it follow that $x \in (a,b)$? (Yes!)

39. Let f be a real-valued function on $[a,b]$ which is differentiable at a point $x \in (a,b)$. Let $(\alpha_n)_{n=1}^{\infty}$ and $(\beta_n)_{n=1}^{\infty}$ be any sequences such that $a < \alpha_n < x < \beta_n < b$, $\alpha_n \to x$, and $\beta_n \to x$ as $n \to \infty$. Show that

$$\lim_{n\to\infty} \frac{f(\beta_n) - f(\alpha_n)}{\beta_n - \alpha_n} = f'(x). \tag{E.1}$$

Solution Let $\lambda_n = (\beta_n - x)/(\beta_n - \alpha_n)$. Then $0 < \lambda_n < 1$ and

$$\frac{f(\beta_n) - f(\alpha_n)}{\beta_n - \alpha_n} - f'(x)$$

$$= \lambda_n \left\{ \frac{f(\beta_n) - f(x)}{\beta_n - x} - f'(x) \right\} + (1 - \lambda_n) \left\{ \frac{f(\alpha_n) - f(x)}{\alpha_n - x} - f'(x) \right\}.$$

The expressions within the braces both tend to zero as $n \to \infty$ and the sequences $(\lambda_n)_{n=1}^{\infty}$ and $(1 - \lambda_n)_{n=1}^{\infty}$ are both bounded. Hence the limit in (E.1) exists and the equality holds. □

40. Let f be a real-valued function on [a,b]. If there are sequences $(\alpha_n)_{n=1}^{\infty}$ and $(\beta_n)_{n=1}^{\infty}$, $a < \alpha_n < x < \beta_n < b$, $\alpha_n \to x$, $\beta_n \to x$, such that

$$\lim_{n\to\infty} \frac{f(\beta_n) - f(\alpha_n)}{\beta_n - \alpha_n}$$

does not exist, show that f is not differentiable at x.

Hint Use Exercise 39.

41. Let

$$f(x) = x \qquad \text{for rational } x, \text{ including } x = 0,$$

$$= x^2 + x \quad \text{for irrational } x.$$

Then f is discontinuous at each $x \neq 0$. Show that f is differentiable at $x = 0$ with $f'(0) = 1$.

42. Let f be continuous on a closed bounded interval [a,b] and α be a step function on [a,b], that is, α is constant on each of the subintervals (a,c_1), (c_1,c_2), ..., (c_m,b), where $a < c_1 < c_2 < \cdots < c_m < b$ with m finite. Show that

$$\int_a^b f(x)\, d\alpha(x) = f(a)[\alpha(a+) - \alpha(a)] + \sum_{k=1}^{m} f(c_k)[\alpha(c_k+) - \alpha(c_k-)]$$
$$+ f(b)[\alpha(b) - \alpha(b-)].$$

Solution Since

$$V_a^b\alpha = |\alpha(a+) - \alpha(a)| + \sum_{k=1}^{m} |\alpha(c_k) - \alpha(c_k-)| + |\alpha(c_k+) - \alpha(c_k)|$$
$$+ |\alpha(b) - \alpha(b-)|,$$

we see that α is of finite variation. Putting $c_0 = a$, $c_{m+1} = b$, we have

$$\int_a^b f(x)\, d\alpha(x) = \sum_{k=0}^{m} \int_{c_k}^{c_{k+1}} f(x)\, d\alpha(x).$$

To compute

$$\int_{c_k}^{c_{k+1}} f(x)\ d\alpha(x),$$

we consider any partition of $[c_k, c_{k+1}]$ and note that the Riemann-Stieltjes sum is

$$f(t_0)[\alpha(c_k+) - \alpha(c_k)] + f(t_{n-1})[\alpha(c_{k+1}) - \alpha(c_{k+1}-)]$$

since all other summands vanish. Passing to the limit, we get

$$\int_{c_k}^{c_{k+1}} f(x)\ d\alpha(x) = f(c_k)[\alpha(c_k+) - \alpha(c_k)]$$
$$+ f(c_{k+1})[\alpha(c_{k+1}) - \alpha(c_{k+1}-)]. \qquad \square$$

<u>43</u>. Let a, b, c, and d be given, with $a > b$ and $c > d$. Define

$$\begin{aligned} f(x) &= ax \quad \text{for } x \ge 0 \text{ and } x \text{ rational}, \\ &= bx \quad \text{for } x > 0 \text{ and } x \text{ irrational}, \\ &= cx \quad \text{for } x \le 0 \text{ and } x \text{ rational}, \\ &= dx \quad \text{for } x < 0 \text{ and } x \text{ irrational}. \end{aligned}$$

Calculate the four Dini derivatives of f at $x = 0$.

<u>44</u>. Let g be continuous and assume that f is monotonically increasing on $[a,b]$. Let A and B be two real numbers such that

$$A \le f(a+) \quad \text{and} \quad B \ge f(b-).$$

Show that there is a point x in $[a,b]$ such that

$$\text{(a)} \quad \int_a^b f(t)\ g(t)\ dt = A \int_a^x g(t)\ dt + B \int_x^b g(t)\ dt.$$

In particular, if $f(t) \ge 0$ for all t in $[a,b]$, we have

$$\text{(b)} \quad \int_a^b f(t)\ g(t)\ dt = B \int_a^x g(t)\ dt.$$

<u>*N.B.:*</u> Part (b) is known as *Bonnet's Theorem*.

<u>*Solution*</u> If $\alpha(t) = \int_a^t g(s)\ ds$, then $\alpha' = g$; Proposition 28

is applicable, and we get

$$\int_a^b f(t)\, g(t)\, dt = f(a) \int_a^x g(t)\, dt + f(b) \int_x^b g(t)\, dt.$$

This shows that part (a) holds whenever $A = f(a)$ and $B = f(b)$. Now if A and B are any two real numbers satisfying $A \le f(a+)$ and $B \ge f(b-)$, we can redefine f at these endpoints a and b to have the values $f(a) = A$ and $f(b) = B$. The modified f is still monotonically increasing on $[a,b]$ and, as we observed in Exercise 26, changing the value of f at a finite number of points does not affect the value of a Riemann integral. Of course, the point x in part (a) will depend on the choice of A and B. By taking $A = 0$, part (b) follows from part (a). □

45. In the proof of Proposition 36 we used the fact that the product of p consecutive integers is divisible by $p!$ Prove the following generalization: If n, n_1, n_2, n_3, etc, are natural numbers such that $n = n_1 + n_2 + n_3 + \cdots$, the quotient

$$\frac{n!}{n_1!\, n_2!\, n_3! \cdots}$$

is an integer.

46. Let g have a continuous derivative g' on an interval $S = [c,d]$. Let f be continuous on $g(S)$ and let $a = g(c)$, $b = g(d)$. Put

$$F(x) = \int_a^x f(t)\, dt$$

for $x \in g(S)$. Prove that for each $x \in S$, the integral

$$\int_c^x f[g(t)]\, g'(t)\, dt$$

exists and has the value $F[g(x)]$. In particular,

$$\int_a^b f(x)\, dx = \int_c^d f[g(t)]\, g'(t)\, dt.$$

Solution The function g' and the composite function f(g) are both continuous on S; hence the integral under consideration exists. Define G on S as follows:

$$G(x) = \int_c^x f[g(t)]\, g'(t)\, dt.$$

We want to show that $G(x) = F[g(x)]$. By Exercise 36 we have $G'(x) = f[g(x)]\, g'(x)$ and by Proposition 3 the derivative of $F[g(x)]$ is also $f[g(x)]\, g'(x)$, since $F'(x) = f(x)$. Thus, $G(x) - F[g(x)]$ is a constant. But $G(c) = 0$ and $F[g(c)] = F(a) = 0$, so that this constant must be zero. Hence $G(x) = F[g(x)]$ for all x in S. In particular, when $x = d$, we have $G(d) = F[g(d)] = F(b)$ and this is the claim made in the last equation of the problem. □

47. Let f be continuous on [a,b] and $f(x) > 0$ for all $x \in [a,b]$. Show that

$$\left(\int_a^b f(x)\, dx\right)\left(\int_a^b \frac{dx}{f(x)}\right) \geq (b - a)^2.$$

48. Let $t > 0$ and define (the continuous function)

$$f(t) = \int_1^t \frac{ds}{s}.$$

Show that, for $x > 0$ and $y > 0$, $f(xy) = f(x) + f(y)$.

N.B.: Compare this result with Proposition 55 of Chapter 2.

49. Let σ and ω be arbitrary positive numbers and f be a bounded function on an interval [a,b] of finite length. Then $f \in \mathcal{R}$ on [a,b] if and only if there is a mode of division of [a,b] into subintervals such that the sum of the lengths of the subintervals in which the oscillation of f is greater than or equal to ω is less than σ.

Solution Let $P = \{a = x_0 < x_1 < x_2 < \cdots < x_n = b\}$ be a partition of [a,b] and consider the sum

$$Z(P,f) = \sum_{k=1}^{n} \omega_k\, \Delta x_k,$$

where $\Delta x_k = x_k - x_{k-1}$ and $\omega_k = M_k - m_k$ with

$$M_k = \sup\{f(x) : x \in [x_{k-1}, x_k]\}, \quad m_k = \inf\{f(x) : x \in [x_{k-1}, x_k]\}$$

(that is, ω_k is the oscillation of f in $[x_{k-1}, x_k]$). We let

$$\Omega = M - m,$$

where

$$M = \sup\{f(x) : x \in [a,b]\}, \quad m = \inf\{f(x) : x \in [a,b]\},$$

and denote the length of the interval [a,b] by L.

We now derive bounds for Z(P,f); incidentally, Z(P,f) equals U(P,f) - L(P,f). Let δ be the sum of the lengths of the subintervals obtained by the partition P in which the oscillation of f is greater than or equal to ω. Then

$$Z(P,f) \geq \delta\omega.$$

But in these subintervals the oscillation of f is less than or equal to Ω and in the remaining subintervals (the sum of whose lengths is $L - \delta$) the oscillation of f is less than ω. Thus

$$Z(P,f) \leq \delta\Omega + (L - \delta)\omega.$$

Since $L - \delta < L$, we see that

$$\delta\omega \leq Z(P,f) \leq \delta\Omega + L\Omega.$$

If f is Riemann integrable on [a,b], then by Proposition 16 there exists a partition P such that for any preassigned positive numbers σ and ω we have

$$Z(P,f) < \omega\sigma.$$

But then $\delta\omega < \omega\sigma$, that is, $\delta < \sigma$.

Conversely, if there exists a partition P for which $\delta < \sigma$, we choose

$$\omega = \frac{\varepsilon}{2L} \quad \text{and} \quad \delta = \frac{\varepsilon}{2\Omega}.$$

Then $Z(P,f) \leq \delta\Omega + L\Omega < \varepsilon/2 + \varepsilon/2 = \varepsilon$. □

50. Using the result in Exercise 49, give another proof for Proposition 20 under the assumption that $\alpha(x) = x$.

51. Read the section entitled "Fermat's Principle and its Consequences" in L. A. Lyusternik's book "Shortest Paths Variational Problems" (Pergamon Press, Oxford, 1964).

52. Given a right circular cone, it is required to inscribe in it a cylinder with greatest area.

Hint Let the radius of the base of the cone be R and its height H. Consider the cases $H > 2R$ and $H \leq 2R$ separately.

53. A point light source is situated on the line connecting the centers of two spheres of different radii. Where should the light source be situated so that the sum of the illuminated parts on the surfaces of the two spheres will be a maximum?

54. Through a point in the interior of an angle we wish to construct a straight line such that the resulting triangle has minimal area. Show that the point has to bisect the segment of the straight line inside the angular region. Give some consequences of this result.

55. Let c be a real number such that n^c is an integer for every positive integer n. Show that c is a nonnegative integer.

Solution The case $n = 2$ shows that c is nonnegative. If the ordinary Mean-Value Theorem (see Proposition 6) is applied to x^c on the interval $[u, u + 1]$ there is a t with $u < t < u + 1$ such that

$$ct^{c-1} = (u + 1)^c - u^c.$$

For any positive integer u the right-hand side is a positive integer. Now, in case $0 < c < 1$, u could be taken large enough so $u^{c-1} < 1/c$ and so $ct^{c-1} < 1$. Thus the Mean-Value Theorem for the first derivative eliminates all c with $0 < c < 1$.

There is an extension of the Mean-Value Theorem which states that if f(x) is k-times differentiable in [a,b] then there is a t, $a < t < b$, such that

$$h^k f^{(k)}(t) = \Delta^k f(a),$$

where $h = (b - a)/k$ and Δ^k is the k-th difference for intervals spaced h apart. The formal definition of the symbol $\Delta^k f(a)$ is

$$\Delta^k f(a) = \sum_{j=0}^{k} (-1)^{k-j} \binom{k}{j} f(a + jh), \quad \text{where} \quad h = \frac{b - a}{k}.$$

Take k as the unique integer such that $k - 1 \le c < k$ and apply this extension of the Mean-Value Theorem on the interval $[u, u + k]$. There is a t with $u < t < u + k$ such that

$$c(c - 1)(c - 2) \cdots (c - k + 1)\, t^{c-k} = \Delta^k f(u).$$

The right-hand side is an integer, and by taking u sufficiently large t^{c-1} becomes sufficiently small so that the left-hand side, though nonnegative, is less than 1. Hence

$$c(c - 1)(c - 2) \cdots (c - k + 1) = 0$$

and so $c = k - 1$. □

56. Let f be a continuous function on the interval [0,1] and suppose that $\int_0^1 f(x)\,dx = 0$, $\int_0^1 x\,f(x)\,dx = 0$, ..., $\int_0^1 x^{n-1} f(x)\,dx = 0$, $\int_0^1 x^n f(x)\,dx = 1$. Show that $|f(x)| \ge 2^n(n + 1)$ on some part of [0,1].

Solution The conditions imply

$$\int_0^1 \left(x - \tfrac{1}{2}\right)^n f(x)\,dx = 1.$$

Suppose $|f(x)| < 2^n(n + 1)$ everywhere on [0,1]. Then

$$1 = \int_0^1 \left(x - \tfrac{1}{2}\right)^n f(x)\,dx < 2^n(n + 1) \int_0^1 \left|x - \tfrac{1}{2}\right|^n dx = 1.$$

But this is a contradiction. □

57. Suppose that f is a function with continuous derivatives of order two and $f(0) = 0$. Show that the function g, defined by

$$g(0) = f'(0), \quad g(x) = \frac{f(x)}{x} \quad \text{for} \quad x \ne 0,$$

has a continuous derivative.

Solution For $x \ne 0$, $g(x) = f(x)/x = f'(\theta x)$ with $0 < \theta < 1$, and since f' is continuous $\lim_{x \to 0} g(x) = f'(0) = g(0)$. Thus g is continuous at the origin and so at all points. We have

$$\lim_{x \to 0} \frac{g(x) - g(0)}{x} = \lim_{x \to 0} \frac{f(x) - x\, f'(0)}{x^2}$$

$$= \lim_{x \to 0} \frac{f'(x) - f'(0)}{2x} = \frac{f''(0)}{2}$$

by L'Hospital's Rule (see Proposition 12). For $x = 0$,

$$g'(x) = \frac{x\, f'(x) - f(x)}{x^2}$$

and by L'Hospital's Rule the limit as $x \to 0$ is also found to be $f''(0)/2$. Thus g' is continuous at $x = 0$ and at other points. □

58. Suppose that $|f(x)| \le 1$ and $|f''(x)| \le 1$ for all x with $0 \le x \le 2$. Verify that $|f'(x)| \le 2$ on the interval $[0,2]$.

Solution Expanding $f(x)$ by Taylor's formula (see Proposition 13) around the points 0 and 2, and subtracting, we get

$$f(2) - f(0) = 2f'(x) - \frac{1}{2} x^2 f''(t_1) + \frac{1}{2} (2 - x)^2 f''(t_2),$$

where t_1 and t_2 are points inside the interval $[0,2]$. This gives

$$2|f'(x)| \le |f(2)| + |f(0)| + \frac{1}{2} x^2 |f''(t_1)|$$

$$+ \frac{1}{2} (2 - x)^2 |f''(t_2)|$$

$$\le 2 + \frac{1}{2}\left[x^2 + (2 - x)^2\right].$$

The largest value of the right-hand side is attained for $x = 0$, and is 4. □

59. Verify that

$$\lim_{x \to \infty} \left[\frac{1}{x} \frac{a^x - 1}{a - 1}\right]^{1/x} = 1 \quad \text{for } 0 < a < 1$$

$$= 0 \quad \text{for } a > 1.$$

60. Consider a polynomial $f(x)$ with real coefficients having the property $f[g(x)] = g[f(x)]$ for every polynomial $g(x)$ with real coefficients. Determine the nature of $f(x)$.

Solution Let $g(x) = x + h$ so that $f(x + h) = f(x) + h$ and $\{f(x + h) - f(x)\}/h = 1$ for all $h \neq 0$. Therefore $f'(x) = 1$ and $f(x) = x + c$. Letting $g(x)$ be the zero polynomial shows that it is necessary that $c = 0$ and thus $f(x) = x$. It is easily seen that $f(x) = x$ is sufficient. □

61. Let f be a real-valued function on an interval $[a,b]$. We say that f is *absolutely continuous on* $[a,b]$ if for any $\varepsilon > 0$ there is a $\delta > 0$ such that

$$\sum_{k=1}^{n} |f(b_k) - f(a_k)| < \varepsilon$$

for every finite, pairwise disjoint sequence

$$(a_1,b_1), \quad (a_2,b_2), \quad \ldots, \quad (a_n,b_n)$$

of open subintervals of $[a,b]$ for which

$$\sum_{k=1}^{n} (b_k - a_k) < \delta.$$

(a) Let f be Riemann integrable on $[a,b]$ and define

$$F(x) = \int_a^x f(t)\,dt.$$

Show that F is absolutely continuous on $[a,b]$.

(b) Show that an absolutely continuous function on $[a,b]$ has finite total variation on $[a,b]$.

(c) Let c be Cantor's function defined in Section 2 of Chapter 2. Show that c is not absolutely continuous on $[0,1]$.

Solution of part (c) We first extend the definition of the function c as follows: we set $c(x) = 0$ for $x < 0$ and $c(x) = 1$ for $x > 1$. Next, we enclose Cantor's ternary set C in a union $\cup_{k=1}^{\infty} (a_k,b_k)$ of pairwise disjoint open intervals such that $\Sigma_{k=1}^{\infty} (b_k - a_k)$ is arbitrarily small. It is easily seen that $\Sigma_{k=1}^{\infty} [c(b_k) - c(a_k)] = 1$ and $\Sigma_{k=1}^{n} [c(b_k) - c(a_k)] \geq 1/2$ for sufficiently large n, while the sum $\Sigma_{k=1}^{n} (b_k - a_k)$ is arbitrarily small. □

62. Let p, f, and g be continuous functions on [a,b] and suppose that p is positive-valued and f and g are monotonically increasing on [a,b]. Verify the following inequality, due to Chebyshev:

$$\left\{\int_a^b p(x)\ f(x)\ dx\right\}\left\{\int_a^b p(x)\ g(x)\ dx\right\}$$

$$\leq \left\{\int_a^b p(x)\ dx\right\}\left\{\int_a^b p(x)\ f(x)\ g(x)\ dx\right\}.$$

63. Let f be Riemann integrable on [a,b] and g be a positive-valued, bounded, and nonincreasing function. Show that

$$\int_a^b f(x)\ g(x)\ dx = g(a+) \int_a^t f(x)\ dx,$$

where t is some number between a and b.

If g is positive-valued and nondecreasing, the corresponding formula is

$$\int_a^b f(x)\ g(x)\ dx = g(b-) \int_t^b f(x)\ dx,$$

where $a < t < b$.

If g is any monotonic function, there is a number t between a and b such that

$$\int_a^b f(x)\ g(x)\ dx = g(a+) \int_a^t f(x)\ dx + g(b-) \int_t^b f(x)\ dx.$$

Solution Let ε be a positive number less than $g(a+) - g(b-)$. Then there is a point x_1 such that

$$g(a+) - g(x) < \varepsilon \quad (a < x < x_1)$$
$$\geq \varepsilon \quad (x > x_1).$$

Similarly there are points $x_2, x_3, \ldots$ such that

$$g(x_{k-1}+) - g(x) < \varepsilon \quad (x_{k-1} < x < x_k)$$

$$g(x_{k-1}+) - g(x) \geq \varepsilon \quad (x > x_k),$$

so long as $g(x_{k-1}+) - g(b-) > \varepsilon$. Otherwise we take $x_n = b$. The point b is thus reached in a finite number of steps, since the variation of g in each interval (x_{k-1}, x_k) is at least ε.

Let $h(x) = g(x_k+)$ in each interval $x_k \leq x < x_{k+1}$. Then

$$0 \leq h(x) - g(x) < \varepsilon$$

except possibly at the points $a = x_0, x_1, x_2, \ldots, b$, and

$$\int_a^b h(x)\, f(x)\, dx = \sum_{k=0}^{n-1} g(x_k+) \int_{x_k}^{x_{k+1}} f(x)\, dx.$$

Let $F(x) = \int_a^x f(s)\, ds$; then, if m and M are the lower and upper bounds of F, it follows from Abel's Lemma (see Exercise 15, part (a), at the end of Chapter 1) that

$$m\, g(a+) \leq \int_a^b h(x)\, f(x)\, dx \leq M\, g(a+).$$

But

$$\left| \int_a^b h(x)\, f(x)\, dx - \int_a^b g(x)\, f(x)\, dx \right| \leq \varepsilon \int_a^b |f(x)|\, dx,$$

which tends to zero with ε. Hence, making $\varepsilon \to 0$, it follows that

$$m\, g(a+) \leq \int_a^b g(x)\, f(x)\, dx \leq M\, g(a+).$$

Since F is continuous, it takes every value between m and M, and so, at every $x = t$, say, the value

$$\frac{1}{g(a+)} \int_a^b g(x)\, f(x)\, dx.$$

The rest follows easily. □

64. Let f be convex on (A,B). Show that f is absolutely continuous on any closed subinterval [a,b] of (A,B).

Hint See Exercise 10 at the end of Chapter 2.

65. Prove that in any set of ten different two digit numbers one can select two disjoint subsets such that the sum of numbers in each of the subsets is the same.

Solution In the set of ten numbers there are altogether $2^{10} = 1024$ subsets. Excluding the empty set, the set of all ten numbers and the ten sets of nine numbers (which could not give a solution) we are left with 1012 possible sets.

The smallest sum is 10, the largest is $92 + 93 + \cdots + 99 = 764$; hence there are not more than 755 possible sums. Thus there must be two subsets having the same sum. □

Chapter 4

UNIFORM CONVERGENCE

1. *Basic Properties of Uniform Convergence*

<u>*Definition*</u> Let E be a set of real numbers and $(f_n)_{n=1}^{\infty}$ a sequence, each term of which is a real-valued function defined on E. We say that $(f_n)_{n=1}^{\infty}$ *converges uniformly on* E *to a function* f if, for every $\varepsilon > 0$, there exists an integer n_0 (depending only on ε) such that $n \geq n_0$ implies

$$|f_n(x) - f(x)| < \varepsilon \tag{1}$$

for every x in E. We denote this symbolically by writing

$$f_n \to f \quad \text{uniformly on E.}$$

Clearly, uniform convergence on E implies pointwise convergence on E. The difference between the two concepts is this: If $(f_n)_{n=1}^{\infty}$ converges pointwise on E, then there exists a function f such that, for every $\varepsilon > 0$, and for every x in E, there is an integer n_0, depending on ε <u>and</u> on x, such that (1) holds if $n \geq n_0$; if $(f_n)_{n=1}^{\infty}$ converges uniformly on E, we can, for each $\varepsilon > 0$, find <u>one</u> integer n_0 which will do for <u>all</u> x in E.

Since (1) is equivalent to

$$f(x) - \varepsilon < f_n(x) < f(x) + \varepsilon, \tag{2}$$

we see that if (2) is to hold for all $n \geq n_0$ and for all $x \in E$ and thus the graph of f_n (that is, the set $\{(x,y): y = f_n(x),\ x \in E\}$) lies within a "band" of height 2ε situated symmetrically about the graph of f.

An immediate consequence of the definition of uniform convergence is the following result.

Proposition 1 Suppose, for every $x \in E$, $\lim_{n \to \infty} f_n(x) = f(x)$. Let

$$M_n = \sup\{|f_n(x) - f(x)|: x \in E\}.$$

Then $f_n \to f$ uniformly on E if and only if $M_n \to 0$ as $n \to \infty$.

The next proposition states *Cauchy's criterion for uniform convergence*.

Proposition 2 Let $(f_n)_{n=1}^{\infty}$ be a sequence of functions defined on a set E of real numbers. There exists a function f such that

$$f_n \to f \quad \text{uniformly on E}$$

if and only if the following (called the *Cauchy condition*) is satisfied: For every $\varepsilon > 0$ there exists an integer n_0 such that $m \geq n_0$, $n \geq n_0$ implies

$$|f_m(x) - f_n(x)| < \varepsilon$$

for every x in E.

Proof Suppose that $f_n \to f$ uniformly on E. Then, given $\varepsilon > 0$, we can find n_0 such that $n \geq n_0$ implies $|f_n(x) - f(x)| < \varepsilon/2$ for all x in E. Taking $m \geq n_0$, we have $|f_m(x) - f(x)| < \varepsilon/2$, and hence we have $|f_m(x) - f_n(x)| < \varepsilon$ for every x in E.

Conversely, suppose the Cauchy condition is satisfied. Then, for each x in E, the sequence $\{f_n(x)\}_{n=1}^{\infty}$ converges by Proposition 58 of Chapter 1. Let $f(x) = \lim_{n\to\infty} f_n(x)$ if $x \in E$. We must show that $f_n \to f$ uniformly on E. If $\varepsilon > 0$ is given, we choose n_0 such that $n \geq n_0$ implies

$$|f_n(x) - f_{n+k}(x)| < \frac{\varepsilon}{2}$$

for every k = 1, 2, ... and every x in E. Thus

$$\lim_{k\to\infty} |f_n(x) - f_{n+k}(x)| = |f_n(x) - f(x)| \leq \frac{\varepsilon}{2}.$$

Hence, $n \geq n_0$ implies $|f_n(x) - f(x)| < \varepsilon$ for every x in E and we see that $f_n \to f$ uniformly on E. □

Examples:

1. For $x \in [0,1]$, let

$$f_n(x) = \frac{x}{1 + n^2x^2}.$$

It is clear that, for every $x \in [0,1]$, $\lim_{n\to\infty} f_n(x) = 0$. Since

$$0 \leq f_n(x) = \frac{1}{2n}\,\frac{2nx}{1 + n^2x^2} \leq \frac{1}{2n},$$

we see that $f_n \to 0$ uniformly on [0,1].

2. For $x \in [0,1]$, let

$$f_n(x) = \frac{nx}{1 + n^2x^2}.$$

It is clear that $f_n(0) = 0$ and $f_n(x) < 1/nx$ for $x > 0$; thus

$$\lim_{n\to\infty} f_n(x) = 0 \quad \text{for every} \quad x \in [0,1].$$

But

$$f_n\left(\frac{1}{n}\right) = \frac{1}{2}$$

and f_n assumes its largest value (namely, 1/2) at the point 1/n; by Proposition 1, f_n fails to be uniformly convergent to 0 on [0,1].

3. For $x \in [0,1]$, let

$$f_n(x) = x^n.$$

Then

$$\lim_{n\to\infty} f_n(x) = 0 \quad \text{for } 0 \le x < 1,$$
$$= 1 \quad \text{for } x = 1;$$

thus the pointwise limit is discontinuous at $x = 1$. For $0 < x < 1$,

$$x^n < \varepsilon \quad (0 < \varepsilon < 1)$$

is equivalent to

$$n > \frac{\log \varepsilon}{\log x}$$

and f_n therefore cannot converge uniformly to 0 on (0,1).

4. For $x \in (0,1)$, let

$$f_n(x) = \frac{1}{1 + nx}.$$

Evidently, $\lim_{n\to\infty} f_n(x) = 0$ for every $x \in (0,1)$. But, for $0 < x < 1$, $0 < \varepsilon < 1$,

$$\frac{1}{1 + nx} < \varepsilon$$

is equivalent to

$$n > \frac{1}{x}\left(\frac{1}{\varepsilon} - 1\right);$$

thus f_n cannot converge uniformly to 0 on (0,1). Note also that

$$f_n\left(\frac{1}{n}\right) = \frac{1}{2}$$

and that f_n assumes its largest value at the point $1/n$.

5. For $x \in (0,1)$, let

$$f_n(x) = 2n^2 x\, e^{-n^2 x^2}.$$

The pointwise limit of $(f_n)_{n=1}^{\infty}$ on $(0,1)$ is 0. But

$$f_n\left(\frac{1}{n}\right) = \frac{2n}{e}$$

and f_n assumes its largest value at the point $1/n$. Again, f_n fails to be uniformly convergent to 0 on $(0,1)$.

Figure 4 - 1 illustrates the graph of

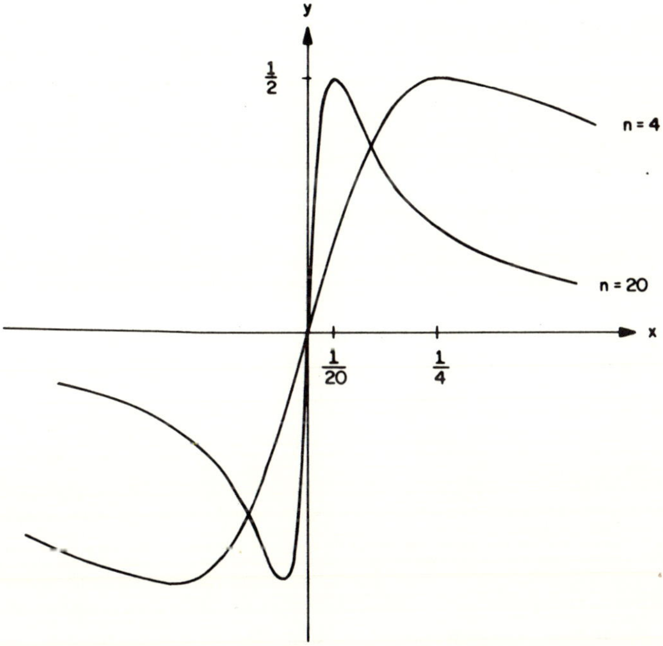

Figure 4 - 1

$$y = f_n(x) = \frac{nx}{1 + n^2x^2}$$

for the cases n = 4, 20.

Remarks In Examples 2, 4, and 5, f_n assumes not only its largest value at 1/n but also has a local maximum there; we may thus speak of a *hump* at 1/n. In Examples 2 and 4 the hump at 1/n has height 1/2, but in Example 5 the hump at 1/n has height 2n/e and so becomes arbitrarily large with increasing n. As we let n take on the values 2, 3, ..., the hump glides toward the point x = 0 in the three examples under consideration.

We shall now discuss the relation between uniform convergence on the one hand and continuity, differentiation, and integration on the other.

Proposition 3 Suppose that $f_n \to f$ uniformly on E. If each f_n is continuous at a point x_0 of E, then the limit function f is also continuous at x_0. If x_0 is a limit point of E, this means that

$$\lim_{x \to x_0} \lim_{n \to \infty} f_n(x) = \lim_{n \to \infty} \lim_{x \to x_0} f_n(x).$$

Proof If x_0 is an isolated point of E, then f is automatically continuous at x_0. Suppose, then, that x_0 is a limit point of E. By assumption, for every $\varepsilon > 0$ there exists an integer n_0 such that $n \geq n_0$ implies

$$|f_n(x) - f(x)| < \frac{\varepsilon}{3}$$

for every x in E. Since f_{n_0} is continuous at x_0, there is a neighborhood $(x_0 - \delta, x_0 + \delta)$ with $\delta > 0$ such that

$$x \in (x_0 - \delta, x_0 + \delta) \cap E$$

implies

$$|f_{n_0}(x) - f_{n_0}(x_0)| < \frac{\varepsilon}{3}.$$

But

$$|f(x) - f(x_0)| \le |f(x) - f_{n_0}(x)| + |f_{n_0}(x) - f_{n_0}(x_0)|$$
$$+ |f_{n_0}(x_0) - f(x_0)|.$$

If $x \in (x_0 - \delta, x_0 + \delta) \cap E$, each term on the right-hand side of the foregoing inequality is less than $\varepsilon/3$ and hence

$$|f(x) - f(x_0)| < \varepsilon. \qquad \square$$

Remark Uniform convergence is therefore sufficient, but not necessary, to transmit continuity from the individual terms of a sequence of functions to the limit function; the foregoing Example 2 illustrates this point. But there is a case in which a partial converse holds and we shall take it up at once.

Proposition 4 (Dini's Theorem) Let E be compact and $(f_n)_{n=1}^{\infty}$ be a sequence of continuous functions on E which converges pointwise to a continuous function f on E. If the sequence $(f_n)_{n=1}^{\infty}$ is monotonic on E, then $f_n \to f$ uniformly on E.

Proof We shall suppose that $f_n(x) \ge f_{n+1}(x)$ for n = 1, 2, ..., and for every $x \in E$. We put

$$g_n(x) = f_n(x) - f(x).$$

Then $g_n \to 0$, and $g_n(x) \ge g_{n+1}(x)$ on E. We have to show that $g_n \to 0$ uniformly on E.

Let $\varepsilon > 0$ be given. For each $x \in E$, there is an integer n_x such that

$$0 \le g_{n_x}(x) < \frac{\varepsilon}{2}.$$

By continuity, and by the monotonicity of the sequence $(g_n)_{n=1}^{\infty}$, there is an open set $J(x)$, which contains x, such that

$$0 \le g_n(t) < \varepsilon \tag{3}$$

if $t \in J(x)$ and $n \ge n_x$.

Since E is compact, there is a finite set of points x_1, x_2, ..., x_m such that

$$E \subset J(x_1) \cup J(x_2) \cup \cdots \cup J(x_m). \tag{4}$$

Putting

$$n_0 = \max\left\{ n_{x_1}, n_{x_2}, \ldots, n_{x_m} \right\},$$

we obtain from (3) and (4) that $0 \le g_n(t) < \varepsilon$ for all $t \in E$, if $n \ge n_0$. This proves the uniform convergence. □

Remark Compactness is essential in Proposition 4, as can readily be seen from Example 4; for $x \in (0,1)$,

$$f_n(x) = \frac{1}{1 + nx}$$

tends monotonically to 0 on $(0,1)$, but the convergence fails to be uniform.

Proposition 5 Let $(f_n)_{n=1}^{\infty}$ be a sequence of functions, differentiable on a bounded closed interval $[a,b]$ and such that $\{f_n(x_0)\}_{n=1}^{\infty}$ converges for some point x_0 of $[a,b]$. If $(f_n')_{n=1}^{\infty}$ converges uniformly on $[a,b]$, then $(f_n)_{n=1}^{\infty}$ converges uniformly on $[a,b]$, to a function f, and, for $a \le x \le b$,

$$f'(x) = \lim_{n \to \infty} f_n'(x). \tag{5}$$

Proof Let $\varepsilon > 0$ be given. Pick n_0 such that $n \ge n_0$, $m \ge n_0$, implies

$$|f_n(x_0) - f_m(x_0)| < \frac{\varepsilon}{2} \tag{6}$$

and, for $a \le t \le b$,

$$|f_n'(t) - f_m'(t)| < \frac{\varepsilon}{2(b - a)}. \tag{7}$$

Applying Proposition 6 of Chapter 3 to the function $f_n - f_m$, (7) shows that

$$|f_n(x) - f_m(x) - f_n(t) + f_m(t)| < \frac{|x - t|\varepsilon}{2(b - a)} < \frac{\varepsilon}{2} \tag{8}$$

for any x and t in $[a,b]$, if $n \ge n_0$, $m \ge n_0$. But

$$|f_n(x) - f_m(x)|$$

$$\le |f_n(x) - f_m(x) - f_n(x_0) + f_m(x_0)| + |f_n(x_0) - f_m(x_0)|$$

implies, by (6) and (8), that

$$|f_n(x) - f_m(x)| < \varepsilon$$

for $a \le x \le b$, $n \ge n_0$, $m \ge n_0$; thus $(f_n)_{n=1}^{\infty}$ converges uniformly on $[a,b]$. Let, for $a \le x \le b$,

$$f(x) = \lim_{n\to\infty} f_n(x).$$

We now fix a point x on $[a,b]$ and define

$$h_n(t) = \frac{f_n(t) - f_n(x)}{t - x}, \quad h(t) = \frac{f(t) - f(x)}{t - x} \tag{9}$$

for $a \le t \le b$, $t \ne x$. Then, for $n = 1, 2, \ldots,$

$$\lim_{t\to x} h_n(t) = f_n'(x). \tag{10}$$

The first inequality in (8) shows that

$$|h_n(t) - h_m(t)| < \frac{\varepsilon}{2(b - a)}$$

for $n \ge n_0$, $m \ge n_0$; thus $(h_n)_{n=1}^{\infty}$ converges uniformly, for $t \ne x$. Since $(f_n)_{n=1}^{\infty}$ converges to f, we get from (9) that

$$\lim_{n \to \infty} h_n(t) = h(t) \tag{11}$$

uniformly for $a \leq t \leq b$, $t \neq x$. Applying Proposition 3 to $(h_n)_{n=1}^{\infty}$, (10) and (11) give

$$\lim_{t \to x} h(t) = \lim_{n \to \infty} f'_n(x);$$

but this is (5), by the definition of $h(t)$. □

Remark Note that the uniform convergence of $(f_n)_{n=1}^{\infty}$ implies nothing about the sequence $(f'_n)_{n=1}^{\infty}$; consider, for example, the sequence

$$f_n(x) = \frac{\sin nx}{\sqrt{n}}$$

for all real x. Then $\lim_{n \to \infty} f_n(x) = f(x) = 0$ for every x and

$$f'_n(x) = \sqrt{n} \cos nx.$$

Hence, $\lim_{n \to \infty} f'_n(x)$ does not exist for any x. The sequence $(f_n)_{n=1}^{\infty}$ converges uniformly on $(-\infty,+\infty)$, but $(f'_n)_{n=1}^{\infty}$ does not even converge pointwise on R^1. For example, $(f'_n)_{n=1}^{\infty}$ diverges since $f'_n(0) = \sqrt{n}$; but $f'(x) = 0$ for all real x.

Proposition 6 Let $[a,b]$ be a bounded closed interval and α a monotonically increasing real-valued function on $[a,b]$. Moreover, suppose that $f_n \in \mathcal{R}(\alpha)$ on $[a,b]$ for $n = 1, 2, \ldots$, and suppose that $f_n \to f$ uniformly on $[a,b]$. Then $f \in \mathcal{R}(\alpha)$ on $[a,b]$,

$$\lim_{n \to \infty} \int_a^b f_n(x)\, d\alpha(x)$$

exists, and

$$\int_a^b f(x)\, d\alpha(x) = \lim_{n \to \infty} \int_a^b f_n(x)\, d\alpha(x). \tag{12}$$

Proof Let $\varepsilon > 0$ be given. Choose $\gamma > 0$ such that

$$\gamma[\alpha(b) - \alpha(a)] < \frac{\varepsilon}{3}. \tag{13}$$

Since $f_n \to f$ uniformly on $[a,b]$, there is an integer n such that

$$|f_n(x) - f(x)| < \gamma$$

for every $x \in [a,b]$. For fixed n_0, we select a partition P of $[a,b]$ such that

$$U(P,f_n,\alpha) - L(P,f_n,\alpha) < \frac{\varepsilon}{3}; \tag{14}$$

this is possible by Proposition 16 of Chapter 3. But

$$f(x) < f_n(x) + \gamma$$

implies, by (13),

$$U(P,f,\alpha) < U(P,f_n,\alpha) + \frac{\varepsilon}{3}, \tag{15}$$

and

$$f(x) > f_n(x) - \gamma$$

implies, by (13),

$$L(P,f,\alpha) > L(P,f_n,\alpha) - \frac{\varepsilon}{3}; \tag{16}$$

combining (14), (15), and (16), we obtain

$$U(P,f,\alpha) - L(P,f,\alpha) < \varepsilon,$$

proving that $f \in \mathcal{R}(\alpha)$ on $[a,b]$.

To verify (12), we pick n_0 such that $n \geq n_0$ implies

$$|f_n(x) - f(x)| < \varepsilon$$

for ever x in $[a,b]$. Then, for $n \geq n_0$,

$$\left|\int_a^b f(x)\ d\alpha(x) - \int_a^b f_n(x)\ d\alpha(x)\right| = \left|\int_a^b [f(x) - f_n(x)]\ d\alpha(x)\right|$$

$$\leq \int_a^b |f(x) - f_n(x)| \, d\alpha(x) < \varepsilon[\alpha(b) - \alpha(a)].$$

Since ε is an arbitrary positive number, (12) holds. □

In connection with Proposition 6 we note the following two examples.

Examples:

1. Let $f_n(x) = nx(1 - x)^n$. It is easy to see that $\lim_{n\to\infty} f_n(x) = 0$ for every x in [0,1]. Also, f_n has a hump at $x = 1/(n + 1)$ and

$$f_n\left(\frac{1}{n+1}\right) = \left(\frac{n}{n+1}\right)^{n+1} \to e^{-1} \quad \text{as} \quad n \to \infty.$$

The sequence $(f_n)_{n=1}^{\infty}$ is not uniformly convergent on [0,1], yet term-by-term integration over [0,1] will lead to the expected result.

2. Let $f_n(x) = nx(1 - x^2)^n$. It is clear that $f_n(0) = 0$. By part (d) of Exercise 14 at the end of Chapter 1, $\lim_{n\to\infty} n\, f_n(x) = 0$ for every x such that $0 < x \leq 1$. Thus $\lim_{n\to\infty} f_n(x) = 0$ for every x in [0,1]. But

$$\int_0^1 x(1 - x^2)^n \, dx = \frac{1}{2n + 2}.$$

Thus

$$\lim_{n\to\infty} \int_0^1 f_n(x) \, dx = \lim_{n\to\infty} \frac{n}{2n + 2} = \frac{1}{2},$$

while

$$\int_0^1 \left(\lim_{n\to\infty} f_n(x)\right) dx = 0.$$

Thus the limit of the integral need not be equal to the integral of the limit in the case of pointwise convergence.

Remark In Exercise 10 at the end of this chapter we shall consider a very important result concerning the interchange of limit and integration signs.

Proposition 7 Let $(\alpha_n)_{n=1}^{\infty}$ be a sequence of functions of finite variation on a closed bounded interval $[a,b]$ such that $\alpha_n(a) = 0$, and suppose there is a function α such that

$$\lim_{n\to\infty} V_a^b(\alpha - \alpha_n) = 0 \tag{17}$$

and $\alpha(a) = 0$. Then, for every f which is continuous on $[a,b]$, we have

$$\lim_{n\to\infty} \int_a^b f(x)\, d\alpha_n(x) = \int_a^b f(x)\, d\alpha(x). \tag{18}$$

Moreover, $\alpha_n \to \alpha$ uniformly on $[a,b]$.

Proof Since $V_a^b\alpha \le V_a^b\alpha_n + V_a^b(\alpha - \alpha_n)$, α is of finite variation over $[a,b]$, so that all integrals in (18) exist. Suppose that $|f(x)| \le M$ on $[a,b]$. Then

$$\left|\int_a^b f(x)\, d\alpha(x) - \int_a^b f(x)\, d\alpha_n(x)\right| = \left|\int_a^b f(x)\, d[\alpha(x) - \alpha_n(x)]\right|$$
$$\le M\, V_a^b(\alpha - \alpha_n),$$

and (18) follows from (17).

Since

$$|\alpha(x) - \alpha_n(x)| \le V_a^b(\alpha - \alpha_n)$$

for $a \le x \le b$, we see that $\alpha_n \to \alpha$ uniformly on $[a,b]$. □

Remark The assumption that $\alpha_n(a) = \alpha(a) = 0$ is only a matter of convenience and serves to simplify the proof of Proposition 7. Addition of constants c_n to α_n leaves the integrals unchanged, and convergence of $\{\alpha_n(x)\}_{n=1}^{\infty}$ for one value of x in $[a,b]$ can readily be shown to imply uniform convergence on [a,b].

Proposition 8 Let f be a continuous function on a closed bounded interval [a,b] and let $(\alpha_n)_{n=1}^{\infty}$ be a sequence of functions which converges pointwise to a real-valued function α on [a,b]. If the total variation satisfies

$$V_a^b \alpha_n \le M$$

for all n, then

$$\lim_{n\to\infty} \int_a^b f(x)\ d\alpha_n(x) = \int_a^b f(x)\ d\alpha(x). \tag{19}$$

Proof For any partition

$$P = \{a = x_0 < x_1 < \cdots < x_m = b\},$$

we have

$$\sum_{k=1}^{m} |\alpha_n(x_k) - \alpha_n(x_{k-1})| \le M$$

for n = 1, 2, ... Passing to the limit as $n \to \infty$, we get

$$\sum_{k=1}^{m} |\alpha(x_k) - \alpha(x_{k-1})| \le M.$$

Since P is an arbitrary partition of [a,b], $V_a^b \alpha \le M$ and so the function α is of finite variation.

To prove (19), we observe that for any partition

$$P = \{a = x_0 < x_1 < \cdots < x_m = b\}$$

of [a,b], we have

$$\left| \int_a^b f(x)\ d\alpha(x) - \int_a^b f(x)\ d\alpha_n(x) \right| \le \mathcal{A} + \mathcal{B} + \mathcal{C},$$

where

$$\mathcal{A} = \sum_{k=1}^{m} \left| \int_{x_{k-1}}^{x_k} [f(x) - f(x_k)] \, d\alpha(x) \right|,$$

$$\mathcal{B} = \sum_{k=1}^{m} |f(x_k)| |\alpha(x_k) - \alpha(x_{k-1}) - \alpha_n(x_k) + \alpha_n(x_{k-1})|,$$

and

$$\mathcal{C} = \sum_{k=1}^{m} \left| \int_{x_{k-1}}^{x_k} [f(x_k) - f(x)] \, d\alpha_n(x) \right|.$$

Observe that $\mathcal{A}$ and $\mathcal{C}$ can be made arbitrarily small by taking the mesh $|P|$ small enough and using the continuity of f. Keeping P fixed, $\mathcal{B}$ tends to zero as $n \to \infty$. □

2. Tests for Uniform Convergence

Definition Let E be a set of real numbers and $(f_n)_{n=1}^{\infty}$ a sequence, each term of which is a real-valued function defined on E. We say that the series $\Sigma_{n=1}^{\infty} f_n$ *converges uniformly* on E if the sequence $(s_n)_{n=1}^{\infty}$ of partial sums defined by

$$s_n(x) = \sum_{j=1}^{n} f_j(x)$$

converges uniformly on E.

Remark Note that the various propositions on sequences studied in Section 1 of this chapter can easily be reformulated in terms of series; we leave the formal details to the student.

Proposition 9 (The M-Test of Weierstrass) Let $(f_n)_{n=1}^{\infty}$ be a sequence of functions defined on E and suppose that

$$|f_n(x)| \le M_n$$

for all $x \in E$ and $n = 1, 2, \ldots$. Then $\Sigma_{n=1}^{\infty} f_n$ converges uniformly if $\Sigma_{n=1}^{\infty} M_n$ converges.

Proof If $\Sigma_{n=1}^{\infty} M_n$ converges, then for arbitrary $\varepsilon > 0$,

$$\left|\sum_{k=n}^{m} f_k(x)\right| \leq \sum_{k=n}^{m} M_k < \varepsilon$$

for all $x \in E$, provided m and n are large enough (see Proposition 64 of Chapter 1). Uniform convergence now follows from Proposition 2.

□

Definition A sequence $(f_n)_{n=1}^{\infty}$ of functions is said to be *uniformly bounded on* E if there is a constant $M > 0$ such that $|f_n(x)| \leq M$ for all x in E and all $n = 1, 2, \ldots$. The number M is called a *uniform bound for* $(f_n)_{n=1}^{\infty}$ *on* E.

Remark If each individual function of a sequence $(f_n)_{n=1}^{\infty}$ is bounded on E, i.e., if $\sup\{|f_n(x)| : x \in E\} \leq M$ for $n = 1, 2, \ldots$, and $f_n \to f$ uniformly on E, then it is easily seen that $(f_n)_{n=1}^{\infty}$ is uniformly bounded on E.

Proposition 10 (Dirichlet's Test for Uniform Convergence) Let $s_n(x)$ denote the n-th partial sum of the series $\Sigma\, a_n(x)$, where a_n is a real-valued function defined on a set E of real numbers. Assume that $(s_n)_{n=1}^{\infty}$ is uniformly bounded on E by a constant M. Let $(b_n)_{n=1}^{\infty}$ be a sequence of real-valued functions which is monotonically decreasing in n for each fixed x in E (that is, $b_{n+1}(x) \leq b_n(x)$ for each x in E and for every $n = 1, 2, \ldots$), and suppose that $b_n \to 0$ uniformly on E. Then the series $\Sigma\, a_n(x)\, b_n(x)$ converges uniformly on E.

Proof Since $b_n \to 0$ uniformly on E, for any $\varepsilon > 0$ there exists an integer j such that for all x in E,

$$|b_n(x)| < \frac{\varepsilon}{2M} \quad \text{when} \quad n \geq j.$$

Moreover, by the Proof in part (b) of Exercise 15 at the end of Chapter 1,

$$|a_{j+1}(x)\ b_{j+1}(x) + \cdots + a_{j+p}(x)\ b_{j+p}(x)| < 2M|b_{j+1}(x)| < \varepsilon$$

whenever p is a positive integer and $x \in E$. The uniform convergence of $\Sigma\ a_n(x)\ b_n(x)$ on E now follows from Proposition 2. □

Proposition 11 (Abel's Test for Uniform Convergence) Let $s_n(x)$ denote the n-th partial sum of the series $\Sigma\ a_n(x)$, where each a_n is a real-valued function defined on a set E of real numbers. Assume that $(s_n)_{n=1}^{\infty}$ converges uniformly on E. Let $(b_n)_{n=1}^{\infty}$ be a sequence of functions which is uniformly bounded on E by a constant K. If $b_n(x)$ is either monotonically increasing in n for each fixed x in E, or monotonically decreasing in n for each fixed x in E, then the series $\Sigma\ a_n(x)\ b_n(x)$ converges uniformly on E.

Proof Let $b_n(x)$ be monotonically increasing in n for each fixed x in E, that is, let $b_n(x) \le b_{n+1}(x)$ for each x in E and for every n = 1, 2, ... Then the pointwise limit of b_n on E, say b, exists. Let

$$c_n(x) = b(x) - b_n(x).$$

Then $c_n(x)$ is positive (or zero) and is monotonically decreasing in n for each fixed x in E. Also,

$$|b(x)| \le K$$

when $x \in E$. Hence, for n = 1, 2, ...,

$$c_n(x) \le 2K$$

when $x \in E$. Since $(s_n)_{n=1}^{\infty}$ is uniformly convergent on E, there is a positive integer j such that

$$|a_{j+1}(x) + \cdots + a_{j+p}(x)| < \frac{\varepsilon}{2K}$$

whenever p is a positive integer and $x \in E$. Hence, by Abel's Lemma [see part (a) of Exercise 15 at the end of Chapter 1],

$$|a_{j+1}(x)\, c_{j+1}(x) + \cdots + a_{j+p}(x)\, c_{j+p}(x)| < \frac{\varepsilon}{2K}\, c_{j+1}(x) < \varepsilon.$$

By Proposition 2, $\Sigma\, a_n(x)\, c_n(x)$ is uniformly convergent on E.

Also, since $|b(x)| \le K$ for all x in E, it easily follows from Proposition 2 that $\Sigma\, b(x)\, a_n(x)$ converges uniformly on E. The uniform convergence of $\Sigma\, a_n(x)\, b_n(x)$ on E is now a consequence of the relation $\Sigma\, a_n(x)\, b_n(x) = \Sigma\, [b(x)\, a_n(x) - a_n(x)\, c_n(x)]$. □

We shall now take up two remarkable applications of Proposition 9.

Proposition 12 There exists a real-valued continuous function on the line R^1 that is nowhere differentiable.

Proof Let

$$h(x) = x \quad \text{for} \quad 0 \le x \le 1,$$

$$= 2 - x \quad \text{for} \quad 1 \le x \le 2,$$

and extend the definition of h(x) to all real x by requiring that

$$h(x + 2) = h(x);$$

in other words, h is periodic with period 2. It is clear that h is continuous on R^1. Define

$$f(x) = \sum_{n=0}^{\infty} \left(\frac{3}{4}\right)^n h(4^n x). \tag{20}$$

Since $0 \le h(x) \le 1$ for all $x \in R^1$, Proposition 9 shows that the series (20) converges uniformly on R^1; hence f is continuous on R^1, by Proposition 3.

We fix a real number x and a positive integer m. There exists an integer k such that

$$k \le 4^m x \le k + 1. \tag{21}$$

We put

$$\alpha_m = 4^{-m}k, \quad \beta_m = 4^{-m}(k + 1),$$

and consider the numbers $4^n \beta_m$ and $4^n \alpha_m$. If $n > m$, their difference is an even integer; if $n = m$, they are integers and their difference is 1; if $n < m$, no integer is situated between them. Hence

$$\begin{aligned} |h(4^n\beta_m) - h(4^n\alpha_m)| &= 0 \quad \text{for} \quad n > m, \\ &= 4^{n-m} \quad \text{for} \quad n \le m. \end{aligned} \tag{22}$$

By (20) and (22),

$$f(\beta_m) - f(\alpha_m) = \sum_{n=0}^{m} \left(\frac{3}{4}\right)^n \left[h(4^n\beta_m) - h(4^n\alpha_m)\right];$$

Thus

$$|f(\beta_m) - f(\alpha_m)| \ge \left(\frac{3}{4}\right)^m - \sum_{n=0}^{m-1} \left(\frac{3}{4}\right)^n 4^{n-m} > \frac{1}{2}\left(\frac{3}{4}\right)^m,$$

or

$$\left|\frac{f(\beta_m) - f(\alpha_m)}{\beta_m - \alpha_m}\right| > \frac{1}{2}\, 3^m. \tag{23}$$

Since $\alpha_m \le x \le \beta_m$ and since $\beta_m - \alpha_m \to 0$ as $m \to \infty$, (23) shows that f is not differentiable at x, by Exercise 40 at the end of Chapter 3.

□

Remarks The example constructed in the foregoing proof is due to B. L. van der Waerden. A general method of constructing nondifferentiable continuous functions was given by Konrad Knopp [see "Einfaches Verfahren zur Bildung stetiger, nirgends differenzierbarer Funktionen," Math. Zeitschrift, 2 (1918), 1 - 26]. The first examples of continuous nondifferentiable functions are due to Weierstrass and to Bolzano; the existence of such functions was suggested by Riemann. As an application of Proposition 45 in Chapter 5 we shall see that the class of continuous nowhere differentiable functions is very large.

In Exercise 21 at the end of Chapter 5 we discuss an example of a function which oscillates everywhere yet has a finite derivative at every point; the construction will also make use of the Weierstrass M-Test and the interested reader can proceed directly to the mentioned exercise in Chapter 5 since the understanding of it requires no additional information to that already gained.

Proposition 13 There exists a continuous curve in the x, y plane that passes through every point of the unit square $[0,1] \times [0,1]$; such curves are known as *space-filling curves*.

Proof Let g be defined on the interval $[0,2]$ by

$$
\begin{aligned}
g(t) &= 0 && \text{for } 0 \le t \le \tfrac{1}{3} \text{ and for } \tfrac{5}{3} \le t \le 2, \\
&= 3t - 1 && \text{for } \tfrac{1}{3} \le t \le \tfrac{2}{3}, \\
&= 1 && \text{for } \tfrac{2}{3} \le t \le \tfrac{4}{3}, \\
&= -3t + 5 && \text{for } \tfrac{4}{3} \le t \le \tfrac{5}{3}.
\end{aligned}
$$

We extend the definition of g to all of R^1 by setting

$$g(t + 2) = g(t).$$

Note that g is continuous on R^1 and has period 2.

We now define two functions x and y by the following equations:

$$x(t) = \sum_{n=1}^{\infty} \frac{g(3^{2n-2}t)}{2^n}, \quad y(t) = \sum_{n=1}^{\infty} \frac{g(3^{2n-1}t)}{2^n}.$$

By Proposition 9, both series converge uniformly on R^1; Proposition 3 shows that both functions x and y are continuous on R^1.

Let $z = (x,y)$ and G denote the image of the interval $[0,1]$ under z. We will show that $G = [0,1] \times [0,1]$.

Evidently, $0 \le x(t) \le 1$ and $0 \le y(t) \le 1$ for each t, since the series $\Sigma_{n=1}^{\infty} 2^{-n} = 1$. Hence, G is a subset of $[0,1] \times [0,1]$. It remains to show that $(a,b) \in G$ whenever $(a,b) \in [0,1] \times [0,1]$. For this purpose we express a and b as p-adic fractions with $p = 2$, that is, we put

$$a = \sum_{n=1}^{\infty} \frac{a_n}{2^n}, \quad b = \sum_{n=1}^{\infty} \frac{b_n}{2^n},$$

where each a_n and each b_n is either 0 or 1 (p-adic fractions were studied in Section 6 of Chapter 1). Let

$$c = 2 \sum_{n=1}^{\infty} \frac{c_n}{3^n},$$

where $c_{2n-1} = a_n$ and $c_{2n} = b_n$, for $n = 1, 2, \ldots.$

Since $2 \Sigma_{n=1}^{\infty} 3^{-n} = 1$, we see that $0 \le c \le 1$. We will show that $x(c) = a$ and $y(c) = b$.

If we can verify that

$$g(3^k c) = c_{k+1} \tag{24}$$

holds for $k = 0, 1, 2, \ldots$, then we will have that

$$g(3^{2n-2}c) = c_{2n-1} = a_n \quad \text{and} \quad g(3^{2n-1}c) = c_{2n} = b_n;$$

this, however, will give us $x(c) = a$ and $y(c) = b$. To prove (24), we put

$$3^k c = 2 \sum_{n=1}^{k} \frac{c_n}{3^{n-k}} + 2 \sum_{n=k+1}^{\infty} \frac{c_n}{3^{n-k}} = (\text{an even integer}) + d_k,$$

where $d_k = 2 \Sigma_{n=1}^{\infty} c_{n+k}/3^n$. Since g has period 2,

$$g(3^k c) = g(d_k).$$

If $c_{k+1} = 0$, we have

$$0 \le d_k \le 2 \sum_{n=2}^{\infty} 3^{-n} = \frac{1}{3},$$

and hence $g(d_k) = 0$; thus (24) is satisfied in this case. The only other case to consider is $c_{k+1} = 1$. But then we get

$$\frac{2}{3} \le d_k \le 1$$

and so $g(d_k) = 1$. Therefore (24) holds in all cases and the proof is finished. □

Remark The construction in the foreging proof is due to I. J. Schoenberg.

N.B.: To gain a better insight into the constructions described in the proofs of Propositions 12 and 13, the student would do well by sketching the graphs of some of the functions involved.

3. *The Approximation Theorems of Stone and Weierstrass*

Definition Let f be a real-valued function defined on the interval [0,1]. The function $B_n(f)$ defined by

$$B_n(f;x) = \sum_{p=0}^{n} \binom{n}{p} f\left(\frac{p}{n}\right) x^p (1-x)^{n-p}, \quad x \in [0,1],$$

is called the *Bernstein polynomial of order* n *of the function* f. $B_n(f)$ is a polynomial of degree $\le n$.

The Bernstein polynomials are linear with respect to the function f, that is, if a_1 and a_2 are constants and $f = a_1 f_1 + a_2 f_2$, then

$$B_n(f) = a_1 B_n(f_1) + a_2 B_n(f_2).$$

Since

$$\binom{n}{p} x^p (1 - x)^{n-p} \geq 0$$

on the interval $0 \leq x \leq 1$ and

$$\sum_{p=0}^{n} \binom{n}{p} x^p (1 - x)^{n-p} = [x + (1 - x)]^n = 1, \tag{25}$$

we have

$$m \leq B_n(f;x) \leq M$$

whenever $m \leq f(x) \leq M$ on $[0,1]$.

Consider the (binomial) identity

$$(x + y)^n = \sum_{p=0}^{n} \binom{n}{p} x^p y^{n-p}.$$

Differentiating with respect to x and multiplying by x, we obtain

$$nx(x + y)^{n-1} = \sum_{p=0}^{n} p \binom{n}{p} x^p y^{n-p}.$$

Similarly, by differentiating the (binomial) identity twice with respect to x and multiplying by x^2, we get

$$n(n - 1)x^2(x + y)^{n-2} = \sum_{p=0}^{n} p(p - 1) \binom{n}{p} x^p y^{n-p}.$$

Thus, if we put

$$r_p(x) = \binom{n}{p} x^p (1 - x)^{n-p},$$

we have

$$\sum_{p=0}^{n} r_p(x) = 1, \quad \sum_{p=0}^{n} p\, r_p(x) = nx, \quad \text{and}$$

$$\sum_{p=0}^{n} p(p - 1)\, r_p(x) = n(n - 1)x^2.$$

Hence

$$\sum_{p=0}^{n} (p - nx)^2 \, r_p(x)$$

$$= n^2x^2 \sum_{p=0}^{n} r_p(x) - 2nx \sum_{p=0}^{n} p \, r_p(x) + \sum_{p=0}^{n} p^2 \, r_p(x)$$

$$= n^2x^2 - 2nx[nx] + \{nx + n[n - 1]x^2\}$$

$$= nx(1 - x).$$

But $4x^2 - 4x + 1 = (2x - 1)^2 \geq 0$; thus $x(1 - x) \leq 1/4$. We have therefore proved the following result:

Lemma For all real x, we have

$$\sum_{p=0}^{n} (p - nx)^2 \binom{n}{p} x^p (1 - x)^{n-p} \leq \frac{n}{4}. \tag{26}$$

Proposition 14 (Bernstein's Theorem) For any continuous function f on [0,1], $B_n(f) \to f$ uniformly on [0,1].

Proof We assume that $|f(x)| \leq M < \infty$ on [0,1]. By the uniform continuity of f, there exists, for any $\varepsilon > 0$, some $\delta > 0$ such that

$$|f(x) - f(x')| < \varepsilon \quad \text{whenever} \quad |x - x'| < \delta.$$

We pick an arbitrary $x \in [0,1]$. By (25) we get

$$f(x) = \sum_{p=0}^{n} f(x) \binom{n}{p} x^p (1 - x)^{n-p}$$

and thus

$$|B_n(f;x) - f(x)| \leq \sum_{p=0}^{n} \left|f\left(\frac{p}{n}\right) - f(x)\right| \binom{n}{p} x^p (1 - x)^{n-p} \tag{27}$$

holds. We partition the numbers p = 0, 1, 2, ..., n into two classes A and B as follows:

$$p \in A \quad \text{if} \quad \left|\frac{p}{n} - x\right| < \delta,$$

$$p \in B \quad \text{if} \quad \left|\frac{p}{n} - x\right| \geq \delta.$$

For $p \in A$ we have

$$\left|f\left(\frac{p}{n}\right) - f(x)\right| < \varepsilon.$$

By (25) we therefore get

$$\sum_{p \in A} \left|f\left(\frac{p}{n}\right) - f(x)\right| \binom{n}{p} x^p (1 - x)^{n-p}$$

$$< \varepsilon \sum_{p \in A} \binom{n}{p} x^p (1 - x)^{n-p}$$

$$\leq \varepsilon \sum_{p=0}^{n} \binom{n}{p} x^p (1 - x)^{n-p} = \varepsilon. \tag{28}$$

For $p \in B$ we have

$$\frac{(p - nx)^2}{n^2\delta^2} \geq 1,$$

and so, by applying the foregoing lemma,

$$\sum_{p \in B} \left|f\left(\frac{p}{n}\right) - f(x)\right| \binom{n}{p} x^p (1 - x)^{n-p}$$

$$\leq \frac{2M}{n^2\delta^2} \sum_{p \in B} (p - nx)^2 \binom{n}{p} x^p (1 - x)^{n-p}$$

$$\leq \frac{2M}{n^2\delta^2} \sum_{p=0}^{n} (p - nx)^2 \binom{n}{p} x^p (1 - x)^{n-p}$$

$$\leq \frac{M}{2n\delta^2}. \tag{29}$$

Combining (27), (28) and (29), we see that

$$|B_n(f;x) - f(x)| < \varepsilon + \frac{M}{2n\delta^2}$$

for any $x \in [0,1]$. This, however, means that

$$n > \frac{M}{2\varepsilon\delta^2}$$

implies $|B_n(f;x) - f(x)| < 2\varepsilon$. □

Proposition 15 (Approximation Theorem of Weierstrass) Let $[a,b]$ be a closed bounded interval and f a continuous function on $[a,b]$. Then, for any $\varepsilon > 0$, there exists a polynomial P such that

$$|f(x) - P(x)| < \varepsilon \quad \text{for all} \quad x \in [a,b] \tag{30}$$

and we say that f *admits uniform approximation by* P.

Proof If $[a,b] = [0,1]$, then the proposition is an immediate consequence of Proposition 14. Let $[a,b] \neq [0,1]$. We consider the following function in y:

$$f[a + y(b - a)].$$

This function is defined and continuous on $[0,1]$. Hence there exists a polynomial $Q(y)$ such that for all $y \in [0,1]$ we have:

$$|f[a + y(b - a)] - Q(y)| < \varepsilon.$$

For $x \in [a,b]$,

$$\frac{x - a}{b - a} \in [0,1]$$

and thus

$$\left| f(x) - Q\left(\frac{x - a}{b - a}\right)\right| < \varepsilon.$$

Therefore the polynomial

$$P(x) = Q\left(\frac{x - a}{b - a}\right)$$

has the required property. □

Remark If $0 < a < b < 1$, Proposition 15 is also true with a polynomial $P(x)$ in (30) which has integral coefficients. For the polynomial

$$P(x) = \sum_{p=1}^{n-1} \left[\binom{n}{p} f\left(\frac{p}{n}\right)\right] x^p (1 - x)^{n-p}$$

(here $[c]$ means the greatest integer $\le c$) has integral coefficients and differs from $B_n(f;x)$ by less than

$$M\{x^n + (1 - x)^n\} + \sum_{p=1}^{n-1} x^p (1 - x)^{n-p}. \tag{31}$$

The last sum does not exceed

$$\frac{1}{n}\sum_{p=1}^{n-1} r_p(x) \le \frac{1}{n};$$

thus (31) converges uniformly to zero on the interval $a \le x \le b$.

We shall now sketch another proof for Proposition 15. This interesting proof, given by H. Lebesgue, is based on the approximation of the special function $f(x) = |x|$.

Lemma 1 For each real number $\varepsilon > 0$ and each closed bounded interval I there exists a polynomial $p(t)$ with no constant term such that

$$|p(t) - |t|| < \varepsilon \quad \text{for all} \quad t \in I.$$

Remark It is of course enough to prove the lemma for an interval of the form $I = [-a,+a]$ and hence, replacing t by at, for the interval $I = [-1,+1]$. Since

$$|t| = \sqrt{t^2},$$

Lemma 1 is a consequence of the following result:

Lemma 2 Let $(p_n)_{n=0}^{\infty}$ be a sequence of polynomials without constant term defined by

$$p_0(t) = 0, \quad p_{n+1}(t) = p_n(t) + \frac{1}{2}\left(t - p_n^2(t)\right), \quad n \geq 0. \tag{32}$$

In the interval [0,1] the sequence $(p_n)_{n=0}^{\infty}$ is increasing and converges uniformly to $\sqrt{t}$.

Proof To prove Lemma 2 it is enough to show that, for all t in [0,1], we have

$$0 \leq \sqrt{t} - p_n(t) \leq \frac{2\sqrt{t}}{2 + n\sqrt{t}}, \tag{33}$$

for (33) implies that $0 \leq \sqrt{t} - p_n(t) \leq 2/n$.

We prove (33) by induction on n. It is true for n = 0. If $n \geq 0$ it follows from the inductive assumption (33) that

$$0 \leq \sqrt{t} - p_n(t) \leq \sqrt{t},$$

hence $0 \leq p_n(t) \leq \sqrt{t}$, and therefore from (32) we have

$$\sqrt{t} - p_{n+1}(t) = \left(\sqrt{t} - p_n(t)\right)\left\{1 - \frac{1}{2}\left(\sqrt{t} + p_n(t)\right)\right\},$$

so that $\sqrt{t} - p_{n+1}(t) \geq 0$, and from (33)

$$\sqrt{t} - p_{n+1}(t) \leq \frac{2\sqrt{t}}{2 + n\sqrt{t}}\left(1 - \frac{\sqrt{t}}{2}\right) \leq \frac{2\sqrt{t}}{2 + n\sqrt{t}}\left(1 - \frac{\sqrt{t}}{2 + (n+1)\sqrt{t}}\right)$$

$$= \frac{2\sqrt{t}}{2 + (n+1)\sqrt{t}}. \qquad \square$$

Returning now to the main line of the argument, we note that the graphical representation of $|x|$ is a polygonal line with one angle. We wish to show that any polygonal function admits uniform approximation by polynomials. First, this is true for $|x|$ on each interval

$[-k,k]$, since $|x| = k|x/k|$ and $|x/k| \le 1$ on $[-k,k]$. Again, each function

$$L_c(x) = 0 \quad \text{for } x \le c,$$
$$= x - c \quad \text{for } x \ge c$$

may be approximated by polynomials on each interval $[a,b]$, since

$$L_c(x) = \frac{1}{2}\Big((x - c) + |x - c|\Big).$$

Suppose now that g is any polygonal function on $[a,b]$ whose graph has angles at the basic points $a = a_0 < a_1 < \cdots < a_n = b$. We shall see that g is a linear combination of the L_c. We put

$$g_0(x) = g(a) + C_0 L_{a_0}(x) + C_1 L_{a_1}(x) + \cdots + C_{n-1} L_{a_{n-1}}(x),$$

and define the constants C_j by the equations

$$g_0(a_j) = g(a_j), \quad j = 0, 1, \ldots, n.$$

The first of these equations is an identity, the second is

$$g(a) + C_0 L_{a_0}(a_1) = g_0(a_1)$$

and defines C_0, the third defines C_1, and so on. The two polygonal functions g and g_0 coincide at all basic points and are therefore identical. This shows that g admits approximation by polynomials.

Finally, if f is continuous on a closed bounded interval $[a,b]$, it is uniformly continuous on it and hence can be approximated uniformly with an arbitrarily small error by a polygonal function on it. Indeed, f is uniformly continuous on a closed bounded interval if and only if f admits uniform approximation by a polygonal function on this interval. This completes Lebesgue's proof of the famous theorem of Weierstrass. □

Remark In Exercise 23 at the end of this chapter we shall discuss E. Landau's proof of Proposition 15.

The next three propositions are due to M. H. Stone and represent a significant generalization of the Approximation Theorem of Weierstrass.

Definition Let f and g be real-valued functions on [a,b]. By max{f,g} we mean the function which at x in [a,b] has the value max{f(x),g(x)}; the function min{f,g} is defined to have the value min{f(x),g(x)} at x.

Proposition 16 Let S be a family of continuous functions s on a closed bounded interval [a,b] having the property that for any two functions s_1, s_2 in S, $\max\{s_1,s_2\}$ and $\min\{s_1,s_2\}$ also belong to S. Then a function f, continuous on [a,b], admits uniform approximation by functions s if and only if for any two points y, z of [a,b] and any $\varepsilon > 0$ there exists a function s such that

$$|f(x) - s(x)| < \varepsilon \quad \text{for} \quad x = y \quad \text{and} \quad x = z. \tag{34}$$

Proof The necessity of the condition is trivial. To prove that the condition is sufficient, we take an arbitrary $\varepsilon > 0$ and denote by s_{yz} a certain $s \in S$ satisfying (34). For fixed y in [a,b] we have

$$s_{yz}(z) < f(z) + \varepsilon,$$

and thus, by continuity,

$$s_{yz}(x) < f(x) + \varepsilon$$

in a certain neighborhood $(z - \delta, z + \delta)$ of z. For all possible z in [a,b], these neighborhoods cover [a,b]; by the Heine-Borel Theorem (see Proposition 10 of Chapter 2) there are a finite number of points $z_1, \ldots, z_n$ whose neighborhoods cover [a,b]. Let

$$s_y(x) = \min\left\{s_{yz_1}(x), \ldots, s_{yz_n}(x)\right\};$$

by assumption, $s_y \in S$. Moreover we have, for $a \le x \le b$,

$$s_y(x) < f(x) + \varepsilon, \tag{35}$$

and

$$s_y(y) > f(y) - \varepsilon. \tag{36}$$

It follows that for each $y \in [a,b]$ there is a neighborhood $(y - \gamma, y + \gamma)$ of y such that $s_y(x) > f(x) - \varepsilon$. Using the Heine-Borel Theorem again, we see that we can select a finite number of points $y_1, \ldots, y_m$ such that for every x,

$$s_{y_k}(x) > f(x) - \varepsilon$$

holds for at least one k. Setting

$$s(x) = \max\left\{s_{y_1}(x), \ldots, s_{y_m}(x)\right\},$$

we have that

$$s(x) > f(x) - \varepsilon$$

for any $x \in [a,b]$. On the other hand, by (35),

$$s(x) < f(x) + \varepsilon,$$

hence

$$|f(x) - s(x)| < \varepsilon$$

for $a \le x \le b$. This proves the proposition since $s \in \mathcal{S}$. □

Remark Note that Dini's Theorem (see Proposition 4) is a corollary to Proposition 16. Indeed, let $(s_n)_{n=1}^{\infty}$ be an increasing sequence of continuous functions on $[a,b]$ which converges pointwise to a continuous function f on $[a,b]$. Then the family $\mathcal{S} = \{s_n : n \in \mathbb{N}\}$ satisfies all conditions of Proposition 16; hence, for each $\varepsilon > 0$, there is an index n_0 with $|f(x) - s_{n_0}(x)| < \varepsilon$. But then we also have $|f(x) - s_n(x)| < \varepsilon$ for $n \ge n_0$, $a \le x \le b$.

Passing to more general families $\mathcal{S}$, we say that the functions s in S *separate the points of* $[a,b]$ if for each pair of distinct points y, z of $[a,b]$ there exists a function s such that $s(y) \neq s(z)$.

Proposition 17 Let a family S have the following properties:

(i) the functions $s \in S$ separate the points of $[a,b]$;

(ii) S contains each finite linear combination of functions of S and each constant function;

(iii) for any $s \in S$, $|s|$ also belongs to S.

Then each function f continuous on $[a,b]$ may be approximated uniformly by a function s of S.

Proof We show that the assumptions of Proposition 16 are satisfied. If $s_1, s_2 \in S$, then

$$\max\{s_1,s_2\} = \frac{1}{2}\left(s_1 + s_2 + |s_1 - s_2|\right)$$

and

$$\min\{s_1,s_2\} = \frac{1}{2}\left(s_1 + s_2 - |s_1 - s_2|\right)$$

also belong to S. Moreover, if y, z are two distinct points of $[a,b]$, there exists an $s \in S$ with $s(y) \neq s(z)$; setting

$$\bar{s}(x) = f(y) + \frac{f(z) - f(y)}{s(z) - s(y)}\left(s(x) - s(y)\right)$$

we obtain a function which belongs to S and satisfies (34) even in the sharper form $\bar{s}(y) = f(y)$, $\bar{s}(z) = f(z)$. □

Definition By a *polynomial with respect to the functions* $s \in S$ we understand a function of the form $p(x) = P\left(s_1(x), \ldots, s_n(x)\right)$, where $s_k \in S$ and $P(s_1, \ldots, s_n)$ is a polynomial, with real coefficients, of the variables $s_1, \ldots, s_n$.

Proposition 18 Let S be a family of continuous functions on $[a,b]$. In order that each continuous function on $[a,b]$ admits uniform approximation by polynomials in the functions belonging to S, it is necessary and sufficient that the functions belonging to S separate the points of $[a,b]$.

Proof The condition is necessary, for if $y \neq z$ and the s do not separate y and z then also the polynomials $p(x)$ in the functions of S do not. This contradicts the fact that any $p(x)$ which is sufficiently close to the function $f(x) = x$ separates y and z.

To verify the sufficiency of the condition, we denote by S_0 the family of all continuous functions admitting approximation by polynomials $p(x)$. To show that S_0 contains all functions continuous on $[a,b]$ it is enough to show that the functions of S_0 approximate any such function. S_0 obviously satisfies conditions (i) and (ii) of Proposition 17. If $s \in S_0$, there is a polynomial

$$p(x) = P\bigl(s_1(x), \ldots, s_n(x)\bigr), \quad s_k \in S,$$

approximating $s(x)$ in $[a,b]$ by less than $\varepsilon/2$. Then $|p(x)|$ approximates $|s(x)|$ by less than $\varepsilon/2$. If $m \le p(x) \le M$ for $x \in [a,b]$, we may find an ordinary polynomial $P_1(t)$ such that

$$\bigl|\,|t| - P_1(t)\bigr| < \frac{\varepsilon}{2}$$

in $m \le t \le M$ (see Lemma 1 following Proposition 15). Then

$$\Bigl|\,|s(x)| - \bar{P}\bigl(s_1(x), \ldots, s_n(x)\bigr)\Bigr| < \varepsilon,$$

where $\bar{P}$ is the polynomial $\bar{P}(s_1, \ldots, s_n) = P_1\bigl(P(s_1, \ldots, s_n)\bigr)$. This establishes (iii) and hence our condition is sufficient, by Proposition 17. □

Remarks We can obtain Proposition 15 by noting that the single function $s(x) = x$ separates the points of any interval. Propositions 16 to 18 and their proofs remain unchanged if the interval $[a,b]$ is replaced by a closed bounded subset A of an n-dimensional Euclidean space or even by a compact metric or topological space (for a compact topological space A, the necessary part of Proposition 18 requires the additional assumption that the points of A are separated by the set of all continuous functions). For example, let A be the unit circle in the plane, a point of A being given by the central angle x. The functions sin x, cos x separate the points of A, and polynomials in sin x, cos x are the trigonometric polynomials

$$T_n(x) = a_0 + \sum_{k=1}^{n} (a_k \cos kx + b_k \sin kx).$$

We obtain, therefore, that any continuous function f on R^1 with period 2π may be approximated uniformly by trigonometric polynomials, a result due to Weierstrass as well. We shall obtain this latter result as a corollary to Fejér's Theorem (see Proposition 39).

We end this section with an application of Proposition 15.

Proposition 19 Let f be a continuous function on a closed bounded interval [a,b] and suppose that

$$\int_a^b x^n f(x)\,dx = 0$$

for n = 0, 1, 2, ... Then f = 0.

Proof By Proposition 15, f admits uniform approximation by a polynomial P; thus, for all $x \in [a,b]$,

$$f(x) = P(x) + \varepsilon h(x),$$

where ε is an arbitrary positive real number and $|h(x)| < 1$ on [a,b]. We see therefore that

$$\int_a^b f^2(x)\,dx = \int_a^b f(x)\left\{P(x) + \varepsilon h(x)\right\} dx.$$

But

$$\int_a^b f(x) P(x)\,dx = 0,$$

by hypothesis. Noting that

$$\int_a^b f^2(x)\ dx < \varepsilon \int_a^b |f(x)|\ dx,$$

we see that we have reached a contradiction, for, if $f \neq 0$, then ε would be bounded below by

$$\frac{\int_a^b f^2(x)\ dx}{\int_a^b |f(x)|\ dx}$$

(see Exercise 25 at the end of Chapter 3). □

4. *The Representation Theorem of F. Riesz*

Let g be a function of finite variation on a closed bounded interval [a,b]. Using this function g, we can associate with every continuous real-valued function f on [a,b] a real number

$$F(f) = \int_a^b f(x)\ dg(x).$$

It is clear that the following conditions are fulfilled:

(i) For any two continuous functions f_1 and f_2 on [a,b] and any real number α, we have

$F(f_1 + f_2) = F(f_1) + F(f_2)$, $F(\alpha f_1) = \alpha F(f_1)$.

(ii) For any continuous function f on [a,b], we have

$|F(f)| \le K\, M(f)$,

where $M(f) = \sup\{|f(x)| : x \in [a,b]\}$ and $K = V_a^b g$.

Definition Let C be the set of all continuous real-valued functions on a closed bounded interval [a,b]. F is said to be a *contin-*

uous linear functional on $\mathcal{C}$ if F is a real-valued function defined on $\mathcal{C}$ and satisfies the following conditions:

(i) For any f_1, $f_2 \in \mathcal{C}$, and any real numbers α and β, we have

$$F(\alpha f_1 + \beta f_2) = \alpha F(f_1) + \beta F(f_2).$$

(ii) There exists a constant K such that

$$|F(f)| \le K\, M(f) \quad \text{for all} \quad f \in \mathcal{C}.$$

The following proposition shows the great significance of the Riemann-Stieltjes integral.

Proposition 20 (Representation Theorem of F. Riesz) Let F be a continuous linear functional on $\mathcal{C}$. Then there exists a function g of finite variation on [a,b] such that for all $f \in \mathcal{C}$ we have

$$F(f) = \int_a^b f(x)\, dg(x). \tag{37}$$

Proof Without loss of generality we may assume that [a,b] = [0,1] (see for example the reasoning in the proof of Proposition 15).

In Section 3 we noted that

$$\sum_{k=0}^{n} \binom{n}{k} x^k (1 - x)^{n-k} = 1$$

and that, for each $x \in [0,1]$, every summand of the foregoing sum is nonnegative. For

$$\varepsilon_k = \pm 1 \qquad (k = 0, 1, 2, \ldots, n)$$

we therefore have that

$$\left| \sum_{k=0}^{n} \varepsilon_k \binom{n}{k} x^k (1 - x)^{n-k} \right| \le 1. \tag{38}$$

Let F be a continuous linear functional on $\mathcal{C}$. By definition there exists a constant K such that

$$|F(f)| \le K\, M(f).$$

From this, together with (38), we get

$$\left|\sum_{k=0}^{n} \varepsilon_k F\left[\binom{n}{k} x^k (1-x)^{n-k}\right]\right| \le K.$$

If we choose the numbers ε_k in such a way that all summands in the foregoing sum are nonnegative, we obtain

$$\sum_{k=0}^{n} \left|F\left[\binom{n}{k} x^k (1-x)^{n-k}\right]\right| \le K. \tag{39}$$

We now define a step function g_n by setting

$$g_n(0) = 0,$$

$$g_n(x) = F\left[\binom{n}{0} x^0 (1-x)^{n-0}\right] \qquad \text{for } 0 < x < \frac{1}{n},$$

$$g_n(x) = F\left[\binom{n}{0} x^0 (1-x)^{n-0}\right] + F\left[\binom{n}{1} x^1 (1-x)^{n-1}\right] \qquad \text{for } \frac{1}{n} \le x < \frac{2}{n},$$

$$\vdots$$

$$g_n(x) = \sum_{k=0}^{n-1} F\left[\binom{n}{k} x^k (1-x)^{n-k}\right] \qquad \text{for } \frac{n-1}{n} \le x < 1,$$

$$g_n(1) = \sum_{k=0}^{n} F\left[\binom{n}{k} x^k (1-x)^{n-k}\right].$$

From (39) we see that there is a constant K which is, for every n, a bound for both the function g_n and for the total variation of g_n. Hence, by Helly's Selection Principle (see Proposition 52 of Chapter 2), we may pick from the sequence $(g_n)_{n=1}^{\infty}$ a subsequence $(g_{n_j})_{j=1}^{\infty}$ which converges at every point of $[0,1]$ to a function g of finite variation.

If $f \in \mathcal{C}$, then, by Exercise 42 at the end of Chapter 3,

$$\int_0^1 f(x)\ dg_n(x) = \sum_{k=0}^{n} f\left(\frac{k}{n}\right) F\left[\binom{n}{k} x^k (1-x)^{n-k}\right].$$

From this we get

$$\int_0^1 f(x)\ dg_n(x) = F\left[B_n(f;x)\right],$$

where

$$B_n(f;x) = \sum_{k=0}^{n} f\left(\frac{k}{n}\right) \left[\binom{n}{k} x^k (1-x)^{n-k}\right]$$

is the Bernstein polynomial of order n of the function f. By Proposition 14, $M(B_n - f) \to 0$ as $n \to \infty$ and, by the definition of continuous linear functional, $|F(B_n) - F(f)| = |F(B_n - f)| \le K\, M(B_n - f)$. Thus, for $n \to \infty$, $F(B_n) \to F(f)$, or

$$\lim_{n\to\infty} \int_0^1 f(x)\ dg_n(x) = F(f).$$

As $n \to \infty$, by taking successively the values $n_1, n_2, \ldots, n_j, \ldots$, we get, by Proposition 8,

$$\lim_{n\to\infty} \int_0^1 f(x)\ dg_n(x) = \int_0^1 f(x)\ dg(x).$$

Thus

$$F(f) = \int_0^1 f(x)\ dg(x),$$

finishing the proof. □

Remark It can be seen from the foregoing proposition that trying to determine the most general form of a continuous linear functional on C would have led to the invention of the Riemann-Stieltjes integral (had it not been discovered earlier already).

5. *Equicontinuity*

Definition Let $(f_n)_{n=1}^{\infty}$ be a sequence of real-valued functions defined on a set E of real numbers. We say that $(f_n)_{n=1}^{\infty}$ is *equicontinuous on* E if for every $\varepsilon > 0$ there exists a $\delta > 0$ such that

$$|f_n(x) - f_n(y)| < \varepsilon$$

whenever $|x - y| < \delta$, $x \in E$, $y \in E$, and $n = 1, 2, \ldots$.

Proposition 21 Let K be a compact set of real numbers. If $(f_n)_{n=1}^{\infty}$ is a uniformly convergent sequence of continuous functions on K, then $(f_n)_{n=1}^{\infty}$ is equicontinuous on K.

Proof Let $\varepsilon > 0$ be given. Then there exists an integer n_0 and a $\delta > 0$ such that, for all $x \in E$ and $n > n_0$,

$$|f_n(x) - f_{n_0}(x)| < \frac{\varepsilon}{3};$$

since continuous functions are uniformly continuous on compact sets (by Proposition 32 of Chapter 2), we have that

$$|f_k(x) - f_k(y)| < \frac{\varepsilon}{3} \tag{40}$$

for $1 \le k \le n_0$ and $|x - y| < \delta$. Thus, if $x \in K$, $y \in K$, $|x - y| < \delta$, and $n > n_0$, we have

$$\begin{aligned} |f_n(x) - f_n(y)| &< |f_n(x) - f_{n_0}(x)| + |f_{n_0}(x) - f_{n_0}(y)| \\ &\quad + |f_{n_0}(y) - f_n(y)| \\ &< \frac{\varepsilon}{3} + \frac{\varepsilon}{3} + \frac{\varepsilon}{3} = \varepsilon. \end{aligned}$$

This together with (40) proves the proposition. □

Definition Let $(f_n)_{n=1}^{\infty}$ be a sequence of real-valued functions defined on a set of real numbers. We say that $(f_n)_{n=1}^{\infty}$ is *pointwise bounded on* E if the sequence $\{f_n(x)\}_{n=1}^{\infty}$ is bounded for every $x \in E$, that is, if there exists a finite-valued function h defined on E such that

$$|f_n(x)| < h(x)$$

for all $x \in E$ and $n = 1, 2, \ldots$.

We recall that $(f_n)_{n=1}^{\infty}$ is said to be *uniformly bounded on* E if there is a real number M such that

$$|f_n(x)| < M$$

for all $x \in E$ and $n = 1, 2, \ldots$.

Proposition 22 Let K be a compact set of real numbers. If $(f_n)_{n=1}^{\infty}$ is pointwise bounded and equicontinuous on K, then $(f_n)_{n=1}^{\infty}$ is uniformly bounded on K.

Proof We define

$$g(x) = \sup\{|f_n(x)|: n = 1, 2, \ldots\}.$$

Given $\varepsilon > 0$, we pick $\delta > 0$ such that $|x - y| < \delta$ implies

$$|f_n(x) - f_n(y)| < \varepsilon$$

for $n = 1, 2, \ldots$. If we fix two points x and y, the inequality

$$|f_n(y)| < |f_n(x)| + \varepsilon$$

implies $g(y) \le g(x) + \varepsilon$ and the inequality

$$|f_n(x)| < |f_n(y)| + \varepsilon$$

implies $g(x) \le g(y) + \varepsilon$; thus

$$|g(y) - g(x)| \le \varepsilon$$

whenever $|x - y| < \delta$ and so g is continuous on K. Since K is compact, g is bounded (by Proposition 29 of Chapter 2). □

Proposition 23 (The Ascoli-Arzelà Theorem) Let K be a compact set of real numbers. If $(f_n)_{n=1}^{\infty}$ is pointwise bounded and equicontinuous on K, then $(f_n)_{n=1}^{\infty}$ contains a uniformly convergent subsequence on the set K.

Proof Let E be the set of all rational points of K. Then E is countable and K is the closure of E.

Let $E = \{x_1, x_2, \ldots\}$. Then $\{f_n(x_1)\}_{n=1}^{\infty}$ is a bounded sequence. Using the Bolzano-Weierstrass Theorem (see Proposition 56 of Chapter 1), we may choose a subsequence $(f_{n,1})_{n=1}^{\infty}$ of $(f_n)_{n=1}^{\infty}$ such that

$$\{f_{n,1}(x_1)\}_{n=1}^{\infty}$$

converges, say,

$$\lim_{n\to\infty} f_{n,1}(x_1) = A_1.$$

We next consider the sequence $\{f_{n,1}(x_2)\}_{n=1}^{\infty}$. But, by the Bolzano-Weierstrass Theorem, we may select a subsequence $(f_{n,2})_{n=1}^{\infty}$ of $(f_{n,1})_{n=1}^{\infty}$ such that

$$\lim_{n\to\infty} f_{n,2}(x_2) = A_2$$

exists. Continuing this process indefinitely, we obtain a subsequence

$$(f_{n,m})_{n=1}^{\infty},$$

for $m = 1, 2, \ldots$, of $(f_n)_{n=1}^{\infty}$ such that

$$\lim_{n\to\infty} f_{n,m}(x_m) = A_m$$

exists, $m = 1, 2, \ldots$. We then consider the diagonal sequence

$$(f_{n,n})_{n=1}^{\infty}.$$

For fixed k, $\{f_{n,n}(x_k)\}_{n\geq k}$ is a subsequence of $\{f_{n,k}(x_k)\}_{n\geq k}$, and hence converges to A_k. Therefore $(f_{n,n})_{n=1}^{\infty}$ converges at every point of E. (The foregoing method of selection is sometimes called *Cantor's Diagonal Process;* we have used this method already in the proof of Proposition 50 in Chapter 2.)

Let $\varepsilon > 0$ be given. Since $(f_n)_{n=1}^{\infty}$ is equicontinuous on K, there exists a $\delta > 0$ such that $|x - y| < \delta$ implies

$$|f_n(x) - f_n(y)| < \frac{\varepsilon}{3}$$

for $n = 1, 2, \ldots$.

Since K is the closure of E and K is compact, there are finitely many points $x_1, \ldots, x_p$ in E such that

$$K \subset J(x_1) \cup \cdots \cup J(x_p),$$

where $J(x_j) = (x_j - \delta, x_j + \delta)$ for $j = 1, \ldots, p$.

Let n_0 be so chosen that

$$|f_{n,n}(x_j) - f_{m,m}(x_j)| < \frac{\varepsilon}{3}$$

for $j = 1, \ldots, p$, whenever $n \geq n_0$ and $m \geq n_0$.

Then, for any $x \in K$, there is a point x_j, with $1 \leq j \leq p$, such that $x \in J(x_j)$; hence $n \geq n_0$, $m \geq n_0$ imply

$$\begin{aligned}
&|f_{n,n}(x) - f_{m,m}(x)| \\
&\leq |f_{n,n}(x) - f_{n,n}(x_j)| + |f_{n,n}(x_j) - f_{m,m}(x_j)| \\
&\quad + |f_{m,m}(x_j) - f_{m,m}(x)| \\
&< \frac{\varepsilon}{3} + \frac{\varepsilon}{3} + \frac{\varepsilon}{3} = \varepsilon.
\end{aligned}$$

Thus $(f_{n,n})_{n=1}^{\infty}$ converges uniformly on K. □

6. Power Series

Definition Given a sequence $(a_n)_{n=0}^{\infty}$ of real numbers, the series

$$\sum_{n=0}^{\infty} a_n x^n \tag{41}$$

is called a *power series;* the numbers are called the *coefficients of* the power series.

Proposition 24 If the power series $\Sigma_{n=0}^{\infty} a_n x^n$ is convergent for $x = x_0$, it is absolutely convergent for every value x such that $|x| < |x_0|$.

Proof Since the power series is convergent for $x = x_0$, there is a positive number M such that $|a_n x_0^n| < M$, when $n \geq 0$. But

$$|a_n x^n| = |a_n x_0^n| \left|\frac{x}{x_0}\right|^n.$$

Therefore, if $|x/x_0| = c < 1$, the terms of the series $\Sigma_{n=0}^{\infty} a_n x^n$ are less than the corresponding terms of the series $M(\Sigma_{n=0}^{\infty} c^n)$. □

Corollary If the power series (41) does not converge for $x = x_0$, it does not converge for any value of x such that $|x| > |x_0|$.

Proof This follows from Proposition 24, since if the series converges for a value of x, such that $|x| > |x_0|$, it must converge for $x = x_0$. □

Remark It follows from Proposition 24 and its corollary that only the following three cases can occur:

(i) The power series (41) converges for x = 0 and no other value of x; for example,

$1 + 1!x + 2!x^2 + 3!x^3 + \cdots, \quad 1 + x + 2^2x^2 + 3^3x^3 + \cdots.$

(ii) The power series (41) converges for all values of x; for example,

$$1 + x + \frac{x^2}{2!} + \frac{x^3}{3!} + \cdots.$$

(iii) There is some positive number R such that, when $|x| < R$, the power series (41) converges, and, when $|x| > R$, the series (41) does not converge; for example

$$x - \frac{1}{2}x^2 + \frac{1}{3}x^3 - \cdots.$$

The interval $-R < x < R$ is called the *interval of convergence* and R is called the *radius of convergence*.

The following proposition is a mere reformulation of Proposition 72 of Chapter 1 and its Corollary.

Proposition 25 Let $(a_n)_{n=0}^{\infty}$ be a sequence of real numbers and let

$$\gamma = \overline{\lim_{n\to\infty}} \sqrt[n]{|a_n|}.$$

Then

(i) If $\gamma = 0$, the power series $\Sigma_{n=0}^{\infty} a_n x^n$ converges absolutely for all real x.

(ii) If $\gamma = L > 0$, then $\Sigma_{n=0}^{\infty} a_n x^n$ converges absolutely for $|x| < 1/L$ and diverges for $|x| > 1/L$.

(iii) If $\gamma = \infty$, then $\Sigma_{n=0}^{\infty} a_n x^n$ converges only for $x = 0$ and diverges for all other x.

(iv) The number $1/\gamma$ is the radius of convergence of $\Sigma_{n=0}^{\infty} a_n x^n$, that is, if $R = 1/\gamma$, then $\Sigma_{n=0}^{\infty} a_n x^n$ converges for every value such that $-R < x < R$.

Proposition 26 Let R be the radius of convergence of the power series $\Sigma_{n=0}^{\infty} a_n x^n$. Then $\Sigma_{n=0}^{\infty} a_n x^n$ is absolutely convergent in the open interval $(-R,R)$.

Proof Let $x_0 \in (-R,R)$. Then between x_0 and the nearer boundary of the interval of convergence $(-R,R)$ there are points x for which the given power series converges, and thus by Proposition 24 it converges absolutely for the value x_0. □

Proposition 27 Suppose that the power series $\Sigma_{n=0}^{\infty} a_n x^n$ converges for $|x| < R$. Then $\Sigma_{n=0}^{\infty} a_n x^n$ converges uniformly on the closed interval $[-R + \delta, R - \delta]$, where δ is any assigned positive number less than R.

Proof Let $\delta > 0$ be given. For $|x| \leq R - \delta$, we have

$$|a_n x^n| \leq |a_n (R - \delta)^n|;$$

and since

$$\sum_{n=0}^{\infty} a_n (R - \delta)^n$$

converges absolutely by Proposition 26, Proposition 9 shows the uniform convergence of $\Sigma_{n=0}^{\infty} a_n x^n$ on $[-R + \delta, R - \delta]$. □

Remark When the interval of convergence extends to infinity, the power series will be absolutely convergent for every value of x, but it need not be uniformly convergent on R^1. However, it will be uniformly convergent in any interval $[-b,b]$, where b is a (finite) real number; a case in point is the *exponential series*

$$1 + x + \frac{x^2}{2!} + \frac{x^3}{3!} + \cdots.$$

Proposition 28 (Abel's Theorem) We have the following:

(i) If $\Sigma_{k=0}^{\infty} a_k$ converges, then $\Sigma_{k=0}^{\infty} a_k x^k$ converges uniformly on $[0,1]$.

(ii) If $\Sigma_{k=0}^{\infty} a_k$ converges to L and $f(x) = \Sigma_{k=0}^{\infty} a_k x^k$, then

$$\lim_{x \to 1-} f(x) = L.$$

Proof Given $\varepsilon > 0$, we may choose n_0 such that

$$\left| \sum_{k=m+1}^{n} a_k \right| < \varepsilon$$

when $m, n \geq n_0$. That is,

$$-\varepsilon < a_{m+1} + a_{m+2} + \cdots + a_n < \varepsilon$$

for $m, n \geq n_0$. If $0 \leq x \leq 1$, then Abel's Lemma [see part (a) of Exercise 15 at the end of Chapter 1], applied to $(a_k)_{k=m+1}^{\infty}$ and $(x^k)_{k=m+1}^{\infty}$, gives

$$-\varepsilon\, x^{m+1} < a_{m+1}x^{m+1} + a_{m+2}x^{m+2} + \cdots + a_n x^n < \varepsilon\, x^{m+1}.$$

Hence

$$\left| \sum_{k=m+1}^{n} a_k x^k \right| < \varepsilon\, x^{m+1} \tag{42}$$

for $m, n \geq n_0$ and $0 \leq x \leq 1$. If $f_n(x) = \Sigma_{k=0}^{n} a_k x^k$, for $0 \leq x \leq 1$, then (42) implies that $|f_n(x) - f_m(x)| < \varepsilon$ for $m, n \geq n_0$, and $0 \leq x \leq 1$. By Proposition 2, the claim under part (i) now follows.

To verify part (ii), we use the fact that $\Sigma_{k=0}^{\infty} a_k x^k$ converges uniformly on [0,1] [by part (i)], and so f is continuous (by Proposition 3) on [0,1]. Thus $\lim_{x \to 1-} f(x) = f(1) = L$. □

Proposition 29 The intervals of convergence of the series

$$a_0 + a_1 x + a_2 x^2 + a_3 x^3 + \cdots \quad \text{and} \quad a_1 + 2a_2 x + 3a_3 x^2 + \cdots$$

are the same.

Proof We are given that

$$|a_0| + |a_1x| + |a_2x^2| + \cdots$$

converges when $|x| < R$.

Take x_0 such that $|x| < x_0 < R$. Then

$$\frac{1}{x_0} + \frac{2}{x_0}\left|\frac{x}{x_0}\right| + \frac{3}{x_0}\left|\frac{x}{x_0}\right|^2 + \cdots$$

is convergent (by Proposition 73 of Chapter 1), because the ratio of the (n + 1)-th term to the n-th term has for its limit $|x/x_0|$, which is less than unity. If we multiply the different terms of this series by the factors

$$|a_1x_0|, \quad |a_2x_0^2|, \quad |a_3x_0^3|, \ \ldots,$$

which form a bounded sequence by Proposition 26, it is clear that the series which we thus obtain, namely

$$|a_1| + 2|a_2x| + 3|a_3x^2| + \cdots,$$

is convergent when $|x| < R$.

We have yet to verify that this last series diverges when $|x| > R$.

But if it converges when $x = |x_0|$, where $|x_0| > R$, the same would hold for the series

$$|a_1x_0| + 2|a_2x_0^2| + 3|a_3x_0^3| + \cdots$$

and also for the series

$$|a_1x_0| + |a_2x_0^2| + |a_3x_0^3| + \cdots,$$

since the terms of the latter are not greater than those of the former. But this is impossible, since we are given that the interval of convergence of the original series is $-R < x < R$.

It follows that the series

$$a_0 + a_1x + a_2x^2 + a_3x^3 + \cdots$$

and the series

$$a_1 + 2a_2x + 3a_3x^2 + \cdots$$

obtained by differentiating the first series term by term have the same interval of convergence. □

Remark To prove the foregoing proposition, it would have been enough to show that

$$\overline{\lim_{n\to\infty}} \sqrt[n]{|a_n|} = \overline{\lim_{n\to\infty}} \sqrt[n]{n|a_n|},$$

by noting that $\sqrt[n]{n} \to 1$, as $n \to \infty$ (the latter limit is considered in part (c) of Exercise 14 at the end of Chapter 1). We shall take up this matter in Exercise 1 at the end of this chapter.

Proposition 30 Let the power series $\Sigma_{n=0}^{\infty} a_n x^n$ have $(-R,R)$ for its interval of convergence and define, for $x \in (-R,R)$,

$$f(x) = \sum_{n=0}^{\infty} a_n x^n.$$

Then

(i) $$\int_{x_0}^{x} f(t)\,dt = a_0(x - x_0) + \sum_{n=1}^{\infty} \frac{a_n}{n+1}\left(x^{n+1} - x_0^{n+1}\right),$$

when $-R < x_0 < x < R$.

(ii) $$f'(x) = a_1 + 2a_2x + 3a_3x^2 + \cdots,$$

where x is any point in the open interval $(-R,R)$.

Moreover, these integrations and differentiations may be repeated any number of times.

Proof Part (i) follows from Propositions 26 and 6; part (ii) is a consequence of Propositions 29 and 5. □

Corollary Let f be as in Proposition 30. Then

$$f^{(k)}(x) = \sum_{n=k}^{\infty} n(n-1) \cdots (n-k+1)a_n x^{n-k}$$

for any $x \in (-R,R)$ and

$$f^{(k)}(0) = k!\, a_k, \tag{43}$$

where $k = 0, 1, 2, \ldots$.

Remarks Formula (43) shows that the coefficients of the power-series development of f are determined by the value of f and of its derivatives at a single point. Also, if the coefficients are given, the values of the derivatives of f at the center of the interval of convergence can be read off immediately from the power series. Note, however, that although a function f may have derivatives of all orders, the series $\Sigma_{n=0}^{\infty} a_n x^n$, where a_n is calculated by (43), need not converge to f(x) for any $x \neq 0$. In this case, f cannot be developed in a power series of the form $f(x) = \Sigma_{n=0}^{\infty} c_n x^n$. For if we had $f(x) = \Sigma_{n=0}^{\infty} c_n x^n$, then

$$n!\, c_n = f^{(n)}(0)$$

would follow, implying that $c_n = a_n$. The function

$$f(x) = e^{-1/x^2} \quad \text{for} \quad x \neq 0,$$
$$= 0 \quad \text{for} \quad x = 0$$

is a case in point; it has derivatives of all orders at $x = 0$ and $f^{(n)}(0) = 0$ for $n = 1, 2, \ldots$.

Lemma Let (a_{ij}), $i = 1, 2, \ldots$, $j = 1, 2, \ldots$, be a double sequence and suppose that, for $i = 1, 2, \ldots$,

$$\sum_{j=1}^{\infty} |a_{ij}| = b_i \tag{44}$$

and $\Sigma_{i=1}^{\infty} b_i$ converges. Then

$$\sum_{i=1}^{\infty} \sum_{j=1}^{\infty} a_{ij} = \sum_{j=1}^{\infty} \sum_{i=1}^{\infty} a_{ij}.$$

Proof Let E be a countable set, consisting of the points x_0, x_1, x_2, ..., and suppose that $x_n \to x_0$ as $n \to \infty$. Let

$$f_i(x_0) = \sum_{j=1}^{\infty} a_{ij} \quad (i = 1, 2, \ldots), \tag{45}$$

$$f_i(x_n) = \sum_{j=1}^{n} a_{ij} \quad (i, n = 1, 2, \ldots), \tag{46}$$

$$h(x) = \sum_{i=1}^{\infty} f_i(x) \quad (x \in E). \tag{47}$$

But from (45), (46), and (44) we see that f_i is continuous at x_0. Since $|f_i(x)| \le b_i$ for $x \in E$, (47) converges uniformly; by Proposition 3, h is continuous at x_0. We therefore have

$$\sum_{i=1}^{\infty} \sum_{j=1}^{\infty} a_{ij} = \sum_{i=1}^{\infty} f_i(x_0) = h(x_0) = \lim_{n \to \infty} h(x_n)$$

$$= \lim_{n \to \infty} \sum_{i=1}^{\infty} f_i(x_n) = \lim_{n \to \infty} \sum_{i=1}^{\infty} \sum_{j=1}^{n} a_{ij}$$

$$= \lim_{n \to \infty} \sum_{j=1}^{n} \sum_{i=1}^{\infty} a_{ij} = \sum_{j=1}^{\infty} \sum_{i=1}^{\infty} a_{ij}$$

and so the order of summation may be interchanged. □

Proposition 31 (Taylor's Expansion Theorem) Suppose that

$$f(x) = \sum_{n=0}^{\infty} c_n x^n$$

with the series converging for $|x| < R$. If $-R < a < R$, then

$$f(x) = \sum_{n=0}^{\infty} \frac{f^{(n)}(a)}{n!} (x - a)^n \tag{48}$$

for all x such that $|x - a| < R - |a|$ and we say that f *can be expanded in a power series about the point* $x = a$ which converges in $|x - a| < R - |a|$.

Proof We observe that

$$f(x) = \sum_{n=0}^{\infty} c_n \left[(x - a) + a\right]^n$$

$$= \sum_{n=0}^{\infty} c_n \sum_{m=0}^{n} \binom{n}{m} a^{n-m} (x - a)^m$$

$$= \sum_{m=0}^{\infty} \left[\sum_{n=m}^{\infty} \binom{n}{m} c_n a^{n-m}\right] (x - a)^m.$$

This is the desired expansion about the point $x = 0$. To see its validity, we have to justify the change which was made in the order of summation. But

$$\sum_{n=0}^{\infty} \sum_{m=0}^{n} \left|c_n \binom{n}{m} a^{n-m} (x - a)^m\right| = \sum_{n=0}^{\infty} |c_n| (|x - a| + |a|)^n$$

and the series on the right-hand side converges if $|x - a| + |a|$ is less than R; hence we may apply the foregoing lemma concerning the interchange of order of summation.

Finally, the form of the coefficients in (48) follows from (43) and the proof is finished. □

Remark The series (48) may also converge in a larger interval than the one given by $|x - a| < R - |a|$.

Proposition 32 (Binomial Expansion Theorem) For any real number α, let

$$\binom{\alpha}{0} = 1 \quad \text{and} \quad \binom{\alpha}{n} = \frac{\alpha(\alpha - 1)(\alpha - 2) \cdots (\alpha - n + 1)}{n!}$$

for n = 1, 2, ... The series

$$\sum_{n=0}^{\infty} \binom{\alpha}{n} x^n \tag{49}$$

converges for all $x \in (-1,1)$. For $\alpha > 0$, the series

$$\sum_{n=0}^{\infty} \left|\binom{\alpha}{n}\right| \tag{50}$$

converges, and the series (49) converges uniformly and absolutely in [-1,1]. Finally, we have

$$\sum_{n=0}^{\infty} \binom{\alpha}{n} x^n = (1 + x)^{\alpha} \tag{51}$$

for $x \in (-1,1)$ for all real α. For $\alpha > 0$, (51) holds for all x in [-1,1].

Proof We shall use Proposition 73 of Chapter 1. If α is a nonnegative integer, then all but finitely many of the numbers $\binom{\alpha}{n}$ are 0, and (49) is seen to converge. Otherwise, for $|x| < 1$ and n = 0, 1, 2, ..., we have

$$\left|\frac{\binom{\alpha}{n+1} x^{n+1}}{\binom{\alpha}{n} x^n}\right| = \frac{|\alpha - n|}{n + 1} |x| \to |x| < 1.$$

This shows that the series (49) converges absolutely for $|x| < 1$ and proves the first claim of the proposition.

Next we consider the second claim of the proposition. The case in which α is a nonnegative integer is again trivial. For α not a nonnegative integer, we let $a_n = \left|\binom{\alpha}{n}\right|$. Then we have

$$\frac{a_{n+1}}{a_n} = \frac{|\alpha - n|}{n + 1} = \frac{n - \alpha}{n + 1},$$

the last equality holding for $n \geq [\alpha] + 1$, where $[\alpha]$ denotes the largest integer not exceeding α. Thus, for $n \geq [\alpha] + 1$, we have

$$(n + 1)a_{n+1} = na_n - \alpha a_n,$$

and so

$$na_n - (n + 1)a_{n+1} = \alpha a_n > 0. \tag{52}$$

This shows that for $n \geq [\alpha] + 1$, (na_n) forms a decreasing sequence which is bounded below; let

$$\lim_{n \to \infty} na_n = \beta \geq 0.$$

Now consider the series $\Sigma_{n=0}^{\infty} \{na_n - (n + 1)a_{n+1}\}$; its p-th partial sum is $-(p + 1)a_{p+1}$, which converges to $-\beta$. Hence the series

$$\sum_{n=0}^{\infty} \{na_n - (n + 1)a_{n+1}\}$$

converges. But, by (52),

$$a_n = \frac{1}{\alpha} \{na_n - (n + 1)a_{n+1}\}$$

for sufficiently large n. It therefore follows that the series $\Sigma_{n=0}^{\infty} a_n$ converges, implying that the series (50) converges. Since

$$\left|\binom{\alpha}{n} x^n\right| < \left|\binom{\alpha}{n}\right|$$

for $|x| \leq 1$, the series (49) converges absolutely and uniformly in $[-1,1]$ and so defines a continuous function on $[-1,1]$ by Proposition 3.

We now show that $\Sigma_{n=0}^{\infty} \binom{\alpha}{n} x^n = (1 + x)^{\alpha}$ holds for $x \in (-1,1)$ and for all real α.

Let

$$f_{\alpha}(x) = \sum_{n=0}^{\infty} \binom{\alpha}{n} x^n$$

for $x \in (-1,1)$ and α real. By Proposition 30,

$$f'_\alpha(x) = \sum_{n=1}^{\infty} n \binom{\alpha}{n} x^{n-1} = \sum_{n=0}^{\infty} (n + 1) \binom{\alpha}{n+1} x^n$$

for $x \in (-1,1)$. Using the identity

$$(n + 1) \binom{\alpha}{n+1} = \alpha \binom{\alpha - 1}{n},$$

we see that

$$f'_\alpha(x) = \alpha \sum_{n=0}^{\infty} \binom{\alpha - 1}{n} x^n = \alpha f_{\alpha-1}(x). \tag{53}$$

We also have

$$\begin{aligned}(1 + x) f_{\alpha-1}(x) &= (1 + x) \sum_{n=0}^{\infty} \binom{\alpha - 1}{n} x^n \\ &= 1 + \sum_{n=1}^{\infty} \left(\binom{\alpha - 1}{n} + \binom{\alpha - 1}{n - 1}\right) x^n \\ &= \sum_{n=0}^{\infty} \binom{\alpha}{n} x^n = f_\alpha(x),\end{aligned}$$

that is,

$$(1 + x) f_{\alpha-1}(x) = f_\alpha(x). \tag{54}$$

Combining (53) and (54), we get

$$(1 + x) f'_\alpha(x) - \alpha f_\alpha(x) = 0$$

for all $x \in (-1,1)$. But

$$\frac{d}{dx}\left(\frac{f_\alpha(x)}{(1 + x)^\alpha}\right) = \frac{1}{(1 + x)^{\alpha+1}} \{(1 + x) f'_\alpha(x) - \alpha f_\alpha(x)\} = 0.$$

Hence

$$\frac{f_\alpha(x)}{(1 + x)^\alpha}$$

is a constant function in $(-1,1)$. For $x = 0$, we find that the constant value of this function is 1; that is, for $x \in (-1,1)$, we get

$$f_\alpha(x) = (1 + x)^\alpha.$$

Finally, if $\alpha > 0$, the series (51) also converges for $x = \pm 1$, and hence (51) holds for all $x \in [-1,1]$ (by either using Proposition 23 or merely applying the result in Exercise 22 at the end of Chapter 2). □

Corollary We have, for all $x \in [-1,1]$,

$$(1 + x)^{1/2} = \sum_{n=0}^{\infty} \binom{1/2}{n} x^n$$

with the series converging absolutely and uniformly for $|x| \le 1$.

Remark For any real number t,

$$|t| = \sqrt{1 + t^2 - 1}.$$

For $|t^2 - 1| \le 1$, that is, for $|t| \le \sqrt{2}$, we have

$$\sqrt{1 + t^2 - 1} = \sum_{n=0}^{\infty} \binom{1/2}{n} (t^2 - 1)^n,$$

and the series converges absolutely and uniformly for all t such that $|t| \le \sqrt{2}$ (by the foregoing corollary). For every positive integer p and for all real t such that $|t| \le \sqrt{2}$, we therefore have

$$\left| |t| - \sum_{n=0}^{p} \binom{1/2}{n} (t^2 - 1)^n \right| \le \sum_{n=p+1}^{\infty} \left| \binom{1/2}{n} \right| \qquad (55)$$

with the right-hand side of (55) being as small as we please for p sufficiently large. In other words, the function $g(t) = |t|$ can be approximated uniformly by a polynomial in t on the closed interval $[-\sqrt{2},\sqrt{2}]$ and therefore in particular on the interval $[-1,1]$. As we have seen in Section 3, both Lebesgue's proof of the Weierstrass Approximation Theorem and Proposition 19 make strong use of this fact.

7. *Fourier Series*

Lemma If $t \neq 2k\pi$, where $k = 0, 1, 2, \ldots,$ then

$$\frac{1}{2} + \cos t + \cos 2t + \cdots + \cos nt = \frac{\sin \frac{2n+1}{2} t}{2 \sin \frac{t}{2}}.$$

Proof Since

$$\sin \frac{2n+1}{2} t = \sin \frac{t}{2} + \left(\sin \frac{3}{2} t - \sin \frac{t}{2}\right) + \left(\sin \frac{5}{2} t - \sin \frac{3}{2} t\right) + \cdots + \left(\sin \frac{2n+1}{2} t - \sin \frac{2n-1}{2} t\right),$$

we get

$$\sin \frac{2n+1}{2} t = \left(\frac{1}{2} + \cos t + \cos 2t + \cdots + \cos nt\right) 2 \sin \frac{t}{2}$$

by use of the identity $\sin A - \sin B = 2 \sin \frac{A-B}{2} \cos \frac{A+B}{2}$. □

For reference we list the following integrals:

$$\int_{-\pi}^{\pi} \cos kx \cos nx \, dx = 0 \quad (k, n, = 0, 1, 2, \ldots; k \neq n);$$

$$\int_{-\pi}^{\pi} \cos^2 nx \, dx = \pi \quad (n = 1, 2, \ldots),$$

$$= 2\pi \quad (n = 0);$$

$$\int_{-\pi}^{\pi} \sin kx \sin nx \, dx = 0 \quad (k, n = 1, 2, \ldots; k \neq n);$$

$$\int_{-\pi}^{\pi} \sin^2 nx \, dx = \pi \quad (n = 1, 2, \ldots);$$

$$\int_{-\pi}^{\pi} \cos kx \sin nx \, dx = 0 \quad (k, n = 0, 1, 2, \ldots).$$

All these values are easily derived by use of the appropriate trigonometric identities and elementary methods of integration.

Definition Let f be a real-valued function defined on the real line R^1 and be periodic with period 2π, that is,

$$f(x + 2\pi) = f(x) \quad \text{for any} \quad x \in R^1.$$

Let us assume, moreover, that f is Riemann integrable over an interval of length 2π. Then the *Fourier series for* f is the series

$$\frac{a_0}{2} + \sum_{k=1}^{\infty} (a_k \cos kx + b_k \sin kx),$$

where

$$a_k = \int_{-\pi}^{\pi} f(x) \cos kx \, dx \quad (k = 0, 1, 2, \ldots),$$

$$b_k = \int_{-\pi}^{\pi} f(x) \sin kx \, dx \quad (k = 1, 2, \ldots).$$

The a_k and b_k are called the *Fourier coefficients for* f. We write

$$f \approx \frac{a_0}{2} + \sum_{k=1}^{\infty} (a_k \cos kx + b_k \sin kx).$$

Note that we write $\approx$ and not =; we have yet to show that the Fourier series for f will converge for some or all values of x to f(x).

Example The function

$$f(x) = 0 \quad (-\pi < x < 0),$$

$$= 1 \quad (0 \le x \le \pi),$$

extended over the real line by making it periodic with period 2π, has the value 1 at x = 0; but

$$f \approx \frac{1}{2} + \frac{2}{\pi}\left(\frac{\sin x}{1} + \frac{\sin 3x}{3} + \frac{\sin 5x}{5} + \cdots\right)$$

and the sum of the Fourier series for f at x = 0 is 1/2.

We shall consider the partial sums of a Fourier series and briefly discuss conditions for convergence. Since all functions appearing in the Fourier series for f have period 2π, it does not matter over what interval of length 2π we integrate to describe the Fourier coefficients.

Let $s_n(x)$ be the sum of the first n + 1 terms of the Fourier series

$$\begin{aligned} s_n(x) &= \frac{a_0}{2} + \sum_{k=1}^{n} (a_k \cos kx + b_k \sin kx) \\ &= \frac{1}{2\pi}\int_0^{2\pi} f(t)\,dt \\ &\quad + \frac{1}{\pi}\sum_{k=1}^{n}\int_0^{2\pi} f(t)\left[\cos kt \cos kx + \sin kt \sin kx\right] dt \\ &= \frac{1}{2\pi}\int_0^{2\pi} f(t)\,dt + \frac{1}{\pi}\sum_{k=1}^{n}\int_0^{2\pi} f(t)\cos k(x - t)\,dt \\ &= \frac{1}{\pi}\int_0^{2\pi} f(t)\left[\frac{1}{2} + \sum_{k=1}^{n}\cos k(x - t)\right] dt. \end{aligned}$$

By the foregoing lemma,

$$s_n(x) = \frac{1}{2\pi}\int_0^{2\pi} \frac{\sin (n + \frac{1}{2})(x - t)}{\sin (x - t)/2}\, f(t)\,dt.$$

On the right-hand side we put t = x + u. Then, on account of periodicity,

$$\begin{aligned} s_n(x) &= \frac{1}{2\pi}\int_{-x}^{2\pi-x} \frac{\sin (n + \frac{1}{2})u}{\sin u/2}\, f(x + u)\,du \\ &= \frac{1}{2\pi}\int_0^{2\pi} \frac{\sin (n + \frac{1}{2})u}{\sin u/2}\, f(x + u)\,du = \frac{1}{2\pi}\left\{\int_0^{\pi} + \int_{\pi}^{2\pi}\right\}. \end{aligned}$$

In the second integral on the right-hand side we put $u = -v$. Then

$$\int_{\pi}^{2\pi} \frac{\sin (n + \frac{1}{2})u}{\sin u/2} f(x + u)\, du = \int_{-2\pi}^{-\pi} \frac{\sin (n + \frac{1}{2})v}{\sin v/2} f(x - v)\, dv.$$

On account of the periodicity the integrand of the right takes on the same values between -2π and $-\pi$ that it does between 0 and π. Hence

$$\int_{0}^{2\pi} \frac{\sin (n + \frac{1}{2})u}{\sin u/2} f(x + u)\, du = \int_{0}^{\pi} \frac{\sin (n + \frac{1}{2})v}{\sin v/2} f(x - v)\, dv,$$

and

$$s_n(x) = \frac{1}{2\pi} \int_{0}^{\pi} \frac{\sin (n + \frac{1}{2})u}{\sin u/2} \Big[f(x + u) + f(x - u)\Big]\, du.$$

If $f(x) = 1$ for all $x \in [0,2\pi]$, then $s_n(x) = 1$ for every n, since $a_k = b_k = 0$, $k > 0$. Hence

$$1 = \frac{1}{2\pi} \int_{0}^{\pi} \frac{\sin (n + \frac{1}{2})u}{\sin u/2}\, 2\, du.$$

It then follows that

$$s_n(x) - \frac{1}{2}\Big[f(x+) + f(x-)\Big] = \frac{1}{2\pi} \int_{0}^{\pi} \frac{\sin (n + \frac{1}{2})u}{\sin u/2} h(u)\, du,$$

where

$$h(u) = \{f(x + u) + f(x - u) - [f(x+) + f(x-)]\}.$$

From this we get the following proposition.

Proposition 33 If x on $[0,2\pi]$ is such that the one-sided limits $f(x+)$ and $f(x-)$ exist, a necessary and sufficient condition that $s_n(x) \to [f(x+) + f(x-)]/2$ is that, as $n \to \infty$,

$$\int_{0}^{\pi} \frac{\sin (n + \frac{1}{2})u}{\sin u/2} h(u)\, du \to 0,$$

where

$$h(u) = \{f(x + u) + f(x - u) - [f(x+) + f(x-)]\}.$$

Proposition 34 (Riemann's Theorem) Let the function f be Riemann integrable on [a,b] and let λ be a positive real number. Then

$$\int_a^b f(x) \sin \lambda x \, dx, \qquad \int_a^b f(x) \cos \lambda x \, dx$$

both tend to zero as $\lambda \to \infty$.

Remark This result has already been discussed in Exercise 30 at the end of Chapter 3.

Proposition 35 (Localization Theorem) The behavior of $s_n(x)$ for large n depends only on the behavior of f(t) for values of t on the interval $[x - \delta, x + \delta]$ for arbitrary small positive δ.

Proof We have

$$s_n(x) = \frac{1}{2\pi}\int_0^\delta \frac{\sin (n + \frac{1}{2})u}{\sin u/2} [f(x + u) + f(x - u)]\, du$$

$$+ \frac{1}{2\pi}\int_\delta^\pi \sin\left(n + \frac{1}{2}\right)u \frac{f(x + u) + f(x - u)}{\sin u/2}\, du.$$

Since $\delta > 0$, the second factor in the integrand of the second integral is an integrable function of u. Hence, by Proposition 34, the second integral tends to zero as $n \to \infty$. It now follows that the behavior of $s_n(x)$ depends, as $n \to \infty$, on the behavior of the first integral. This in turn involves only values of f(t) for $t = x + u$ and $t = x - u$ with $|u| \le \delta$. Hence t is on $[x - \delta, x + \delta]$, completing the proof. □

The next proposition is a consequence of Propositions 33 and 35.

Proposition 36 A necessary and sufficient condition that

$$s_n(x) \to \frac{f(x+) + f(x-)}{2}$$

is that for δ with $0 < \delta < \pi$,

$$\lim_{n\to\infty} \int_0^{\delta} \frac{\sin\,(n + \frac{1}{2})u}{\sin\,u/2}\, h(u)\, du = 0,$$

where

$$h(u) = \{f(x + u) + f(x - u) - [f(x+) + f(x-)]\}.$$

Now consider the possibility of replacing $\sin(u/2)$ by $u/2$ in the integrand of Proposition 36. L'Hospital's Rule (see Proposition 12 of Chapter 3) may be used to show that

$$\frac{1}{\sin(u/2)} - \frac{1}{u/2} = \frac{u - 2\,\sin(u/2)}{u\,\sin(u/2)} \to 0 \quad \text{as} \quad u \to 0.$$

Consequently the product of this function and $h(u)$ is integrable in $[0,\delta]$. It then follows from Proposition 34 that

$$\int_0^{\delta} \sin\left(n + \frac{1}{2}\right)u \left[\frac{1}{\sin(u/2)} - \frac{1}{u/2}\right] h(u)\, du \to 0 \quad \text{as} \quad n \to \infty.$$

As a consequence of this we have the following proposition.

Proposition 37 A necessary and sufficient condition that

$$s_n(x) \to \frac{f(x+) + f(x-)}{2} \quad \text{as} \quad n \to \infty$$

is that for $0 < \delta < \pi$

$$\int_0^{\delta} \frac{\sin\,(n + \frac{1}{2})u}{u}\, h(u)\, du \to 0 \quad \text{as} \quad n \to \infty,$$

where

$$h(u) = \{f(x + u) + f(x - u) - [f(x+) + f(x-)]\}.$$

Proposition 38 (Dirichlet-Jordan Convergence Condition) Let f be of finite variation over $[0,2\pi]$. Then for x_0 on $(0,2\pi)$

$$s_n(x_0) \to \frac{f(x_0+) + f(x_0-)}{2} \quad \text{as} \quad n \to \infty.$$

Proof The function

$$h(u) = \{f(x_0 + u) + f(x_0 - u) - [f(x_0+) + f(x_0-)]\}$$

is of finite variation on $0 \le u \le \delta$, provided δ is sufficiently small to ensure that $x_0 + u$ and $x_0 - u$ are on $[0,2\pi]$. Hence

$$h(u) = h_1(u) - h_2(u)$$

(by Proposition 46 of Chapter 2), where $h_1(u)$ and $h_2(u)$ are positive and h_1 and h_2 are monotonically increasing functions. For u close to zero, $f(x_0 + u)$, $f(x_0 - u)$ are respectively close to $f(x_0+)$, $f(x_0-)$, both of which exist because f is of finite variation on $[0,2\pi]$. Hence $h(u) \to 0$ as $u \to 0$ and the same is true of $h_1(u)$ and $h_2(u)$.

$$\int_0^\delta \frac{\sin (n + \frac{1}{2})u}{u} h(u)\, du$$

$$= \int_0^\delta \frac{\sin (n + \frac{1}{2})u}{u} h_1(u)\, du - \int_0^\delta \frac{\sin (n + \frac{1}{2})u}{u} h_2(u)\, du$$

$$= I_1 - I_2.$$

Let $\varepsilon > 0$ be given. Fix δ, with $0 < \delta < \pi$, so that $h_1(u) < \varepsilon$ for $0 < u < \delta$. For γ, $0 < \gamma < \delta$, by Bonnet's Theorem (see Exercise 44 at the end of Chapter 3),

$$\left|\int_0^\gamma \frac{\sin (n + \frac{1}{2})u}{u} h_1(u)\, du\right|$$

$$= \left|h_1(\gamma) \int_\sigma^\gamma \frac{\sin (n + \frac{1}{2})u}{u}\, du\right| \qquad (0 \le \sigma < \gamma)$$

$$= \left| h_1(\gamma) \int_{(n+\frac{1}{2})\sigma}^{(n+\frac{1}{2})\gamma} \frac{\sin v}{v} \, dv \right|$$

$$< \varepsilon |A(\sigma,n,\gamma)|,$$

where $A(\sigma,n,\gamma)$ is bounded for all σ, n, γ (as can readily be seen from the graph of the function $y = (\sin v)/v$ given in Fig. 4 - 2.)

With γ fixed it follows from Proposition 34 that as $n \to \infty$

$$\int_{\gamma}^{\delta} \frac{\sin (n + \frac{1}{2})u}{u} h_1(u) \, du \to 0.$$

We can use Proposition 34 because h_1 is Riemann integrable (and so is $1/u$ because $0 < \gamma < \delta$). It now follows that $I_1 \to 0$ as $n \to \infty$, and it may be shown in a similar way that $I_2 \to 0$ as $n \to \infty$. Thus the condition for Proposition 37 is fulfilled and we may conclude that

$$s_n(x_0) \to \frac{f(x_0+) + f(x_0-)}{2} \quad \text{as} \quad n \to \infty.$$

If x_0 is a point of continuity of f, then $s_n(x_0) \to f(x_0)$ as $n \to \infty$, and since f is of finite variation, $s_n(x) \to f(x)$ as $n \to \infty$, for all values of x except possibly at the points of discontinuity of f. At a point of discontinuity $s_n(x) \to f(x)$ as $n \to \infty$ only if

$$f(x) = \frac{f(x+) + f(x-)}{2},$$

that is, if $f(x)$ is the arithmetical average of $f(x+)$ and $f(x-)$. □

We shall now consider the function

$$\frac{\sin x}{x}$$

in greater detail. The graph of the function, shown in Fig. 4 - 2, is clearly symmetrical about the y axis. For $x \geq 0$ the successive waves have the same breadth and diminishing amplitude, and the area between 0 and π is greater than that between π and 2π in absolute value; that between π and 2π is greater than between 2π and 3π, and so on; since $\sin x$ goes through the same set of values in each case,

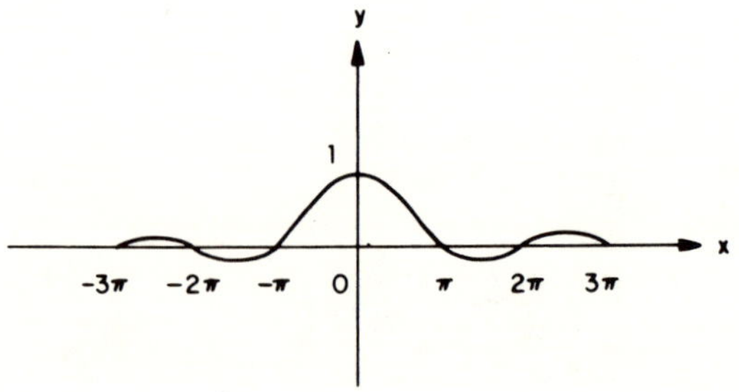

Figure 4 - 2

and 1/x diminishes as x increases. Thus

$$\int_0^t \frac{\sin x}{x}\,dx \le \int_0^\pi \frac{\sin x}{x}\,dx < \pi$$

whatever positive value t may have. Also

$$\int_c^d \frac{\sin x}{x}\,dx = \int_0^d \frac{\sin x}{x}\,dx - \int_0^c \frac{\sin x}{x}\,dx,$$

and each of these integrals on the right-hand side is positive and less than π for $0 < c < d$. Therefore, for $0 \le c < d$,

$$\int_c^d \frac{\sin x}{x}\,dx < \pi.$$

Using Proposition 28 of Chapter 3, we get

$$\int_a^b \frac{\sin x}{x}\,dx = \frac{1}{a}\int_a^t \sin x\,dx + \frac{1}{b}\int_t^b \sin x\,dx,$$

where $0 < a \le t \le b$. Hence

$$\left|\int_a^b \frac{\sin x}{x}\,dx\right| \le 2\left(\frac{1}{a} + \frac{1}{b}\right)$$

(this is so because $\left|\int_\alpha^\beta \sin x\,dx\right| \le 2$ for any α and β) and so

$$\left|\int_a^b \frac{\sin x}{x}\,dx\right| < \varepsilon$$

when $b > a \geq A$, provided $A > 4/\varepsilon$. This however means that

$$I = \lim_{A\to\infty} \int_0^A \frac{\sin x}{x}\,dx = \int_0^\infty \frac{\sin x}{x}\,dx$$

exists (that is, there is a real number I with the property: given any $\varepsilon > 0$, there is a positive number X such that

$$\left|I - \int_0^A \frac{\sin x}{x}\,dx\right| < \varepsilon$$

provided that $A \geq X$).

In Exercise 11 at the end of this chapter we shall give a method for calculating

$$I = \int_0^\infty \frac{\sin x}{x}\,dx;$$

the method to be discussed is due to the famous geometer N. I. Lobatshewski who was particularly annoyed with the variety of imprecise methods used to calculate this improper integral.

Example Let $f(x) = x^2$ for $x \in [-\pi,\pi]$ and extend the domain of definition of f to all real x by letting $f(x + 2\pi) = f(x)$. Fig. 4 - 3 shows the graph of f. We see that

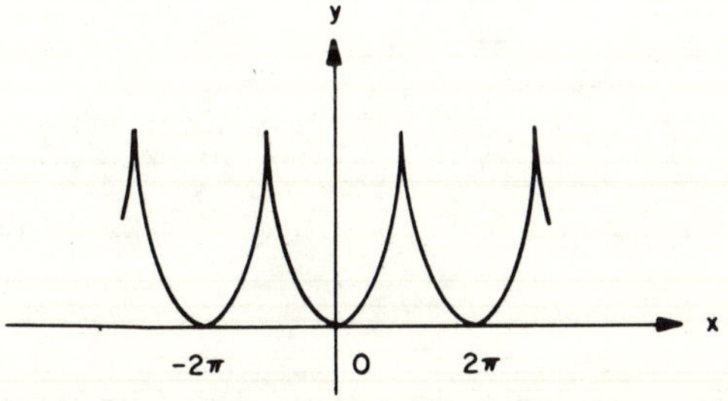

Figure 4 - 3

$$\frac{1}{2} a_0 = \frac{1}{2\pi} \int_0^{2\pi} x^2 \, dx = \frac{\pi^2}{3}$$

and, using integration by parts,

$$a_k = (-1)^k \frac{4}{k^2} \quad (k > 0).$$

Since f is an even function, that is, $f(-x) = f(x)$, we see that

$$b_k = 0 \quad \text{for} \quad k = 1, 2, \ldots.$$

By Proposition 38 we have, for $-\pi \le x \le \pi$,

$$x^2 = \frac{\pi^2}{3} + \sum_{k=1}^{\infty} (-1)^k \frac{\cos kx}{k^2}.$$

For $x = \pi$ and $x = 0$ we get the interesting expansions

$$\frac{\pi^2}{6} = \sum_{k=1}^{\infty} \frac{1}{k^2} \quad \text{and} \quad \frac{\pi^2}{12} = \sum_{k=1}^{\infty} \frac{(-1)^{k-1}}{k^2},$$

respectively. The first of these two series was discussed in Exercise 40 at the end of Chapter 1. Note also that one may directly derive one series from the other in the case of the foregoing two series.

Substituting $u = 2t$ into the formula

$$s_n(x) = \frac{1}{2\pi} \int_0^{\pi} \frac{\sin (n + \frac{1}{2})u}{\sin u/2} [f(x + u) + f(x - u)] \, du,$$

we get

$$s_n(x) = \frac{1}{\pi} \int_0^{\pi/2} \frac{\sin (2n + 1)t}{\sin t} [f(x + 2t) + f(x - 2t)] \, dt.$$

The foregoing expression for $s_n(x)$ is sometimes called a *Dirichlet integral*.

Proposition 39 (Fejér's Theorem) Let f be a continuous function of period 2π and suppose that s_n is the sum of the first $n + 1$ terms of the Fourier series for f. We put

$$\sigma_n(x) = \frac{s_0(x) + s_1(x) + \cdots + s_{n-1}(x)}{n}.$$

Then the *Fejér sums* σ_n converge to f uniformly on R^1.

Proof We consider the Dirichlet integrals

$$s_k(x) = \frac{1}{\pi}\int_0^{\pi/2} \frac{f(x + 2t) + f(x - 2t)}{\sin t} [\sin(2k + 1)t] \, dt.$$

It follows that

$$\sigma_n(x) = \frac{1}{n\pi}\int_0^{\pi/2} \frac{f(x + 2t) + f(x - 2t)}{\sin t} \sum_{k=0}^{n-1} \sin(2k + 1)t \, dt. \tag{56}$$

But

$$\sum_{k=0}^{n-1} \sin(2k + 1)t = \frac{\sin^2 nt}{\sin t}$$

because

$$2 \sin A \sin B = \cos(A - B) - \cos(A + B);$$

and so

$$2 \sin t \sum_{k=0}^{n-1} \sin(2k + 1)t = \sum_{k=0}^{n-1} [\cos 2kt - \cos(2k + 2)t]$$

$$= 1 - \cos 2nt = 2 \sin^2 nt.$$

Thus

$$\sigma_n(x) = \frac{1}{n\pi}\int_0^{\pi/2} [f(x + 2t) + f(x - 2t)] \left(\frac{\sin nt}{\sin t}\right)^2 dt.$$

The foregoing integral is sometimes called a *Fejér integral*. We see that a Fejér sum of a function can be represented by the use of a Fejér integral. This is true in particular of the constant function f = 1, whose Fourier series is

$$1 + \sum_{k=1}^{\infty} 0;$$

for this function all sums s_n and therefore all Fejér sums σ_n are identically equal to 1. If we use formula (56) with the constant function $f = 1$, we get

$$1 = \frac{1}{n\pi} \int_0^{\pi/2} 2\left(\frac{\sin nt}{\sin t}\right)^2 dt. \tag{57}$$

Multiplying (57) by f and then subtracting the result from the expression in (56), we obtain

$$\begin{aligned} &\sigma_n(x) - f(x) \\ &= \frac{1}{n\pi} \int_0^{\pi/2} [f(x + 2t) + f(x - 2t) - 2f(x)] \left(\frac{\sin nt}{\sin t}\right)^2 dt. \end{aligned} \tag{58}$$

Since f is uniformly continuous, for every $\varepsilon > 0$ we can find a $\delta > 0$ such that $|x'' - x'| < 2\delta$ implies $|f(x'') - f(x')| < \varepsilon/2$. If we decompose the integral in (58) into integrals over $[0,\delta]$ and $[\delta,\pi/2]$, then on the first interval we get

$$|f(x + 2t) + f(x - 2t) - 2f(x)| < \varepsilon;$$

by (57), therefore,

$$\begin{aligned} &\frac{1}{n\pi} \int_0^{\delta} |f(x + 2t) + f(x - 2t) - 2f(x)| \left(\frac{\sin nt}{\sin t}\right)^2 dt \\ &< \frac{\varepsilon}{n\pi} \int_0^{\pi/2} \left(\frac{\sin nt}{\sin t}\right)^2 dt = \frac{\varepsilon}{2}. \end{aligned} \tag{59}$$

On the second interval we get

$$|f(x + 2t) + f(x - 2t) - 2f(x)| \le 4M,$$

where M is the largest value of $|f(x)|$ on $[-\pi,\pi]$; moreover,

$$\left(\frac{\sin nt}{\sin t}\right)^2 \le \frac{1}{\sin^2 \delta},$$

and therefore

$$\left|\frac{1}{n\pi}\int_{\delta}^{\pi/2} [f(x + 2t) + f(x - 2t) - 2f(x)] \left(\frac{\sin nt}{\sin t}\right)^2 dt\right|$$
$$\le \frac{1}{n\pi}\left(\frac{4M}{\sin^2 \delta}\right)\frac{\pi}{2} = \frac{2M}{n \sin^2 \delta}. \tag{60}$$

By (59) and (60),

$$|\sigma_n(x) - f(x)| < \frac{\varepsilon}{2} + \frac{2M}{n \sin^2 \delta}.$$

But, for sufficiently large n, we have

$$\frac{2M}{n \sin^2 \delta} < \frac{\varepsilon}{2}$$

and therefore

$$|\sigma_n(x) - f(x)| < \varepsilon. \qquad \square$$

Corollary 1 (Second Weierstrass Approximation Theorem) Every real-valued continuous function on R^1 having period 2π can be approximated uniformly by trigonometric polynomials.

Remark We noted this result in the Remark following Proposition 18; by Fejér's Theorem we see that the Fejér sums will effect this uniform approximation.

Corollary 2 If all Fourier coefficients of a continuous function f of period 2π vanish, then $f = 0$.

Proof If all Fourier coefficients of f vanish, then the Fejér sums $\sigma_n(x)$ vanish as well and therefore

$$f(x) = \lim_{n\to\infty} \sigma_n(x) = 0$$

for all real x. $\square$

Corollary 3 If $|f(x)| \le M$ for all $x \in [0,2\pi]$, then

$$|\sigma_n(x)| \le \frac{2M}{n\pi}\int_0^{\pi/2}\left(\frac{\sin nt}{\sin t}\right)^2 dt$$

and so

$$|\sigma_n(x)| \le M$$

[by (57)].

Exercises

1. Let $(u_n)_{n=1}^{\infty}$ and $(v_n)_{n=1}^{\infty}$ be sequences of positive real numbers, the first sequence being bounded above, and in the second $v_n \to 1$ as $n \to \infty$. Show that

$$\overline{\lim_{n\to\infty}}\ u_n v_n = \overline{\lim_{n\to\infty}}\ u_n.$$

Solution It is clear that the sequence $(u_n v_n)_{n=1}^{\infty}$ is a positive sequence bounded above. If possible, let

$$\overline{\lim_{n\to\infty}}\ u_n v_n = s' > s = \overline{\lim_{n\to\infty}}\ u_n.$$

Take $2\varepsilon = s' - s$, and let

$$K = \sup\{u_n : n = 1, 2, \ldots\}.$$

Since $v_n \to 1$, as $n \to \infty$, there is a positive integer k_1 such that

$$|v_n - 1| < \frac{\varepsilon}{2K} \quad \text{when} \quad n \ge k_1.$$

Thus

$$|u_n v_n - u_n| = |u_n||v_n - 1| < K\,\frac{\varepsilon}{2K} = \frac{\varepsilon}{2} \quad \text{when} \quad n \ge k_1.$$

Therefore

$$u_n v_n < u_n + \frac{\varepsilon}{2} \quad \text{when} \quad n \ge k_1.$$

But, since

$$\overline{\lim_{n\to\infty}}\ u_n = s,$$

there is a positive integer k_2 such that

$$u_n < s + \frac{\varepsilon}{2} \quad \text{when} \quad n \ge k_2.$$

Therefore $u_n v_n < s + \varepsilon$ when $n \ge k$, where k is the larger of the integers k_1 and k_2. But, since

$$\overline{\lim_{n\to\infty}}\ u_n v_n = s',$$

we know that $u_n v_n > s' - \varepsilon$, for an infinite number of values of n. Therefore s' cannot be greater than s.

Similarly it can be shown that s' is not less than s; hence s' = s. □

2. Consider the series

$$\sum_{n=1}^{\infty} \tau(n)\, x^n,$$

where x is positive and $\tau(n)$ denotes the number of divisors of the positive integer n. Show that

$$x \le \sqrt[n]{\tau(n)}\, x \le \sqrt[n]{n}\, x$$

and deduce that the series converges for $0 \le x < 1$ and diverges for $x \ge 1$.

3. Show that the function, for $x > 1$,

$$\zeta(x) = \sum_{n=1}^{\infty} \frac{1}{n^x}$$

has derivatives of any order.

4. For $0 \le x \le 1$, let

$$f(x) = \sum_{n=1}^{\infty} \frac{x^n}{n^2}.$$

Show that, for $0 < x < 1$,

$$f(x) + f(1 - x) + (\log x)[\log(1 - x)] = C,$$

where

$$C = \lim_{x \to 1-} f(x) = f(1) = \sum_{n=1}^{\infty} \frac{1}{n^2}.$$

Hint Note that

$$f'(x) = \sum_{n=1}^{\infty} \frac{x^n}{n^2} = -\frac{1}{x} \log(1 - x)$$

and

$$f'(1 - x) = -\frac{1}{1 - x} \log x;$$

thus

$$f'(x) - f'(1 - x) + \frac{1}{x} \log(1 - x) - \frac{1}{1 - x} \log x = 0.$$

5. Let f be a real-valued function on R^1 with period 2π and have finite total variation v over $(0,2\pi)$. Show that the Fourier coefficients a_k and b_k, for $k = 1, 2, \ldots$, satisfy the inequalities

$$|a_k| \le \frac{v}{k} \quad \text{and} \quad |b_k| \le \frac{v}{k}.$$

6. Let f be as in Proposition 38 and assume moreover that f is continuous on $(0,2\pi)$. Let $[a,b]$ be a closed subinterval of $(0,2\pi)$. Is it true that $s_n \to f$ uniformly on $[a,b]$?

7. If $t \neq 2k\pi$, where $k = 0, 1, 2, \ldots$, show that

$$\sum_{k=1}^{n} \sin kt = \frac{\sin (n + \frac{1}{2})t - \sin t/2}{2 \sin t/2}.$$

8. Let $p, q \ge 0$ and either p or q be larger than 1. Show that the function

$$f(x) = \sum_{n=1}^{\infty} \frac{x}{n^p + x^2 n^q}$$

is continuous for any $x \in R^1$, except $p < 1$, $q > 1$ but $p + q \le 2$; in the latter case f is not continuous at $x = 0$.

9. Let (x) denote the positive or negative excess of x above the nearest integer to it, and if x is halfway between two successive integers, let (x) = 0. Let a function f be defined for the interval (0,1) as the limit of

$$\frac{(x)}{1} + \frac{(2x)}{4} + \frac{(3x)}{9} + \cdots + \frac{(nx)}{n^2}$$

as $n \to \infty$. Observe that the right- and left-hand limits of f at $x = p/(2n)$, with p and 2n relatively prime, satisfy

$$f(x+) = f(x) - \frac{1}{2n^2}\left(1 + \frac{1}{9} + \frac{1}{25} + \cdots\right) = f(x) - \frac{\pi^2}{16n^2},$$

$$f(x-) = f(x) + \frac{1}{2n^2}\left(1 + \frac{1}{9} + \frac{1}{25} + \cdots\right) = f(x) + \frac{\pi^2}{16n^2},$$

and that everywhere else $f(x+) = f(x)$, $f(x-) = f(x)$. Note that the set K of points at which the jump is $\ge k$ is finite for each positive value of k; the number of points of K is the number of irreducible proper fractions having even denominators 2n, such that $\pi^2/8n^2 \ge k$. The set of all points of discontinuity of f is such that every open subinterval of (0,1) contains discontinuities of f.

Show that f is Riemann integrable over (0,1).

10. *Threorem of Arzelà:* Let $(f_n)_{n=1}^{\infty}$ be a uniformly bounded sequence of functions on a closed bounded interval [a,b] with pointwise limit f on [a,b]. If f and all f_n (n = 1, 2, ...) are Riemann integrable over [a,b], then

$$\lim_{n \to \infty} \int_a^b f_n(x)\, dx = \int_a^b f(x)\, dx.$$

The following elementary proof of the Theorem of Arzelà is due to F. Hausdorff [Math. Zeitschrift, 26 (1927)]. The special feature of this proof is that it stays completely within Riemann's integration theory and does not make use of Lebesgue's integration theory.

We divide the proof into five steps.

I. Let the functions $(g_n)_{n=1}^{\infty}$ be continuous on [a,b] and for each x in [a,b], let $g_n(x) \geq g_{n+1}(x) \to 0$. Then $g_n \to 0$ uniformly on [a,b].

Proof Note that $M_n = \sup\{g_n(x) : x \in [a,b]\}$ is assumed on at least one point x_n in [a,b] and we have

$$M_{n+1} = g_{n+1}(x_{n+1}) \leq g_n(x_{n+1}) \leq M_n.$$

The sequence of numbers $M_n \geq 0$ is therefore monotonically decreasing and thus $M_n \to M$ with $M \geq 0$. Let t be a limit point of the set of all x_n. Then there is a subsequence $x_n' \to t$. Then for $m < n'$

$$0 \leq M \leq M_{n'} = g_{n'}(x_{n'}) \leq g_m(x_{n'}),$$

so that, letting $n' \to \infty$, we find for each m,

$$0 \leq M \leq g_m(t).$$

Since $g_m(t) \to 0$ as $m \to \infty$, it follows that $M = 0$. This proves I. □

II. Let h be a bounded function on [a,b], but not necessarily Riemann integrable. Then

$$\underline{J}(h) = \underline{\int_a^b} h(x)\, dx$$

is equal to the supremum of all

$$J(g) = \int_a^b g(x)\, dx,$$

where g is continuous on [a,b] and

$$\gamma \leq g(x) \leq h(x).$$

Here γ denotes the infimum of h(x) on [a,b].

Proof It is easy to see that $J(g) \leq \underline{J}(h)$. On the other hand, let $\varepsilon > 0$ be given and consider the partition

$$P = \{a = x_0 < x_1 < \cdots < x_n = b\}$$

of the interval [a,b]. We first form the "step function"

$$h^*(x) = m_k(h) = \inf\{h(x) : x \in [x_{k-1}, x_k]\}$$

in the subintervals (x_{k-1}, x_k). If the mesh of the partition P is sufficiently small, we can approximate this step function from below so closely by a continuous function $g(x) \le h(x)$ that

$$J(g) \ge \underline{J}(h) - \varepsilon.$$

We can obviously choose $g(x) \ge \gamma$, which proves II. □

III. Let h_1 and h_2 be bounded functions on $[a,b]$ and let

$$h(x) = \min\{h_1(x), h_2(x)\},$$

$$H(x) = \max\{h_1(x), h_2(x)\},$$

so that $h_1(x) + h_2(x) = h(x) + H(x)$. Then

$$\underline{J}(h_1) + \underline{J}(h_2) \le \underline{J}(h) + \underline{J}(H).$$

Proof Indeed, for any subinterval (x_{k-1}, x_k) of $[a,b]$ we have

$$m_k(h) = \min\{m_k(h_1), m_k(h_2)\},$$

$$m_k(H) \ge \max\{m_k(h_1), m_k(h_2)\};$$

here $m_k(h_j)$ denotes the infimum of h_j on the subinterval $[x_{k-1}, x_k]$ for $j = 1, 2$. From this follows III. □

IV. Let $(h_n)_{n=1}^{\infty}$ be a sequence of bounded functions on $[a,b]$, and furthermore for each x in $[a,b]$, let $h_n(x) \ge h_{n+1}(x) \to 0$. Then

$$\underline{J}(h_n) \to 0 \quad \text{as} \quad n \to \infty.$$

Proof Let $\varepsilon > 0$ be given. It suffices to determine a sequence of continuous functions $(g_n)_{n=1}^{\infty}$ for which

$$g_n(x) \le h_n(x), \quad g_n(x) \ge g_{n+1}(x) \to 0,$$
$$J(g_n) > \underline{J}(h_n) - \varepsilon \tag{E.1}$$

in $[a,b]$. For the g_n are then uniformly convergent to 0 (by I), so that $J(g_n) \to 0$ and hence

$$\overline{\lim_{n\to\infty}}\ \underline{J}(h_n) \le \varepsilon.$$

On the other hand, each $\underline{J}(h_n) \ge 0$. The sequence $(g_n)_{n=1}^{\infty}$ can now be determined inductively. First let a $g_1(x) \ge 0$ be given satisfying (E.1). This is possible by II. Next, let us suppose that $g_k(x) \ge 0$ has been determined for all $k \le n$, and we set

$$\begin{aligned} h(x) &= \min\{h_{n+1}(x),\ g_n(x)\}, \\ H(x) &= \max\{h_{n+1}(x),\ g_n(x)\}. \end{aligned} \tag{E.2}$$

Then, using III and $H(x) \le \max\{h_n(x),\ g_n(x)\} = h_n(x)$, we have

$$\begin{aligned} \underline{J}(h) + \underline{J}(h_n) &\ge \underline{J}(h) + \underline{J}(H) \ge \underline{J}(h_{n+1}) + \underline{J}(g_n) \\ &> \underline{J}(h_{n+1}) + \underline{J}(h_n) - \varepsilon, \\ \underline{J}(h) &> \underline{J}(h_{n+1}) - \varepsilon. \end{aligned} \tag{E.3}$$

We now select a continuous function $g_{n+1}(x) \ge 0$ for which

$$g_{n+1}(x) \le h(x), \quad J(g_{n+1}) > \underline{J}(h_{n+1}) - \varepsilon. \tag{E.4}$$

This is possible by (E.3) and II. For this g_{n+1}, by the definition of h_n, we have in fact

$$0 \le g_{n+1}(x) \le h_{n+1}(x), \quad g_{n+1}(x) \le g_n(x).$$

The first inequality implies that $g_{n+1}(x) \to 0$, so that condition (E.1) is satisfied. This proves IV. □

V. Proof of the Theorem of Arzelà: We can assume that $f = 0$, since otherwise we need only consider the sequence $(f_n - f)_{n=1}^{\infty}$. In $[a,b]$ we let

$$h_n(x) = \sup\{f_n(x),\ f_{n+1}(x),\ \ldots\}.$$

Since $(f_n)_{n=1}^{\infty}$ is uniformly bounded on $[a,b]$, h_n is bounded on $[a,b]$, and obviously

$$h_n(x) \ge h_{n+1}(x).$$

Since $f_n \to f$ pointwise on $[a,b]$ and $f = 0$, we see that $h_n(x) \to 0$ on $[a,b]$. Thus, by IV, we conclude that $\underline{J}(h_n) \to 0$. Since $f_n(x) \le h_n(x)$, we have

$$J(f_n) \le \underline{J}(h_n)$$

so that

$$\overline{\lim_{n\to\infty}}\; J(f_n) \le 0.$$

Since the sequence $(-f_n)_{n=1}^{\infty}$ satisfies the same hypothesis, we have

$$\overline{\lim_{n\to\infty}}\; J(-f_n) = -\underline{\lim_{n\to\infty}}\; J(f_n) \le 0;$$

$$\underline{\lim_{n\to\infty}}\; J(f_n) \ge 0.$$

However, this proves the Theorem of Arzelà. □

11. (a) Let a be a real number, but not an integer, and consider the function cos ax in $[-\pi,\pi]$. Noting that

$$\frac{1}{\pi}\int_0^{\pi} \cos ax\, dx = \frac{\sin a\pi}{a\pi}$$

and, for $n > 0$,

$$\frac{1}{\pi}\int_0^{\pi} [\cos(a+n)x + \cos(a-n)x]\, dx = (-1)^n \frac{2a}{a^2 - n^2}\,\frac{\sin a\pi}{\pi}$$

we get, for $-\pi \le x \le \pi$,

$$\frac{\pi}{2}\,\frac{\cos ax}{\sin a\pi} = \frac{1}{a\pi} + \sum_{n=1}^{\infty} \frac{a \cos nx}{a^2 - n^2}.$$

Setting $x = 0$, this gives

$$\frac{1}{\sin a\pi} = \frac{1}{a\pi} + \sum_{n=1}^{\infty} \frac{2a\pi}{(a\pi)^2 - (n\pi)^2};$$

substituting $t = a\pi$, we obtain what is known as the *partial fraction decomposition of* $(\sin t)^{-1}$, namely,

$$\frac{1}{\sin t} = \frac{1}{t} + \sum_{n=1}^{\infty} (-1)^n \frac{2t}{t^2 - (n\pi)^2}$$

$$= \frac{1}{t} + \sum_{n=1}^{\infty} (-1)^n \left(\frac{1}{t - n\pi} + \frac{1}{t + n\pi}\right),$$

where t is not a multiple of π.

(b) We now consider the integral

$$I = \int_0^{\infty} \frac{\sin x}{x}\, dx$$

and write

$$I = \sum_{k=0}^{\infty} \int_{k\pi/2}^{(k+1)\pi/2} \frac{\sin x}{x}\, dx.$$

For $k = 2m$ we consider the substitution $x = m\pi + t$ and for $k = 2m - 1$ we consider the substitution $x = m\pi - t$. This leads to

$$\int_{2m\pi/2}^{(2m+1)\pi/2} \frac{\sin x}{x}\, dx = (-1)^m \int_0^{\pi/2} \frac{\sin t}{m\pi + t}\, dt$$

and

$$\int_{(2m-1)\pi/2}^{2m\pi/2} \frac{\sin x}{x}\, dx = (-1)^{m-1} \int_0^{\pi/2} \frac{\sin t}{m\pi - t}\, dt.$$

It follows that

$$I = \int_0^{\pi/2} \frac{\sin t}{t}\, dt + \sum_{m=1}^{\infty} \int_0^{\pi/2} (-1)^m \left(\frac{1}{t - m\pi} + \frac{1}{t + m\pi}\right) \sin t\, dt.$$

But the series

$$\sum_{m=1}^{\infty} (-1)^m \left(\frac{1}{t - m\pi} + \frac{1}{t + m\pi}\right) \sin t$$

converges uniformly in the interval $0 \le t \le \pi/2$ because it is majorized by the convergent series

$$\frac{1}{\pi}\sum_{m=1}^{\infty}\frac{1}{m^2 - \frac{1}{4}}.$$

We therefore have

$$I = \int_0^{\pi/2} \sin t \left\{\frac{1}{t} + \sum_{m=1}^{\infty} (-1)^m \left(\frac{1}{t - m\pi} + \frac{1}{t + m\pi}\right)\right\} dt.$$

By part (a),

$$I = \int_0^{\pi/2} \sin t \frac{1}{\sin t} dt = \int_0^{\pi/2} dt = \frac{\pi}{2}.$$

(This elegant calculation of I is due to N. I. Lobatshewski.)

12. Let $(f_n)_{n=1}^{\infty}$ be a differentiable sequence of functions on [0,1] such that the sequence $\{f_n'(x_0)\}_{n=1}^{\infty}$ converges for some x_0 in [0,1]. Show that if the derived sequence $(f_n')_{n=1}^{\infty}$ is uniformly bounded on [0,1], then $(f_n)_{n=1}^{\infty}$ contains a uniformly convergent subsequence on [0,1].

13. Let $(f_n)_{n=1}^{\infty}$ be a uniformly bounded sequence of functions which are Riemann integrable on [a,b], and let

$$F_n(x) = \int_a^x f_n(t)\, dt.$$

Show that there exists a subsequence $(F_{n_k})_{k=1}^{\infty}$ of $(F_n)_{n=1}^{\infty}$ which converges uniformly on [a,b].

14. Let

$$f_n(x) = \frac{x^2}{x^2 + (1 - nx)^2}$$

for $0 \le x \le 1$ and n = 1, 2, ... Show that $(f_n)_{n=1}^{\infty}$ contains no subsequence that converges uniformly on [0,1].

Hint Note that $\lim_{n\to\infty} f_n(x) = 0$ for $0 \le x \le 1$, but

$$f_n\left(\frac{1}{n}\right) = 1 \quad \text{for} \quad n = 1, 2, \ldots.$$

15. Show that a uniformly convergent sequence of bounded functions is uniformly bounded.

16. If $(f_n)_{n=1}^{\infty}$ and $(g_n)_{n=1}^{\infty}$ converge uniformly on E, show that $(f_n + g_n)_{n=1}^{\infty}$ converges uniformly on E. If, moreover, $(f_n)_{n=1}^{\infty}$ and $(g_n)_{n=1}^{\infty}$ are sequences of bounded functions, show that $(f_n g_n)_{n=1}^{\infty}$ converges uniformly on E.

17. Let $(f_n)_{n=1}^{\infty}$ be a sequence of monotonic functions on $[0,1]$, and $(f_n)_{n=1}^{\infty}$ converge pointwise to a continuous function f on $[0,1]$. Verify that the convergence is uniform on $[0,1]$.

18. Suppose $(f_n)_{n=1}^{\infty}$ is an equicontinuous sequence of functions on $[0,1]$, and $(f_n)_{n=1}^{\infty}$ converges pointwise on $[0,1]$. Verify that $(f_n)_{n=1}^{\infty}$ converges uniformly on $[0,1]$.

19. Let $(f_n)_{n=1}^{\infty}$ be a sequence of continuous functions which converges uniformly to a function f on a set E of real numbers. Show that $\lim_{n\to\infty} f_n(x_n) = f(x)$ for every sequence of points $x_n \in E$ such that $x_n \to x$ as $n \to \infty$, and $x \in E$. Is the converse of this true?

20. Show that the series

$$\sum_{n=1}^{\infty} (-1)^n \frac{x^2 + n}{n^2}$$

converges uniformly in every bounded interval, but does not converge absolutely for any value of x.

21. Let

$$f(x) = \sum_{n=1}^{\infty} \frac{1}{1 + n^2 x}.$$

(a) For what values of x does the series converge absolutely?

(b) On what intervals does it converge uniformly?

(c) On what intervals does it fail to converge uniformly?

(d) Is f continuous wherever the series converges?

(e) Is f bounded?

22. A sequence (c_n), $n = 0, \pm 1, \pm 2, \ldots$ of complex numbers (be the sequence finite or not) is called *positive definite* if for any finite set of complex numbers $\lambda_1, \ldots, \lambda_p$ and every subset of p elements of the sequence (c_n) the following inequality holds:

$$\sum_{j=1}^{p} \sum_{k=1}^{p} c_{j-k} \lambda_j \bar{\lambda}_k \geq 0$$

(here $\bar{\lambda}_k$ denotes the complex conjugate of λ_k).

From the foregoing definition one can deduce the following basic relations for any positive definite sequence (c_n):

(i) $c_0 \geq 0$;

(ii) $c_{-n} = \bar{c}_n$;

(iii) $|c_n| \leq c_0$;

(iv) $|c_n - c_{n+m}|^2 \leq 2c_0(c_0 - \mathrm{Re}\{c_m\})$

(here $\mathrm{Re}\{c_m\}$ denotes the real part of c_m).

The student is invited to follow the subsequent proof of a theorem, due to G. Herglotz, concerning the Fourier-Stieltjes integral representation for positive definite sequences.

Theorem A sequence (c_n) is positive definite if and only if it has the representation

$$c_n = \int_{-\pi}^{\pi} e^{inx} \, dg(x) \quad (n = 0, \pm 1, \pm 2, \ldots),$$

where g is a monotonically increasing function of finite variation.

Proof Let the sequence (c_n) be positive definite. Then we have in particular that

$$\sum_{j=1}^{p} \sum_{k=1}^{p} c_{j-k} \, e^{-i(j-k)x} \geq 0.$$

We now observe that

$$\sum_{j=1}^{p} \sum_{k=1}^{p} c_{j-k} e^{-i(j-k)x} = p \sum_{n=-p+1}^{p-1} \left(1 - \frac{|n|}{p}\right) c_n e^{-inx}.$$

We let

$$Q_p(x) = \sum_{n=-p+1}^{p-1} \left(1 - \frac{|n|}{p}\right) c_n e^{-inx}.$$

Since

$$\int_{-\pi}^{\pi} e^{-(n-m)x} dx = 0 \quad \text{for } n \neq m,$$

$$= 2\pi \quad \text{for } n = m,$$

we see that

$$\frac{1}{2\pi} \int_{-\pi}^{\pi} Q_p(x) e^{inx} dx = \left(1 - \frac{|n|}{p}\right) c_n.$$

We put

$$\frac{1}{2\pi} \int_{-\pi}^{\pi} Q_p(x) e^{inx} dx = \int_{-\pi}^{\pi} e^{inx} dg_p(x).$$

Then

$$g_p(x) = \frac{1}{2\pi} \int_{-\pi}^{x} Q_p(y) dy;$$

g_p is a monotonically increasing function and its total variation over $[-\pi,\pi]$ is

$$\int_{-\pi}^{\pi} dg_p(x) dx = c_0.$$

By Helly's Selection Principle (see Proposition 52 of Chapter 3), we can find a sequence $p_k \to \infty$ (for $k \to \infty$) such that $(g_{p_k})_{k=1}^{\infty}$ converges to a limit function. We denote this limit function by g. The function g is also monotonically increasing and of finite variation. Using Proposition 8, we obtain

$$\lim_{k\to\infty} \int_{-\pi}^{\pi} e^{inx}\, dg_{p_k}(x) = \int_{-\pi}^{\pi} e^{inx}\, dg(x).$$

Conversely, if a sequence permits the type of representation under consideration, then

$$\begin{aligned}\sum_{j=1}^{p}\sum_{k=1}^{p} c_{j-k}\,\lambda_j\,\overline{\lambda}_k &= \sum_{j=1}^{p}\sum_{k=1}^{p}\left[\int_{-\pi}^{\pi} e^{i(j-k)x}\, dg(x)\right]\lambda_j\,\overline{\lambda}_k \\ &= \int_{-\pi}^{\pi}\left\{\sum_{j=1}^{p} e^{ijx}\,\lambda_j\right\}\left\{\sum_{k=1}^{p} e^{-ikx}\,\overline{\lambda}_k\right\} dg(x) \\ &= \int_{-\pi}^{\pi}\left|\sum_{j=1}^{p} e^{ijx}\,\lambda_j\right|^2 dg(x) \\ &\ge 0,\end{aligned}$$

and the proof is complete. □

23. We now take up Landau's proof of the Weierstrass Approximation Theorem (see Proposition 15).

Without loss of generality, we may assume that $[a,b] = [0,1]$. We also assume that $f(0) = f(1) = 0$; for, if the theorem is proved for this case, consider

$$g(x) = f(x) - f(0) - x\{f(1) - f(0)\}$$

for $0 \le x \le 1$. Here $g(0) = g(1) = 0$, and if g admits uniform approximation by polynomials, so does f, because $f - g$ is a polynomial.

Furthermore, we assume that $f(x) = 0$ when $x \notin [0,1]$. Then f is uniformly continuous on R^1.

We put, for $n = 1, 2, \ldots,$

$$Q_n(x) = c_n(1 - x^2)^n,$$

where c_n is so chosen that

$$\int_{-1}^{1} Q_n(x)\, dx = 1 \tag{E.5}$$

for n = 1, 2, ... But $Q_n(-x) = Q_n(x)$ and $(1 - x^2)^n \geq 1 - nx^2$; the latter can be seen from the fact that

$$(1 - x^2)^n + nx^2 - 1$$

is zero at x = 0 and has a positive derivative throughout (0,1). Thus

$$\int_{-1}^{1} (1 - x^2)^n\, dx = 2\int_{0}^{1} (1 - x^2)^n\, dx \geq 2\int_{0}^{1/\sqrt{n}} (1 - x^2)^n\, dx$$

$$\geq 2\int_{0}^{1/\sqrt{n}} (1 - nx^2)\, dx = \frac{4}{3\sqrt{n}} > \frac{1}{\sqrt{n}}.$$

It therefore follows from (E.5) that

$$c_n < \sqrt{n}.$$

For any $\delta > 0$, this implies

$$Q_n(x) \leq \sqrt{n}(1 - \delta^2)^n \tag{E.6}$$

when $\delta \leq |x| \leq 1$. But $\Sigma_{n=1}^{\infty} \sqrt{n}(1 - \delta^2)^n$ converges by Proposition 73 of Chapter 1 for $0 < \delta \leq 1$; hence $n(1 - \delta^2)^n \to 0$. Thus $Q_n \to 0$ uniformly in $\delta \leq |x| \leq 1$.

We put, for $x \in [0,1]$,

$$P_n(x) = \int_{-1}^{1} f(x + t)\, Q_n(t)\, dt.$$

Our assumptions about f show, by a simple change of variable, that

$$P_n(x) = \int_{-x}^{1-x} f(x + t)\, Q_n(t)\, dt = \int_{0}^{1} f(t)\, Q_n(t - x)\, dt,$$

where the last integral is evidently a polynomial in x. Thus $(P_n)_{n=1}^{\infty}$ is a sequence of polynomials.

Given $\varepsilon > 0$, we pick $\delta > 0$ such that $|y - x| < \delta$ implies

$$|f(y) - f(x)| < \frac{\varepsilon}{2};$$

this is possible since f is uniformly continuous on R^1. Let

$$M = \sup\{|f(x)| : x \in R^1\}.$$

Since $Q_n(x) \geq 0$ and because of (E.5) and (E.6), we obtain, for $0 \leq x \leq 1$.

$$\begin{aligned}|P_n(x) - f(x)| &= \left|\int_{-1}^{1} [f(x + t) - f(x)]\, Q_n(t)\, dt\right| \\ &\leq \int_{-1}^{1} |f(x + t) - f(x)|\, Q_n(t)\, dt \\ &\leq 2M \int_{-1}^{-\delta} Q_n(t)\, dt + \frac{\varepsilon}{2}\int_{-\delta}^{\delta} Q_n(t)\, dt \\ &\quad + 2M \int_{\delta}^{1} Q_n(t)\, dt \\ &\leq 4M\sqrt{n}(1 - \delta^2)^n + \frac{\varepsilon}{2} < \varepsilon\end{aligned}$$

for all sufficiently large n, finishing what we set out to do. □

Remark Another way of establishing the inequality

$$\int_0^1 (1 - x^2)^n\, dx \geq \frac{2}{3\sqrt{n}} \quad (n = 1, 2, \ldots)$$

is to use that

$$\begin{aligned}\int_0^1 (1 - x^2)^{n+1}\, dx &= \int_0^1 (1 - x^2)(1 - x^2)^n\, dx \\ &= \frac{2n + 2}{2n + 3}\int_0^1 (1 - x^2)^n\, dx \\ &= \int_0^1 (1 - x^2)^n\, dx - \int_0^1 x\{x(1 - x^2)^n\}\, dx\end{aligned}$$

(which follows from integration by parts); we leave the details as an exercise.

24. Let $\Sigma_{n=0}^{\infty} a_n x^n$ and $\Sigma_{n=0}^{\infty} b_n x^n$ be two power series representing the functions f(x) and g(x), respectively, within their intervals of convergence (-A,A) and (-B,B). Show that if

$$c_n = \sum_{k=0}^{n} a_k b_{n-k} \quad (n = 0, 1, 2, \ldots),$$

then the power series $\Sigma_{n=0}^{\infty} c_n x^n$ represents f(x) g(x) for all points x interior to both intervals of convergence (-A,A) and (-B,B).

25. Let $\Sigma_{n=0}^{\infty} a_{nm} x^n$, with m = 0, 1, 2, ..., be a given sequence of power series. We form the double series

$$\sum_{m=0}^{\infty} \left(\sum_{n=0}^{\infty} a_{nm} x^n \right), \tag{E.7}$$

Show: If for a certain value of x the double series formed by taking the absolute value is convergent, then the series (E.7) converges also and its sum A(x) is given by

$$A(x) = \sum_{n=0}^{\infty} A_n x^n,$$

where

$$A_n = \sum_{m=0}^{\infty} a_{nm} \quad (n = 0, 1, 2, \ldots).$$

Hint Use the lemma preceding Proposition 31.

26. Consider the expansion of the function

$$\frac{1}{\sqrt{1 - 2x\alpha + \alpha^2}} = [1 + (\alpha^2 - 2x\alpha)]^{-1/2}$$

into powers of α with x being arbitrary but fixed. Such an expansion is possible when

$$|\alpha|^2 + 2|x||\alpha| < 1.$$

The coefficient of α^n $(n \geq 1)$ is a certain polynomial $P_n = P_n(x)$ of degree n, so that

$$\frac{1}{\sqrt{1 - 2x\alpha + \alpha^2}} = 1 + P_1\alpha + P_2\alpha^2 + \cdots + P_n\alpha^n + \cdots. \qquad \text{(E.8)}$$

To determine the coefficients we differentiate with respect to α:

$$\frac{x - \alpha}{\left(\sqrt{1 - 2x\alpha + \alpha^2}\right)^3} = P_1 + 2P_2\alpha + \cdots + nP_n\alpha^{n-1} + \cdots.$$

Comparison with (E.8) gives

$$(1 - 2x\alpha + \alpha^2)(P_1 + P_2\alpha + \cdots + nP_n\alpha^{n-1} + \cdots)$$

$$= (x - \alpha)(1 + P_1\alpha + P_2\alpha^2 + \cdots + P_n\alpha^n + \cdots).$$

Comparing coefficients of like powers of α, we get

$$P_1 = x, \quad 2P_2 - 2xP_1 = -1 + xP_1,$$

that is,

$$P_2 = \frac{3x^2 - 1}{2}.$$

In general, we obtain

$$(n + 1)P_{n+1} - 2nxP_n + (n - 1)P_{n-1} = xP_n - P_{n-1}$$

or

$$(n + 1)P_{n+1} - (2n + 1)xP_n + nP_{n-1} = 0.$$

We see that

$$\frac{1}{\sqrt{1 - 2x\alpha + \alpha^2}}$$

is the *generating function* for the Legendre polynomials

$$P_n(x) = c_n \frac{d^n(x^2 - 1)}{dx^n}, \quad \text{where} \quad c_n = \frac{1}{2^n n!}.$$

Note that these polynomials were already encountered in Exercises 19 and 24 at the end of Chapter 3.

The details of justifying the various steps are left to the student.

27. Show that

$$\frac{\sin x}{1} + \frac{\sin 2x}{2} + \frac{\sin 3x}{3} + \cdots$$

converges uniformly on $[\delta, \pi/2]$ for any $\delta > 0$.

28. Let $b_1 \geq b_2 \geq \cdots \geq b_n \geq 0$. Show that a necessary and sufficient condition that the series $\Sigma_{n=1}^{\infty} b_n \sin nx$ should be uniformly convergent throughout any interval is that $nb_n \to 0$ as $n \to \infty$.

Solution To see that the condition is necessary, observe that, if $x = \pi/(2p)$, and $n = [\frac{1}{2}p + 1]$, where $[x]$ means the integral part of x,

$$b_n \sin nx + b_{n+1} \sin(n+1)x + \cdots + b_p \sin px$$

$$> b_p(\sin nx + \cdots + \sin px) > b_p(\tfrac{1}{2}p - 1)\sin \tfrac{1}{4}\pi,$$

since there are at least $\frac{1}{2}p - 1$ terms in the bracket, in each of which $mx > \frac{1}{4}\pi$. Since the given series is uniformly convergent in an interval including the origin, the left-hand side of the above inequality tends to zero as $p \to \infty$. Hence $pb_p \to 0$.

Next we show the sufficiency of the condition. It will be enough to consider the interval $[0,\pi]$, since each term of the series is odd and has the period 2π. Consider the sum

$$s_{n,p} = b_n \sin nx + \cdots + b_p \sin px,$$

where now n and p are unconnected. Let $\mu_n = \sup\{mb_m : m \geq n\}$, so that $\mu_n \to 0$. If $x \geq \pi/n$, we apply Abel's Lemma (see Exercise 15, part (a), at the end of Chapter 1). We have

$$|\sin nx + \cdots + \sin rx| = \left|\frac{\cos(n - \frac{1}{2})x - \cos(r + \frac{1}{2})x}{2 \sin \frac{1}{2}x}\right| \leq \frac{1}{\sin \frac{1}{2}x}$$

for all values of n and r, and, since $(\sin \theta)/\theta$ is steadily decreasing for $0 < \theta < \frac{1}{2}\pi$,

$$\frac{1}{\sin \frac{1}{2}x} \le \frac{\pi}{x},$$

and we deduce that

$$|s_{n,p}| \le \frac{b_n \pi}{x} \le nb_n \le \mu_n.$$

If $x \le \pi/p$, we have, since $\sin \theta < \theta$,

$$|s_{n,p}| \le b_n nx + \cdots + b_p px \le p\mu_n x \le \pi\mu_n.$$

If $\pi/p < x < \pi/n$, we combine the two arguments. We have

$$|s_{n,p}| \le |s_{n,k}| + |s_{k+1,p}|,$$

and, applying Abel's Lemma to the second part, and the other method to the first part, obtain

$$|s_{n,p}| \le k\mu_n x + b_{k+1}\pi/x \le \mu_n[kx + \pi/\{(k + 1)x\}].$$

Taking $k = [\pi/x]$, we have

$$|s_{n,p}| \le \mu_n(\pi + 1).$$

Hence in any case

$$|s_{n,p}| \le A\mu_n,$$

and, since $\mu_n \to 0$, the proof is complete. □

29. Justify the following calculations:

(i) $$\int_0^1 \frac{\log x}{1 - x}\, dx = \sum_{n=0}^{\infty} \int_0^1 x^n \log x \, dx = -\sum_{n=0}^{\infty} \frac{1}{(n + 1)^2} = -\frac{\pi^2}{6}.$$

(ii) $$\int_0^1 \log \frac{1 + x}{1 - x} \frac{dx}{x} = 2 \sum_{n=0}^{\infty} \frac{1}{(2n - 1)^2} = \frac{\pi^2}{4}.$$

30. Let f be a continuous function on a closed bounded interval [a,b]. Show that f permits uniform approximation by a polynomial with rational coefficients.

Solution Let $\varepsilon > 0$ be given. By Proposition 15 there exists a polynomial p such that

$$\sup\{|f(t) - p(t)|: t \in [a,b]\} < \frac{\varepsilon}{2},$$

where

$$p(t) = a_0 + a_1 t + \cdots + a_n t^n$$

and $a_0, a_1, \ldots, a_n$ are real numbers. Let $c = \max\{|a|,|b|\}$; for each j, $0 \le j \le n$, choose a rational number b_j such that

$$|b_j - a_j| < \frac{\varepsilon}{2(n+1)c^j}$$

and let

$$q(t) = b_0 + b_1 t + \cdots + b_n t^n.$$

Then

$$|p(t) - q(t)| = \left|\sum_{j=0}^{n} (b_j - a_j)t^j\right| < \frac{\varepsilon}{2}$$

and so

$$\sup\{|p(t) - q(t)|: t \in [a,b]\} < \frac{\varepsilon}{2}.$$

Hence $\sup\{|f(t) - q(t)|: t \in [a,b]\} < \varepsilon$. □

31. Let a be any real number; put $(a)_0 = 1$ and

$$(a)_k = a(a-1)(a-2)\cdots(a-k+1)$$

for $k = 1, 2, \ldots$. For $x \in (-1,1)$, the series

$$\sum_{k=0}^{\infty} (a)_k \frac{x^k}{k!} \tag{E.9}$$

converges absolutely for any real number a. Fix x in $(-1,1)$ and denote (E.9) by $g(a)$. Show that g is a continuous function on the real line R^1 and satisfies the functional equation

$$g(a)\,g(b) = g(a+b).$$

(By Proposition 54 of Chapter 2 we therefore have that $g(a) = c^a$, where $c = g(1) = 1 + x$, and so

$$(1 + x)^a = \sum_{k=0}^{\infty} (a)_k \frac{x^k}{k!} = \sum_{k=0}^{\infty} \binom{a}{k} x^k.)$$

32. The Möbius function μ is defined as follows:

$\mu(n) = 1$ if $n = 1$,

$\mu(n) = (-1)^k$ if n is the product of k distinct primes,

$\mu(n) = 0$ if n is divisible by a square number > 1.

For example: $\mu(1) = 1$, $\mu(2) = -1$, $\mu(3) = -1$, $\mu(4) = 0$, $\mu(5) = -1$, $\mu(6) = 1, \ldots$.

(a) Consider the sum of all $\mu(d)$ for which d is a divisor of n:

$$\sum_{d|n} \mu(d).$$

Show that this sum is 1 for $n = 1$ and 0 for $n > 1$.

(b) Investigate the relations

$$\sum_{n=1}^{[x]} \mu(n) \left[\frac{x}{n}\right] = 1 \quad \text{for} \quad x \geq 1,$$

and

$$\sum_{n=1}^{\infty} \frac{\mu(n)}{n} = 0.$$

(c) Show that

$$x = \sum_{n=1}^{\infty} \mu(n) \frac{x^n}{1 - x^n} \quad \text{for} \quad |x| < 1.$$

33. *Tauber's Theorem:* Let $f(x) = \Sigma_{n=0}^{\infty} a_n x^n$ for $-1 < x < 1$, and suppose that $na_n \to 0$ as $n \to \infty$. If $f(x) \to S$ as $x \to 1-$, then $\Sigma_{n=0}^{\infty} a_n$ converges and has sum S.

Proof Let $n\sigma_n = \Sigma_{k=0}^{n} k|a_k|$. Then $\sigma_n \to 0$ as $n \to \infty$. Also $f(x_n) \to S$ as $n \to \infty$ for $x_n = 1 - 1/n$. Hence, given $\varepsilon > 0$, we can choose n_0 such that $n \geq n_0$ implies

$$|f(x_n) - S| < \frac{\varepsilon}{3}, \quad \sigma_n < \frac{\varepsilon}{3}, \quad n|a_n| < \frac{\varepsilon}{3}.$$

Now let $s_n = \Sigma_{k=0}^{n} a_k$. Then, for $-1 < x < 1$, we can write

$$s_n - S = f(x) - S + \sum_{k=0}^{n} a_k(1 - x^k) - \sum_{k=n+1}^{\infty} a_k x^k.$$

Now keep x in (0,1). Then

$$(1 - x^k) = (1 - x)(1 + x + \cdots + x^{k-1}) \le k(1 - x)$$

for each k. Thus, if $n \ge n_0$ and $0 < x < 1$, we have

$$|s_n - S| \le |f(x) - S| + (1 - x) \sum_{k=0}^{n} k|a_k| + \frac{\varepsilon}{3n(1 - x)}.$$

Taking $x = x_n = 1 - 1/n$, we find

$$|s_n - S| < \frac{\varepsilon}{3} + \frac{\varepsilon}{3} + \frac{\varepsilon}{3} = \varepsilon. \qquad \square$$

34. Let $f(x) = \Sigma_{n=0}^{\infty} (-1)^n x^n$ for $-1 < x < 1$. Then $f(x) = 1/(1 + x)$ for $-1 < x < 1$ and $f(x) \to \frac{1}{2}$ as $x \to 1-$. However, $\Sigma_{n=0}^{\infty} (-1)^n$ diverges. Does this show that the converse of part (ii) of Proposition 28 is false in general?

35. Let $\Sigma_{n=1}^{\infty} u_n(x)$ be a series of functions and let the sequence of arithmetic means $\sigma_n(x)$ converge uniformly to $\sigma(x)$ in an interval (a,b). Then, if either the partial sums $s_n(x)$ of $\Sigma_{n=1}^{\infty} u_n(x)$ satisfy

$$n[s_n(x) - s_{n+1}(x)] < K \quad \text{or} \quad n[s_{n+1}(x) - s_n(x)] < K,$$

when $n \ge n_0$, where K is independent of x and n, and the same n_0 serves for all values of x in (a,b), $s_n(x) \to \sigma(x)$ uniformly.

Hint See Exercise 44 at the end of Chapter 1.

36. Using the result in Exercise 33, show the following: Let $\Sigma_{n=0}^{\infty} a_n x^n$ converge for $0 < x < 1$ and suppose that

$$\lim_{x \to 1-} \sum_{n=0}^{\infty} a_n x^n = A.$$

If

$$\lim_{n \to \infty} \frac{a_1 + 2a_2 + \cdots + na_n}{n} = 0,$$

then $\Sigma_{n=0}^{\infty} a_n = A$.

Solution Put $v_0 = 0$ and $v_n = a_1 + 2a_2 + \cdots + na_n$ for $n \geq 1$. Then $a_n = (1/n)(v_n - v_{n-1})$ for $n \geq 1$ and hence

$$\sum_{n=0}^{\infty} a_n x^n = a_0 + \sum_{n=1}^{\infty} \frac{v_n}{n} x^n - \sum_{n=1}^{\infty} \frac{v_{n-1}}{n} x^n$$

$$= a_0 + (1 - x) \sum_{n=1}^{\infty} \frac{v_n}{n} x^n + \sum_{n=1}^{\infty} \frac{v_n}{n(n+1)} x^{n+1}.$$

But $(v_n)/n \to 0$ as $n \to \infty$ and so

$$\lim_{x \to 1-} (1 - x) \sum_{n=1}^{\infty} \frac{v_n}{n} x^n = 0$$

because we may break up the series

$$(1 - x) \sum_{n=1}^{\infty} \frac{v_n}{n} x^n$$

into

$$(1 - x) \sum_{n=1}^{\nu-1} \frac{v_n}{n} x^n + (1 - x) \sum_{n=\nu}^{\infty} \frac{v_n}{n} x^n$$

and choose ν such that in the second series all factors $(v_n)/n$ have absolute value less than a given $\varepsilon > 0$; the first series has only a finite number of terms and we can render its absolute value arbitrarily small by letting x tend to 1. It therefore follows that

$$\lim_{x \to 1-} (1 - x) \sum_{n=1}^{\infty} \frac{v_n}{n(n+1)} x^{n+1} = A - a_0.$$

By Exercise 33,

$$\sum_{n=1}^{\infty} \frac{v_n}{n(n+1)} = A - a_0.$$

But

$$\sum_{m=1}^{n} \frac{v_m}{m(m+1)} = \sum_{m=1}^{n} \frac{v_m}{m} - \sum_{m=1}^{n+1} \frac{v_{m-1}}{m} = -\frac{v_n}{n+1} + \sum_{m=1}^{n} a_m.$$

Hence

$$\lim_{n\to\infty} \sum_{m=1}^{n} a_m = A - a_0. \qquad \square$$

37. Consider the series $\Sigma_{k=0}^{\infty} a_k$ and let $A_0 = a_0$ and

$$A_n = a_0 + a_1 + \cdots + a_n \quad \text{for} \quad n \geq 1.$$

Let $(p_n)_{n=0}^{\infty}$ be a sequence of positive numbers and put $P_0 = p_0$ and

$$P_n = p_0 + p_1 + \cdots + p_n \quad \text{for} \quad n \geq 1.$$

Finally, let

$$w_n = \frac{p_n A_0 + p_{n-1} A_1 + \cdots + p_0 A_n}{P_n}.$$

Show that if $(p_n)/P_n \to 0$ as $n \to \infty$, then $A_n \to A$ implies $w_n \to A$.

Solution Recall the result in Exercise 26 at the end of Chapter 1; putting $x_n = A_n$ and $t_{nm} = p_{n-m}/P_n$, we see that

$$t_{nm} = \frac{p_{n-m}}{P_n} < \frac{p_{n-m}}{P_{n-m}} \to 0 \quad (\text{as } n \to \infty)$$

and condition (i) is satisfied. Since

$$\sum_{m=0}^{n} |t_{nm}| = \sum_{m=0}^{n} t_{nm} = 1,$$

we have that conditions (ii) and (iii) are satisfied as well. Thus $w_n \to A$. $\square$

38. In this exercise

$$s_n = a_0 + \sum_{r=1}^{n} (a_r \cos rx + b_r \sin rx),$$

$$\sigma_n = \frac{1}{n}(s_0 + s_1 + \cdots + s_{n-1}),$$

and $a_0, a_1, \ldots, b_1, \ldots$ are the Fourier coefficients of the function f.

Let f be a Riemann integrable function on $(-\pi,\pi)$. Show that

$$2a_0^2 + \sum_{n=1}^{\infty} (a_n^2 + b_n^2) = \frac{1}{\pi}\int_{-\pi}^{\pi} f^2(x)\, dx,$$

a result usually called *Parseval's Equality*.

Solution First we show that

$$2a_0^2 + \sum_{n=1}^{\infty} (a_n^2 + b_n^2) \le \frac{1}{\pi}\int_{-\pi}^{\pi} f^2(x)\, dx,$$

a result known as *Bessel's Inequality*.

We note that

$$\int_{-\pi}^{\pi} [f(x) - s_n]^2\, dx = \int_{-\pi}^{\pi} f^2(x)\, dx - 2\int_{-\pi}^{\pi} f(x)\, s_n\, dx + \int_{-\pi}^{\pi} s_n^2\, dx.$$

Substituting for s_n on the right-hand side, we get at once

$$\int_{-\pi}^{\pi} [f(x) - s_n]^2\, dx = \int_{-\pi}^{\pi} f^2(x)\, dx - 2\pi\left[2a_0^2 + \sum_{r=1}^{n} (a_r^2 + b_r^2)\right]$$

$$+ \pi\left[2a_0^2 + \sum_{r=1}^{n} (a_r^2 + b_r^2)\right].$$

It follows that

$$\frac{1}{\pi}\int_{-\pi}^{\pi} [f(x) - s_n]^2\, dx = \frac{1}{\pi}\int_{-\pi}^{\pi} f^2(x)\, dx - \left[2a_0^2 + \sum_{r=1}^{n} (a_r^2 + b_r^2)\right].$$

Thus

$$2a_0^2 + \sum_{r=1}^{n} (a_r^2 + b_r^2) \le \frac{1}{\pi}\int_{-\pi}^{\pi} f^2(x)\,dx \tag{E.10}$$

and therefore

$$2a_0^2 + \sum_{r=1}^{\infty} (a_r^2 + b_r^2) \le \frac{1}{\pi}\int_{-\pi}^{\pi} f^2(x)\,dx. \tag{E.11}$$

This proves Bessel's Inequality.

Next we prove Parseval's Equality. Since

$$s_n = \frac{1}{2\pi}\int_{-\pi}^{\pi} f(x')\left[1 + 2\sum_{r=1}^{n} \cos r(x' - x)\right] dx',$$

we have

$$\sigma_n = \frac{1}{2n\pi}\int_{-\pi}^{\pi} f(x')\,\frac{\sin^2 \frac{1}{2}n(x' - x)}{\sin^2 \frac{1}{2}(x' - x)}\,dx'.$$

Also we know that

$$\int_{-\pi}^{\pi} \frac{\sin^2 \frac{1}{2}n(x' - x)}{\sin^2 \frac{1}{2}(x' - x)}\,dx' = 2n\pi.$$

Therefore

$$\sigma_n - f(x) = \frac{1}{2n\pi}\int_{-\pi}^{\pi} [f(x') - f(x)]\,\frac{\sin^2 \frac{1}{2}n(x' - x)}{\sin^2 \frac{1}{2}(x' - x)}\,dx' \tag{E.12}$$

and

$$|\sigma_n - f(x)| \le M - m, \tag{E.13}$$

where M, m are the upper and lower bounds of f(x) on $(-\pi,\pi)$ and x is any point in this interval.

Since f is Riemann integrable on $(-\pi,\pi)$ we see that for any pair α, β of positive numbers there is a mode of division of $(-\pi,\pi)$ such that the sum of the lengths of the partial intervals in which the oscillation (that is, the difference of the upper and the lower bound of the function) is greater than or equal to β is less than α; this is both a necessary and a sufficient condition for the Riemann integrability of f over the interval $(-\pi,\pi)$. (This characterization of Riemann integrability was discussed in Exercise 49 at the end of Chapter 3.) Let Δ denote the set of partial intervals in this mode of division in which the oscillation of f is greater than or equal to β, and δ the other partial intervals.

Cut off from the intervals of δ at each of their ends a part, so that the sum of the lengths of these segments cut off is less than α. Let δ'' denote the segment cut off, and δ' the inner parts of δ which remain. Then

$$\int_{-\pi}^{\pi} [\sigma_n - f(x)]^2\, dx$$

$$= \left\{\sum_{\delta'} \int_{\delta'} + \sum_{\delta''} \int_{\delta''} + \sum_{\Delta} \int_{\Delta}\right\} [\sigma_n - f(x)]^2\, dx, \tag{E.14}$$

where, by this notation, we mean that the integrals are taken respectively over the intervals of δ', δ'', and Δ.

Now let (a,b) be one of the intervals of δ, and (a',b') the corresponding interval of δ'. Also let x be a point of (a',b'). Then from (E.12),

$$\sigma_n - f(x)$$

$$= \frac{1}{2n\pi}\left[\int_{-\pi}^{a} + \int_{a}^{b} + \int_{b}^{\pi}\right] [f(x') - f(x)]\, \frac{\sin^2 \frac{1}{2}n(x' - x)}{\sin^2 \frac{1}{2}(x' - x)}\, dx'. \tag{E.15}$$

Taking these three integrals separately, we have

$$\left|\frac{1}{2n\pi}\int_{-\pi}^{a} [f(x') - f(x)]\, \frac{\sin^2 \frac{1}{2}n(x' - x)}{\sin^2 \frac{1}{2}(x' - x)}\, dx'\right|$$

$$< \frac{M - m}{2n\pi}\int_{-\pi}^{a} \csc^2 \tfrac{1}{2}(x' - x)\, dx' < \frac{K}{n},$$

where K is a positive constant depending upon the position of (a,b) and (a',b'). Similarly we find that

$$\left|\frac{1}{2n\pi}\int_b^\pi [f(x') - f(x)]\,\frac{\sin^2 \frac{1}{2}n(x' - x)}{\sin^2 \frac{1}{2}(x' - x)}\,dx'\right|$$

$$< \frac{M - m}{2n\pi}\int_b^\pi \csc^2 \tfrac{1}{2}(x' - x)\,dx' < \frac{K}{n},$$

where it is clear that we may take the same value for K in both, and replace it later by any larger value we please. Finally,

$$\left|\frac{1}{2n\pi}\int_a^b [f(x') - f(x)]\,\frac{\sin^2 \frac{1}{2}n(x' - x)}{\sin^2 \frac{1}{2}(x' - x)}\,dx'\right|$$

$$< \frac{\beta}{2n\pi}\int_a^b \frac{\sin^2 \frac{1}{2}n(x' - x)}{\sin^2 \frac{1}{2}(x' - x)}\,dx'$$

$$< \frac{\beta}{2n\pi}\int_{-\pi}^\pi \frac{\sin^2 \frac{1}{2}n(x' - x)}{\sin^2 \frac{1}{2}(x' - x)}\,dx' < \beta.$$

Thus, from (E.15)

$$|\sigma_n - f(x)| \le \left(\beta + 2\,\frac{K}{n}\right). \tag{E.16}$$

But the sum of the lengths of the intervals of δ' does not exceed 2π. Therefore in (E.14), we have

$$\sum_{\delta'}\int_{\delta'} [\sigma_n - f(x)]^2\,dx \le 2\pi\left(\beta + 2\,\frac{K}{n}\right)^2. \tag{E.17}$$

Also

$$\sum_{\delta''}\int_{\delta''} [\sigma_n - f(x)]^2\,dx \le \alpha(M - m)^2, \tag{E.18}$$

by (E.13), since the lengths of the sum of the intervals δ'' is less than α. And

$$\sum_{\Delta}\int_{\Delta} [\sigma_n - f(x)]^2 \le \alpha(M - m)^2 \tag{E.19}$$

for a similar reason.

It follows from (E.14), (E.17), (E.18), and (E.19) that

$$\lim_{n\to\infty} \int_{-\pi}^{\pi} [\sigma_n - f(x)]^2 \, dx = 0, \tag{E.20}$$

because β and α are arbitrary positive numbers which can be chosen as small as we please.

But

$$\sigma_n = a_0 + \sum_{r=1}^{n-1} \left(\frac{n-r}{n}\right)(a_r \cos rx + b_r \sin rx).$$

Therefore, as in the proof of Bessel's Inequality, we have

$$\frac{1}{\pi}\int_{-\pi}^{\pi} [\sigma_n - f(x)]^2 \, dx$$

$$= \frac{1}{\pi}\int_{-\pi}^{\pi} f^2(x) \, dx - \left[2a_0^2 + \sum_{r=1}^{n-1} \left(\frac{n^2 - r^2}{r^2}\right)(a_r^2 + b_r^2)\right]$$

$$= \left\{\frac{1}{\pi}\int_{-\pi}^{\pi} f^2(x) \, dx - \left[2a_0^2 + \sum_{r=1}^{n-1} (a_r^2 + b_r^2)\right]\right\}$$

$$+ \frac{1}{n^2}\sum_{r=1}^{n-1} r^2(a_r^2 + b_r^2). \tag{E.21}$$

But we know from (E.10) that

$$\frac{1}{\pi}\int_{-\pi}^{\pi} f^2(x) \, dx - \left[2a_0^2 + \sum_{r=1}^{n-1} (a_r^2 + b_r^2)\right] \ge 0.$$

Therefore from (E.20) and (E.21) we see that if f is Riemann integrable, then

$$\frac{1}{\pi}\int_{-\pi}^{\pi} f^2(x) \, dx = 2a_0^2 + \sum_{r=1}^{\infty} (a_r^2 + b_r^2).$$

Also

$$\lim_{n\to\infty} \left[\frac{1}{n^2}\sum_{r=1}^{n-1} r^2(a_r^2 + b_r^2)\right] = 0. \qquad \square$$

(The foregoing proof of Parseval's Equality is due to A. Hurwitz.)

39. Let f and g be Riemann integrable in $(-\pi,\pi)$ and let a_n, b_n and α_n, β_n be the Fourier coefficients of f and g respectively. Then

$$\frac{1}{\pi}\int_{-\pi}^{\pi} f(x)\ g(x)\ dx = 2a_0\alpha_0 + \sum_{n=1}^{\infty} (a_n\alpha_n + b_n\beta_n).$$

Hint We have from Exercise 38,

$$\frac{1}{\pi}\int_{-\pi}^{\pi} [f(x) \pm g(x)]^2\ dx$$

$$= 2(a_0 \pm \alpha_0) \sum_{n=1}^{\infty}\left[(a_n \pm \alpha_n)^2 + (b_n \pm \beta_n)^2\right].$$

But $(A + B)^2 - (A - B)^2 = 4AB$.

40. The *classical isoperimetric problem* is: To determine, among all closed rectifiable Jordan arcs that are free of double points and have a fixed length L, that one which encloses the largest area I.

This problem was historically one of the first to arise in the calculus of variations (Euler). Steiner (1839) showed by simple geometrical considerations that no curve other than the circle can possibly possess the desired maximum property. The problem of whether or not the circle actually has the isoperimetric property remained unsolved for the time being. Refined arguments of the Steiner type, however, completely dispose of the problem. We now look at a famous analytic proof, which A. Hurwitz gave in 1902.

The type of Jordan arc C in question is given by a set of points P(x,y) of the Cartesian x, y plane of the form $P_t = P[x(t),y(t)]$, where x(t) and y(t) are continuous functions, of bounded variation, of the parameter t in the interval $[t_0,t_1]$ and for which $P_{t_0} = P_{t_1}$ while otherwise $P_t \neq P_{t'}$ with $t \neq t'$. It is convenient to employ $t = 2\pi s/L$ as the parameter, where s is the arc length calculated from some fixed point. Thus for x(t) and y(t) we may choose two arbitrary continuous functions of bounded variation and of period 2π, for which $P_t = P_{t'}$ if and only if $t - t'$ is a multiple of 2π.

We first assume that $x'(t)$ and $y'(t)$ exist except at most at a finite number of points, and that they are Riemann integrable. Let α_k, β_k and α_k', β_k' be the Fourier coefficients of $x(t)$ and $y(t)$ respectively. Then

$$x'(t) \approx \sum_{k=1}^{\infty} k(\beta_k \cos kt - \alpha_k \sin kt),$$

$$y'(t) \approx \sum_{k=1}^{\infty} k(\beta_k' \cos kt - \alpha_k' \sin kt).$$

Since $[x'(t)]^2 + [y'(t)]^2 = [s'(t)]^2 = (L/2\pi)^2$, we have by Exercise 38

$$\frac{1}{\pi}\int_0^{2\pi} \left\{[x'(t)]^2 + [y'(t)]^2\right\} dt = \frac{L^2}{2\pi^2}$$

$$= \sum_{k=1}^{\infty} k^2(\alpha_k^2 + \alpha_k'^2 + \beta_k^2 + \beta_k'^2).$$

On the other hand, by Exercise 39,

$$I = \int_0^{2\pi} x(t)\ y'(t)\ dt = \pi \sum_{k=1}^{\infty} k(\alpha_k\beta_k' - \beta_k\alpha_k'),$$

$$L^2 - 4\pi I$$

$$= 2\pi^2 \sum_{k=1}^{\infty} \left[(k\alpha_k - \beta_k')^2 + (k\alpha_k' + \beta_k)^2 + (k^2 - 1)(\beta_k^2 + \beta_k'^2)\right]. \tag{E.22}$$

Since the right-hand side of the foregoing equation is not negative, we have $I \le L^2/4\pi$, the equality sign holding if and only if

$$\alpha_1 = \beta_1', \quad \alpha_1' = -\beta_1, \quad \alpha_k = \alpha_k' = \beta_k = \beta_k' = 0 \quad \text{for} \quad k > 1.$$

But this means that the curve is a circle of circumference L.

In the general case we can approximate the curve C by a sequence of curves $C^{(r)}$ for which the foregoing proof is valid (for example, by polygons), and for which the $L^{(r)} \to L$, $I^{(r)} \to I$ while the Fourier coefficients for the $C^{(r)}$ converge to the corresponding coefficients of the given curve, C. We shall not go into details, but assuming such a sequence of approximating curves has been determined, we conclude from (E.22) that for each m,

$$[L^{(r)}]^2 - 4\pi I^{(r)}$$

$$\geq 2\pi^2 \sum_{k=1}^{m} \Big[(k\alpha_k^{(r)} - \beta_k'^{(r)})^2$$

$$+ (k\alpha_k'^{(r)} + \beta_k^{(r)})^2 + (k^2 - 1)(\beta_k^{(r)\,2} + \beta_k'^{(r)\,2}) \Big];$$

thus, letting $r \to \infty$,

$$L^2 - 4\pi I$$

$$\geq 2\pi^2 \sum_{k=1}^{m} \Big[(k\alpha_k - \beta_k')^2 + (k\alpha_k' + \beta_k)^2 + (k^2 - 1)(\beta_k^2 + \beta_k'^2) \Big].$$

This easily yields the result as above.

Remark The problem of differentiating monotonic functions will be discussed in Exercise 34 at the end of Chapter 5; Lebesgue's characterization of Riemann integrability will be taken up in Exercise 32 at the end of Chapter 5.

41. Let [a,b] be a closed bounded interval. The series

$$\sum_{k=1}^{\infty} u_k(x)$$

is said to *converge quasiuniformly on* [a,b] to f(x) if for any $\varepsilon > 0$ and any index n' the interval [a,b] may be covered by a finite number of open intervals

$$(a_1,b_1),\ (a_2,b_2),\ \ldots,\ (a_i,b_i),\ \ldots,\ (a_m,b_m),$$

and there exist m indices

$$n_1,\ n_2,\ \ldots,\ n_i,\ \ldots,\ n_m \quad (> n')$$

such that for all x situated in (a_i, b_i), $i = 1, 2, \ldots, m$, we have simultaneously the inequality

$$\left| f(x) - \sum_{k=1}^{n_i} u_k(x) \right| < \varepsilon.$$

Prove the following result due to C. Arzelà: Let the functions $u_k(x)$ be continuous on a compact interval $[a,b]$ and suppose that the series $\Sigma_{k=1}^{\infty} u_k(x)$ converges (pointwise) to $f(x)$ on $[a,b]$. Then f is continuous on $[a,b]$ if and only if the series $\Sigma_{k=1}^{\infty} u_k(x)$ converges quasi-uniformly on $[a,b]$.

42. Use the result in the foregoing exercise to deduce the following theorem of U. Dini: Let the functions $u_k(x)$ be continuous and nonnegative on a compact interval $[a,b]$. If the (pointwise) limit $f(x)$ of the series $\Sigma_{k=1}^{\infty} u_k(x)$ is a continuous function on $[a,b]$, then the series $\Sigma_{k=1}^{\infty} u_k(x)$ converges uniformly to f on $[a,b]$.

N.B.: We have met the foregoing result in Proposition 4.

43. What is the smallest amount that may be invested at interest rate i, compounded annually, in order that one may withdraw 1 dollar at the end of the first year, 4 dollars at the end of the second year, ..., n^2 dollars at the end of the n-th year, in perpetuity?

Solution The present value of n^2 dollars n years from now at rate i per year is $n^2(1 + i)^{-n}$. Thus the required sum is

$$\sum_{n=1}^{\infty} \frac{n^2}{(1 + i)^n}$$

Since $(1 - x)^{-1} = \Sigma_{n=0}^{\infty} x^n$, it follows by differentiation that

$$x(1 - x)^{-2} = \sum_{n=1}^{\infty} nx^n, \quad (x + x^2)(1 - x)^{-3} = \sum_{n=1}^{\infty} n^2x^n,$$

all series being convergent for $-1 < x < 1$. Taking $x = (1 + i)^{-1}$, the required sum is found to be

$$\frac{(1 + i)(2 + i)}{i^3}. \qquad \square$$

44. The lengths of successive segments of a broken line are represented by the successive terms of the harmonic progression

$$1, \frac{1}{2}, \ldots, \frac{1}{n}, \ldots.$$

Each segment makes with the preceding segment a given angle θ. What is the distance and what is the direction of the limiting point (if there is one) from the initial point of the first segment?

45. Show that

$$\frac{1}{r_1} + \frac{1}{r_2} + \cdots + \frac{1}{r_n} = \frac{n}{p},$$

where $r_1, r_2, \ldots, r_n$ are the lengths of consecutive radius vectors separated by an angle of $2\pi/n$ and emanating from the focal point of the parabola $y^2 = 2px$.

46. Show that

$$\frac{1}{r_1^2} + \frac{1}{r_2^2} + \cdots + \frac{1}{r_n^2} = \frac{n}{2}\left(\frac{1}{a^2} + \frac{1}{b^2}\right),$$

where $r_1, r_2, \ldots, r_n$ are the lengths of consecutive radius vectors separated by an angle of π/n and emanating from the center of the ellipse

$$\frac{x^2}{a^2} + \frac{y^2}{b^2} = 1.$$

47. Suppose that the series $\Sigma_{n=0}^{\infty} a_n x^n$ and $\Sigma_{n=0}^{\infty} b_n x^n$ converge in the same segment $S = (-R,R)$. Let E be the set of all $x \in S$ at which

$$\sum_{n=0}^{\infty} a_n x^n = \sum_{n=0}^{\infty} b_n x^n. \qquad \text{(E.22)}$$

If E has a limit point in S, then $a_n = b_n$ for $n = 0, 1, 2, \ldots$. Hence (E.22) holds for all $x \in S$.

Proof Let $c_n = a_n - b_n$ and, for $x \in S$,

$$f(x) = \sum_{n=0}^{\infty} c_n x^n.$$

Then $f(x) = 0$ on E.

Let A be the set of all limit points of E in S, and let B consist of all other points of S. Then B is an open set. By assumption, A is not empty. Also, $S = A \cup B$, $A \cap B = \emptyset$, and S is connected. Thus B must be empty, hence $A = S$. Since f is continuous in S, $A \subset E$. Thus $E = S$, and (43) shows that $c_n = 0$ for $n = 0, 1, 2, \ldots$, which is what we wish to show.

Thus we have to prove that A is open. If $x_0 \in A$, Proposition 31 shows that

$$f(x) = \sum_{n=0}^{\infty} d_n (x - x_0)^n \qquad (|x - x_0| < R - |x_0|). \tag{E.23}$$

We claim that $d_n = 0$ for all n. If not, let k be the smallest natural number such that $d_k \neq 0$. Then

$$f(x) = (x - x_0)^k \, g(x) \qquad (|x - x_0| < R - |x_0|), \tag{E.24}$$

where

$$g(x) = \sum_{m=0}^{\infty} d_{k+m} (x - x_0)^m.$$

Since g is continuous at x_0 and

$$g(x_0) = d_k \neq 0,$$

there exists a $\delta > 0$ such that $g(x) \neq 0$ if $|x - x_0| < \delta$. It follows from (E.24) that $f(x) \neq 0$ if $0 < |x - x_0| < \delta$. But this contradicts the fact that x_0 is a limit point of E.

Thus $d_n = 0$ for all n, so that $f(x) = 0$ for all x for which (E.23) holds, i.e., in a neighborhood of x_0. This shows that A is open, and the proof is finished. □

<u>48</u>. A polygon is said to be convex if it contains the line segments connecting any two of its points. Prove the following theorem: Any convex polygon K (in the complex plane) which contains all the zeros of a polynomial P(x) also contains all the zeros of the deriv-

ative $P'(x)$, regardless of whether x is real or complex. Hence any convex polygon which contains all the zeros of a polynomial P also contains all the zeros of its derivatives.

Remark From Rolle's Theorem (see Proposition 5 in Chapter 3) we know that any interval on the real line which contains all the zeros of a real-valued polynomial P also contains all the zeros of the derivative P'; the result in Exercise 48 generalizes this fact.

Hint Let P be a polynomial of degree n and have the zeros $\xi_1, \ldots, \xi_n$. Then

$$P(x) = a_0(x - \xi_1)(x - \xi_2) \cdots (x - \xi_n).$$

Thus

$$\frac{P'(x)}{P(x)} = \sum_{k=1}^{n} \frac{1}{x - \xi_k}.$$

Let K be the smallest convex polygon containing $\xi_1, \ldots, \xi_n$. If η is a zero of P' and coincides with one of the ξ_k, then there is nothing to prove; if η is different from all ξ_k, then

$$\sum_{k=1}^{n} \frac{1}{\eta - \xi_k} = 0$$

holds and it is sufficient to show that the foregoing equation cannot be satisfied for any point η outside of K.

49. Let $a > 0$ and f' be a continuous function on $[0,a]$. Show that

$$|f(0)| \le \frac{1}{a}\int_0^a |f(x)|\,dx + \int_0^a |f'(x)|\,dx.$$

50. Let $X = (0,1) \cup (1,2) \cup A \cup B$, where A denotes the set of all rational points in the open interval $(2,3)$ and B denotes the set containing only the point $x = 4$. How many new sets can be formed from X by using complementation and closure?

Chapter 5

METRIC SPACES

1. Basic Notions

Definition Let X be a nonempty set and suppose that with any two elements x and y of X there is associated a real number d(x,y), called the *distance from* x *to* y, such that

(i) $d(x,y) > 0$ if $x \neq y$; $d(x,x) = 0$;

(ii) $d(x,y) = d(y,x)$;

(iii) $d(x,y) \leq d(x,z) + d(z,y)$ for any $z \in X$.

The pair (X,d) is called a *metric space* and the function d is called the *metric of* (X,d).

The elements of the set X are also referred to as the *points of* the metric space (X,d) and the axioms (i), (ii), and (iii) are called the *metric axioms;* in particular, (ii) is the *axiom of symmetry,* and (iii) is the *triangle inequality axiom.*

Proposition 1 Let (X,d) be a metric space. Then, for all x, x', y, and y' in X, we have

$$|d(x,x') - d(y,y')| \le d(x,y) + d(x',y').$$

Proof By the triangle inequality axiom,

$$d(x,x') \le d(x,y) + d(y,y') + d(y',x')$$

and so

$$d(x,x') - d(y,y') \le d(x,y) + d(x',y').$$

In the same way we show that

$$d(y,y') - d(x,x') \le d(x,y) + d(x',y').$$ □

Definition A sequence $(x_n)_{n=1}^{\infty}$ of elements of a metric space (X,d) is said to *converge in* (X,d) if there exists x in X such that

$$d(x_n,x) \to 0 \quad \text{as} \quad n \to \infty.$$

We then write $x_n \to x$ as $n \to \infty$ or

$$\lim_{n\to\infty} x_n = x,$$

and call x the *limit of* the sequence $(x_n)_{n=1}^{\infty}$

Proposition 2 We have:

(i) If $x_n \to x$ and $y_n \to y$ as $n \to \infty$ in the metric space (X,d), then $d(x_n,y_n) \to d(x,y)$ as $n \to \infty$.

(ii) If the sequence $(x_n)_{n=1}^{\infty}$ converges in the metric space (X,d), then its limit is unique.

Proof To prove part (i), we use Proposition 1. Since

$$|d(x_n,y_n) - d(x,y)| \le d(x_n,x) + d(y_n,y),$$

it is clear that $d(x_n,y_n) \to d(x,y)$ as $n \to \infty$, because $x_n \to x$ and $y_n \to y$ as $n \to \infty$.

To verify the claim under (ii), we use the triangle inequality axiom. Suppose that $x_n \to x$ and $x_n \to x'$ as $n \to \infty$ for some elements x, x' of X. By the triangle inequality

$$0 \le d(x,x') \le d(x,x_n) + d(x_n,x');$$

letting $n \to \infty$, it follows that $d(x,x') = 0$; thus $x = x'$ by the metric axiom (i). □

<u>*Definition*</u> A sequence $(x_n)_{n=1}^{\infty}$ of points in a metric space (X,d) is called a *Cauchy sequence in* (X,d) if, given any $\varepsilon > 0$, there is a positive integer n_0 such that $d(x_m,x_n) < \varepsilon$ for all $m, n \ge n_0$.

<u>*Definition*</u> A metric space (X,d) is said to be *complete* if every Cauchy sequence in (X,d) is convergent in (X,d).

<u>*Examples*</u> We shall now consider examples illustrating the foregoing concepts; some of the details are left as exercises for the reader.

1. Let X be any nonempty set, and let d be defined by

$$d(x,y) = 0 \quad \text{if} \quad x = y,$$
$$= 1 \quad \text{if} \quad x \ne y.$$

This d is called the *discrete metric of* X.

2. The most important examples of metric spaces, from our standpoint, are the *Euclidean spaces* R^n, especially R^1 (the real line); R^n is the set of all ordered n-tuples

$$x = (x_1,x_2, \ldots,x_n),$$

where $x_1, x_2, \ldots, x_n$ are real number, called the *coordinates of* x. If

$$y = (y_1,y_2, \ldots,y_n)$$

is another ordered n-tuple of real numbers, we define the *Euclidean distance in* R^n by

$$d(x,y) = \left[\sum_{j=1}^{n} (x_j - y_j)^2\right]^{1/2}, \quad (x, y \in R^n).$$

In case n = 1, we have

$$d(x,y) = |x - y|.$$

It is clear that we can define various other distance functions in the set of all ordered n-tuples of real numbers; for example,

$$d_\infty(x,y) = \max\{|x_j - y_j|: 0 \le j \le n\},$$

$$d_p(x,y) = \left[\sum_{j=1}^{n} |x_j - y_j|^p\right]^{1/p}, \quad (1 \le p < \infty).$$

The fact that d_p is a metric in the set of all ordered n-tuples of real numbers is a consequence of Minkowski's Inequality (see Proposition 49 in Chapter 1). The notation d_∞ is motivated by the fact that

$$\lim_{p\to\infty} \left[\sum_{j=1}^{n} |x_j - y_j|^p\right]^{1/p} = \max\{|x_j - y_j|: 0 \le j \le n\}.$$

The set of all ordered n-tuples of real numbers is a complete metric space with respect to the Euclidean metric ; the same is true with respect to the metrics d_p and d_∞. Of course, endowing the same basic set X with two different metrics, say d_1 and d_2, gives rise to two distinct metric spaces (X,d_1) and (X,d_2).

3. The set of all sequences of real numbers provide natural extensions of the foregoing example. The space ℓ_p, for $1 \le p \le \infty$, is the set of all sequences of real numbers

$$x = (x_1, x_2, \ldots)$$

satisfying

$$\sum_{j=1}^{\infty} |x_j|^p < \infty$$

with the metric

$$d_p(x,y) = \left[\sum_{j=1}^{\infty} |x_j - y_j|^p\right]^{1/p}, \quad [y = (y_1,y_2, \ldots)].$$

(Here we need the extended Minkowski Inequality; see Exercise 13 at the end of Chapter 1.)

The space ℓ_∞ is defined to be the set of all bounded sequences of real numbers, that is, all $x = (x_1,x_2, \ldots)$ for which $\sup\{|x_j|: 1 \le j < \infty\} < \infty$, with the metric

$$d_\infty(x,y) = \sup_{1\le j<\infty} |x_j - y_j|.$$

The spaces (ℓ_p,d_p) and (ℓ_∞,d_∞) are complete metric spaces. Metric convergence in both of these spaces implies coordinatewise convergence; in fact, metric convergence in (ℓ_∞,d_∞) means uniform convergence with respect to the index j.

4. Given any metric space (X,d), we can construct a new metric space (X,d_1) by simply setting

$$d_1(x,y) = \frac{d(x,y)}{1 + d(x,y)}.$$

To verify the triangle inequality axiom for d_1, we only need to observe that the function

$$h(x) = \frac{x}{1 + x}$$

is an increasing function for $x \ge 0$.

This observation also helps us in the consideration of the set S of all sequences of real numbers. Let

$$x = (x_1,x_2, \ldots) \quad \text{and} \quad y = (y_1,y_2, \ldots)$$

belong to S. We define the distance between x and y by

$$d(x,y) = \sum_{j=1}^{\infty} \frac{1}{2^j} \frac{|x_j - y_j|}{1 + |x_j - y_j|}.$$

It is easy to see that (S,d) is a metric space and that

$$x^{(n)} = (x_1^{(n)}, x_2^{(n)}, \ldots)$$

tends to

$$x = (x_1, x_2, \ldots)$$

as $n \to \infty$ if and only if $\lim_{n\to\infty} x_k^{(n)} = x_k$ for $k = 1, 2, \ldots$ (that is, convergence in this metric is coordinatewise convergence). Moreover, (S,d) is a complete metric space.

5. Another important example of a complete metric space is the set $C[a,b]$ of all continuous real-valued functions on a given closed bounded interval $[a,b]$. If $f, g \in C[a,b]$, we define a distance in $C[a,b]$ by

$$d(f,g) = \sup\{|f(t) - g(t)|: t \in [a,b]\}.$$

The resulting metric space is complete and convergence in the sense of this metric is uniform convergence; for the completeness of this space, consult Proposition 3 of Chapter 4.

The set of all continuous real-valued functions on $[a,b]$ can also be endowed with the metric

$$d_1(f,g) = \int_a^b |f(t) - g(t)|\, dt.$$

However, under this new metric the resulting metric space is not complete. Indeed, let $[a,b] = [0,1]$ and let

$$f_n(t) = \min\left\{1, \frac{1}{2} - n\left(x - \frac{1}{2}\right)\right\} \quad \text{on } \left[0, \frac{1}{2}\right],$$

$$= \max\left\{0, \frac{1}{2} - n\left(x - \frac{1}{2}\right)\right\} \quad \text{on } \left[\frac{1}{2}, 1\right].$$

From the graph of this function it is clear that

$$d_1(f_n, f_m) = \frac{1}{4}\left|\frac{1}{n} - \frac{1}{m}\right|.$$

Hence $(f_n)_{n=1}^{\infty}$ is a Cauchy sequence. Suppose that $d_1(f_n, f) \to 0$ as $n \to \infty$ for some continuous function f on $[0,1]$. Then, for $\alpha_n = \frac{1}{2} - 1/2n$ and $\beta_n = \frac{1}{2} + 1/2n$,

$$d_1(f_n,f) \geq \int_0^{\alpha_n} |1 - f(t)|\, dt + \int_{\beta_n}^{1} |f(t)|\, dt.$$

Letting $n \to \infty$, we see that

$$\int_0^{\frac{1}{2}} |1 - f(t)|\, dt = \int_{\frac{1}{2}}^{1} |f(t)|\, dt = 0.$$

Since f is continuous, we must have $f(t) = 1$ on $[0,\frac{1}{2}]$ and $f(t) = 0$ on $[\frac{1}{2},1]$, which is impossible.

6. Consider the set V of all functions of finite variation on a bounded closed interval $[a,b]$. Define, for $f, g \in V$,

$$d(f,g) = |f(a) - g(a)| + V_a^b(|f| - |g|).$$

Then (V,d) is a complete metric space.

7. Consider the set $\mathcal{R}$ of all Riemann integrable functions on a closed bounded interval $[a,b]$, and let d_1 be defined as in 5. We note that d_1 satisfies all conditions of a metric except that $d_1(f,g) = 0$ need not imply $f = g$. This situation can however be remedied by identifying two functions f and g in $\mathcal{R}$ whenever

$$\int_a^b |f(t) - g(t)|\, dt = 0. \tag{1}$$

Thus, instead of considering the functions of $\mathcal{R}$ as elements of the space, we consider classes of functions forming the elements of the space; here f and g belong to the same class if and only if (1) is satisfied. Let E_{f_1} denote the class of all functions u such that

$$\int_a^b |f_1(t) - u(t)|\, dt = 0,$$

where u is of course assumed to be Riemann integrable on $[a,b]$, and let E_{f_2} be the class of all functions v which are Riemann integrable on $[a,b]$ and such that

$$\int_a^b |f_2(t) - v(t)|\, dt = 0.$$

We define

$$\sigma(E_{f_1}, E_{f_2}) = \int_a^b |f_1(t) - f_2(t)|\, dt$$

and note that σ is well defined, that is, σ is independent of the choice of which elements are chosen. Indeed. If $g_i \in E_{f_i}$ $(i = 1, 2)$, then

$$|d_1(f_1,f_2) - d_1(g_1,g_2)| \le d_1(f_1,g_1) + d_1(f_2,g_2)$$

because

$$||a - b| - |a' - b'|| \le |a - a'| + |b - b'|$$

holds for any real numbers a, a', b, b' (by Proposition 1). Thus

$$\int_a^b |f_1(t) - f_2(t)|\, dt = \int_a^b |g_1(t) - g_2(t)|\, dt.$$

It is now an easy matter to establish that σ satisfies all the requirements of a metric; we leave the details to the reader.

We shall call two Riemann integrable functions f and g *equivalent* if

$$\int_a^b |f(t) - g(t)|\, dt = 0$$

and consider σ on the classes of functions so determined, that is, the distance of two classes is $\sigma(f,g)$ where f and g are, respectively, any representatives of these classes. The resulting metric space is incomplete, as we shall now show. For convenience we let [a,b] = [0,1].

We define a sequence $(S_n)_{n=0}^{\infty}$ of sets as follows: Let

$S_0 = [0,1]$,

$S_1 = S_0$ with the middle open interval of length 1/4 removed,

$S_2 = S_1$ with the middle open intervals of the component intervals of S_1 removed, each of length $1/4^2$.

Then, by induction, having already defined S_n so as to consist of 2^n disjoint closed intervals, of equal length, let

$S_{n+1} = S_n$ with the middle open intervals of the component intervals of S_n removed, each of length $1/4^{n+1}$.

For each n = 1, 2, ..., the sum of the length of the component open intervals of S_n is given by

$$|S_n| = \sum_{j=0}^{n-1} \frac{1}{4^{j+1}} 2^j = \frac{1}{2} + \frac{1}{2^{n+1}}.$$

For every n = 1, 2, ..., let f_n be defined by

$$f_n(t) = 1 \quad \text{if} \quad t \in S_n,$$
$$= 0 \quad \text{if} \quad t \notin S_n.$$

Let m > n. Then

$$\int_0^1 |f_n(t) - f_m(t)| \, dt = |S_n| - |S_m| = \frac{1}{2^{n+1}} - \frac{1}{2^{m+1}}.$$

It therefore follows that $(f_n)_{n=1}^{\infty}$ is a Cauchy sequence.

We now show that $(f_n)_{n=1}^{\infty}$ does not converge to any Riemann integrable function. Let f be a Riemann integrable function and suppose

$$\lim_{n\to\infty} \int_0^1 |f(t) - f_n(t)| \, dt = 0. \tag{2}$$

Let I_1 be the open interval removed in forming S_1; I_1, I_2, I_3 the open intervals removed in forming S_2, etc. For each k = 1, 2, ... there is an n_0 so that $n > n_0$ implies $f_n(t) = 0$, $t \in I_k$. It follows that f is equivalent to a function which is identically zero on

$$T = \bigcup_{k=1}^{\infty} I_k.$$

Evidently, the lower Riemann integral of such a function is zero. Since f is assumed integrable,

$$\int_0^1 f(t)\ dt = 0.$$

But

$$\int_0^1 f_n(t)\ dt > \frac{1}{2}, \quad (n = 1, 2, \ldots).$$

This contradicts (2) and the proof is complete.

8. Let $\mathcal{B}$ denote the set of all sequences $(x_j)_{j=1}^{\infty}$ of positive integers. For

$$x = (x_1, x_2, \ldots) \quad \text{and} \quad y = (y_1, y_2, \ldots)$$

in $\mathcal{B}$, define

$$d(x,x) = 0 \quad \text{and} \quad d(x,y) = \frac{1}{k},$$

where k is the first integer for which $x_k \neq y_k$. The function d is a metric and the triangle inequality assumes here the sharper form

$$d(x,z) \leq \max\left\{ d(x,y),\ d(y,z) \right\}.$$

The metric space $(\mathcal{B},d)$ is called the *null space of Baire* and d is called *Baire's metric*. For

$$x = (x_1, x_2, \ldots) \quad \text{and} \quad x_n = (x_1^{(n)}, x_2^{(n)}, \ldots)$$

in $(\mathcal{B},d)$, $\lim_{n\to\infty} x_n = x$ if and only if for every j there is an n_j such that $x_j^{(n)} = x_j$ for $n \geq n_j$.

Indeed. If $\lim_{n\to\infty} x_n = x$, then $d(x_n,x) \to 0$; hence there is an n_j such that $d(x_n,x) < 1/j$ for $n \geq n_j$; but then $x_j^{(n)} = x_j$ for $n \geq n_j$. Conversely, putting

$$\bar{n}_j = \max\{n_1, n_2, \ldots, n_j\},$$

we have, for $n \geq \bar{n}_j$,

$$x_1^{(n)} = x_1 \quad x_2^{(n)} = x_2, \ \ldots, \quad x_j^{(n)} = x_j;$$

thus $d(x_n,x) < 1/j$ for $n \geq \bar{n}_j$, hence $d(x_n,x) \to 0$, that is,

$$\lim_{n\to\infty} x_n = x.$$

To see that the metric space $(\mathcal{B},d)$ is complete, we let

$$x_n = (x_1^{(n)}, x_2^{(n)}, \ldots)$$

be the n-th term of a Cauchy sequence $(x_n)_{n=1}^{\infty}$ in $(\mathcal{B},d)$. Then, for every j, there is an n_j and a natural number x_j such that

$$x_j^{(n)} = x_j \quad \text{for} \quad n \geq n_j.$$

We let

$$x = (x_1, x_2, \ldots)$$

and note that $x \in (\mathcal{B},d)$. From what we showed above, it is clear that $\lim_{n\to\infty} x_n = x$.

Remark Note that if in a metric space (X,d) the triangle inequality axiom can be replaced by formula (3), then a sequence $(x_n)_{n=1}^{\infty}$ of elements of (X,d) is Cauchy if and only if $d(x_n,x_{n+1}) \to 0$ as $n \to \infty$.

9. Let J denote the set of all irrational numbers between 0 and 1 and let $x \in J$ be expressed as an infinite decimal fraction

$$x = .x_1x_2 \ldots.$$

Then (J,d), where d is taken to be Baire's metric, is a complete metric space.

Remarks For the set I of all rational numbers between 0 and 1 there is no metric ρ such that (I,ρ) is a complete metric space. The set J of all irrationals between 0 and 1 is of course an incomplete metric space under the Euclidean metric $d_e(x,y) = |x - y|$. However, d_e and Baire's metric d are *equivalent metrics on* J, that is, for each $x \in J$ and $\varepsilon > 0$ the following two conditions hold:

(i) There exists $\delta_1 = \delta_1(x,\varepsilon)$ such that

$$d(x,y) < \delta_1 \quad \text{implies} \quad d_e(x,y) < \varepsilon.$$

(ii) There exists $\delta_2 = \delta_2(x,\varepsilon)$ such that

$$d_e(x,y) < \delta_2 \quad \text{implies} \quad d(x,y) < \varepsilon.$$

The null space of Baire was introduced by R. L. Baire; see *Acta Mathematica*, 32, 105 (1909). The more advanced reader might be interested to prove that the null space of Baire is homeomorphic to the set of all irrational points of R^1.

2. *Topological Concepts*

Definition Let (X,d) be a metric space. Given any x_0 in X and any $r > 0$, the set of all points x in X such that $d(x_0,x) < r$ is called an *open sphere in* (X,d), and we denote it by $S(x_0,r)$. Correspondingly, the set of all points x in X such that $d(x_0,x) \le r$ is called a *closed sphere in* (X,d), and we denote it by $\bar{S}(x_0,r)$. In either case, the point x_0 is called the *center of the sphere* and r the *radius of the sphere*.

Definition Let (X,d) be a metric space. A *neighborhood of a point* x_0 in X is any open sphere in (X,d) with center x_0.

Proposition 3 Let (X,d) be a metric space and $x \in X$. If $S(x,\varepsilon)$ is any neighborhood of x and $y \in S(x,\varepsilon)$, then there is a neighborhood $S(y,\varepsilon')$ of y such that $S(y,\varepsilon') \subset S(x,\varepsilon)$.

Proof It is clear that $d(x,y) < \varepsilon$. We let $\varepsilon' = \varepsilon - d(x,y)$; then it is easily seen that $S(y,\varepsilon') \subset S(x,\varepsilon)$. □

Definition Let (X,d) be a metric space. If $Y \subset X$, then a point x_0 in Y is called an *interior point of* Y if there exists a neighborhood of x_0 lying entirely in Y.

Remark It is clear that every point of X is an interior point of X in a metric space (X,d).

Definition Let (X,d) be a metric space and $Y \subset X$. A point x_0 in X is said to be a *limit point of* Y if every neighborhood of x_0 contains at least one point $(\neq x_0)$ of Y.

Hence x_0 is a limit point of Y if $[S(x_0,\varepsilon) - \{x_0\}] \cap Y \neq \emptyset$ for all $\varepsilon > 0$; here $\{x_0\}$ denotes the set containing the single element x_0 and $\emptyset$ denotes the empty set.

The existence of one point different from x_0 as specified in the definition of limit point implies the existence of an infinite number of such points. Indeed, if ε is the radius of an arbitrary neighborhood of x_0, which contains a point x_1 (say) of Y other than x_0, let $\varepsilon_1 = d(x_0,x_1) < \varepsilon$. Then the neighborhood $S(x_0,\varepsilon_1)$ of x_0 also contains a point x_2 (say) of Y other than x_0, and since x_1 is not in $S(x_0,\varepsilon_1)$, we have that $x_1 \neq x_2$; we let $\varepsilon_2 = d(x_0,x_2)$. By next considering the neighborhood $S(x_0,\varepsilon_2)$ we obtain a third point x_3 distinct from x_0, x_1, x_2; and so forth. Thus in the definition of limit point the phrase "at least one point $(\neq x_0)$ of Y" could be replaced equivalently by "an infinite number of points of Y".

It is obvious that no finite subset of a metric space has a limit point. It is also clear that a limit point of a set does not necessarily belong to that set.

Definition A subset Y of a metric space (X,d) may contain points which are not limit points of Y; any such point is called an *isolated point of* Y. A point x_0 of Y is an isolated point of Y if and only if there exists an $\varepsilon > 0$ such that

$$[S(x_0,\varepsilon) - \{x_0\}] \cap Y = \emptyset.$$

Proposition 4 Let (X,d) be a metric space and $Y \subset X$. Then x_0 is a limit point of Y if and only if there exists a sequence $(x_n)_{n=1}^{\infty}$ of distinct points of Y such that $x_n \to x_0$ as $n \to \infty$.

Proof Let x_0 be a limit point of Y. Given any $\varepsilon_0 > 0$, there is a point x_1 (say) of Y in $S(x_0,\varepsilon_0)$ with $x_1 \neq x_0$. Suppose that $\varepsilon_1 = \frac{1}{2} d(x_0,x_1) < \frac{1}{2}\varepsilon_0$; then there is a point x_2 (say) of Y in $S(x_0,\varepsilon_1)$ with $x_2 \neq x_0$. Obviously $x_2 \neq x_1$. Let $\varepsilon_2 = \frac{1}{2} d(x_0,x_2) < \frac{1}{2}\varepsilon_1$. Proceeding in this way, we obtain an infinite sequence of distinct points such that

$$d(x_0,x_n) < \varepsilon_{n-1} < \frac{1}{2}\varepsilon_{n-2} < \frac{1}{2^{n-1}}\varepsilon_0.$$

Then $d(x_0,x_n) \to 0$ as $n \to \infty$, that is, $x_n \to x_0$ as $n \to \infty$.

The proof of the converse is trivial. □

Remark Note that a limit is a point associated with a sequence, whereas a limit point is a point associated with a set. Thus the concepts of a limit and a limit point are distinct and apply in different situations. We can, however, establish some connection between these concepts, when given a sequence, by considering the limit points (if any) of the set whose elements are the terms of a sequence.

We leave the verification of the following result as an exercise.

Proposition 5 Let (X,d) be a metric space containing a sequence which converges to a point x in X. Then x is a limit point of the set whose elements are the terms of the sequence if and only if the sequence has infinitely many distinct terms.

Definition Let (X,d) be a metric space and $Y \subset X$. Then Y is said to be *open in* (X,d) if every point of Y is an interior point of Y.

Definition Let (X,d) be a metric space and $Y \subset X$. Then Y is said to be *closed in* (X,d) if its complement $X - Y$ is open in (X,d); the set $X - Y$ consists of all points x in X which do not belong to Y.

If no confusion arises, we will only say that Y is *open* or Y is *closed*, respectively.

Remarks A set Y may neither be open nor closed in (X,d); for example, the half-open interval $(a,b]$ in (R^1,d_e) with $d_e(x,y) = |x - y|$ is neither open nor closed. The sets $\emptyset$ and X are both open and closed in a metric space (X,d). Indeed, since a set Y is open if all its points are interior points of Y, $\emptyset$ certainly contains no points which are not interior points, so $\emptyset$ is open in (X,d). To see that X is open in (X,d), we only need to recall that every point of X is an interior point of (X,d) (see the remark following the definition of interior point, which follows Proposition 3). From the definition of closed sets it is also clear that $\emptyset$ and X are closed sets in (X,d).

Proposition 6 Let (X,d) be a metric space and $Y \subset X$. Then Y is closed if and only if Y contains all its limit points.

Proof Let Y be closed; then $X - Y$ is open. Let $x \in X - Y$ so there is a neighborhood $S = S(x,\varepsilon)$ of x such that $S \subset X - Y$. Thus $S \cap Y = \emptyset$ and hence $(S - \{x\}) \cap Y = \emptyset$. Hence x is not a limit point of Y showing that Y contains all its limit points.

Conversely, assume that Y contains all its limit points; we verify that $X - Y$ is open. Let $x \in X - Y$, so $x \notin Y$. Thus x is not a limit point of Y; so there is a neighborhood $S = S(x,\varepsilon)$ of x such that $(S - \{x\}) \cap Y = \emptyset$; since $x \notin Y$, $S \cap Y = \emptyset$, and so $S \subset X - Y$. Thus x is an interior point of $X - Y$; hence $X - Y$ is open. □

Definition Let (X,d) be a metric space and $Y \subset X$. The set consisting of Y together with all its limit points is called the *closure of* Y, and is denoted by $\overline{Y}$.

It is easy to see that

(i) $Y \subset \overline{Y}$ for all sets $Y \subset X$;

(ii) $Y = \overline{Y}$ if and only if Y is closed.

The following result is an easy consequence of the corresponding definitions; the verification is left as an exercise.

Proposition 7 Let (X,d) be a metric space and $Y \subset X$. Then the following statements are equivalent:

(i) $x \in \overline{Y}$;

(ii) $S(x,\varepsilon) \cap Y \neq \emptyset$ for every neighborhood $S(x,\varepsilon)$ of x;

(iii) there exists an infinite sequence $(x_n)_{n=1}^{\infty}$ of points (not necessarily distinct) of Y such that $x_n \to x$ as $n \to \infty$.

Proposition 8 Let (X,d) be a metric space and $Y \subset X$. Let Y' denote the set of all limit points of Y. Then

(i) $\overline{Y}$ is closed (so $\overline{\overline{Y}} = \overline{Y}$);

(ii) Y' is closed.

Proof

(i) Let x be a limit point of $\overline{Y}$ and let $S = S(x,\varepsilon)$ be any neighborhood of x. Then S contains a point y of $\overline{Y}$. By Proposition 3, S contains some neighborhood of y, and therefore some point of Y (by Proposition 7). It follows that $S \cap Y \neq \emptyset$ for every neighborhood of x, and thus $x \in \overline{Y}$; hence $\overline{Y}$ is closed.

(ii) Let x be a limit point of Y' and proceed as before. Thus any neighborhood S of x contains a point y of Y' and hence an infinite number of points of Y (see the discussion following the definition of limit points, that follows Proposition 3). Therefore $x \in Y'$ and so Y' is closed. □

Proposition 9 Let (X,d) be a metric space. Then

(i) $X, \emptyset$ are open in (X,d);

(ii) the union of any collection of open sets is open;

(iii) the intersection of any finite collection of open sets is open.

Proof

(i) This was noted in the remark following the definition of open sets, which follows Proposition 5.

(ii) Let $\{Y_t : t \in T\}$ be a collection of open sets in (X,d), and denote the union of all the sets Y_t by Y; let $x \in Y$. Then there exists some t in T (say t_0) such that $x \in Y_{t_0}$. Since Y_{t_0} is open in (X,d), there is some neighborhood $S = S(x,\varepsilon)$ of x such that $S \subset Y_{t_0}$; hence $S \subset Y$ and Y is seen to be open in (X,d).

(iii) We need only show that the intersection of two open sets is open; by induction it follows that the intersection of any finite collection of open sets is open. Hence, assume that Y_1 and Y_2 are open. We may further assume that Y_1 and Y_2 are not disjoint [for if $Y_1 \cap Y_2 = \emptyset$, there is nothing to prove, by part (i)]. Let $x \in Y_1 \cap Y_2$, so $x \in Y_1$ and $x \in Y_2$. Then there exist neighborhoods $S(x,\varepsilon_1)$, $S(x,\varepsilon_2)$ of x such that $S(x,\varepsilon_j) \subset Y_j$ for $j = 1, 2$. Let $\varepsilon = \min\{\varepsilon_1,\varepsilon_2\}$; then $S(x,\varepsilon) \subset Y_1 \cap Y_2$ and hence $Y_1 \cap Y_2$ is open. □

Remark To see that the intersection of an infinite collection of open sets is not necessarily open, consider the open intervals $(0, 1 + 1/n)$ for $n = 1, 2, \ldots$. Then

$$\bigcap_{n=1}^{\infty} \left(0,1 + \frac{1}{n}\right] = (0,1].$$

Proposition 10 Let (X,d) be a metric space. Then

(i) $X, \emptyset$ are closed in (X,d);

(ii) the intersection of any collection of closed sets is closed;

(iii) the union of any finite collection of closed sets is closed.

Proof Note that $Y \subset X$ is closed if and only if $X - Y$ is open. Hence the present proposition is the mere dual statement of Proposition 9 and follows from Proposition 9 by use of De Morgan's rules (see Section 1 of Chapter 2). □

Proposition 11 Let (X,d) be a metric space.

(i) If $Y \subset Z \subset X$, then $\overline{Y} \subset \overline{Z}$.

(ii) If $Y \subset X$, then $\overline{Y}$ is the smallest closed set which contains Y (that is, if Z is any closed set such that $Y \subset Z$, then $\overline{Y} \subset Z$).

(iii) If $Y, Z \subset X$, then $\overline{Y \cup Z} = \overline{Y} \cup \overline{Z}$, but $\overline{Y \cap Z} \subset \overline{Y} \cap \overline{Z}$.

Proof

(i) Let x be a limit point of Y and S be any neighborhood of x. Then $(S - \{x\}) \cap Y \neq \emptyset$ and since $Y \subset Z$, $(S - \{x\}) \cap Z \neq \emptyset$, so $x \in \overline{Z}$; hence $\overline{Y} \subset \overline{Z}$.

(ii) This claim follows immediately from part (i). Moreover, $\overline{Y}$ is seen to be the intersection of all closed sets containing Y.

(iii) Since $Y \subset \overline{Y}$ and $Z \subset \overline{Z}$, it follows that $Y \cup Z \subset \overline{Y} \cup \overline{Z}$; hence

$$\overline{Y \cup Z} \subset \overline{\overline{Y} \cup \overline{Z}} = \overline{Y} \cup \overline{Z}$$

by (i) and because $\overline{Y} \cup \overline{Z}$ is closed. Thus $\overline{Y \cup Z} \subset \overline{Y} \cup \overline{Z}$.

To obtain the inclusion in the other direction, we note that $\overline{Y} \subset \overline{Y \cup Z}$ and $\overline{Z} \subset \overline{Y \cup Z}$ (because $Y, Z \subset Y \cup Z$); thus $\overline{Y} \cup \overline{Z} \subset \overline{Y \cup Z}$.

To obtain the second claim in part (iii), we note that $Y \cap Z \subset Y$ and $Y \cap Z \subset Z$; thus $\overline{Y \cap Z} \subset \overline{Y}$ and $\overline{Y \cap Z} \subset \overline{Z}$. This, however, implies that $\overline{Y \cap Z} \subset \overline{Y} \cap \overline{Z}$.

Finally, to see that strict inclusion is possible, take $Y = (0,1)$ and $Z = (1,2)$ in (R^1, d_e) with $d_e(x,y) = |x - y|$. Then

$$\overline{Y \cap Z} = \emptyset \quad \text{and} \quad \overline{Y} \cap \overline{Z} = \{1\}. \qquad \square$$

Definition Let (X,d) be a metric space and Y be a nonempty subset of X. Then the restriction d_Y of the distance function d to the set of all pairs (x,y) with $x, y \in Y$ will serve as a metric for Y; thus (Y,d_Y) is also a metric space. If $z \in Y$, we will denote the neighborhoods of z with radius r in (Y,d_Y), (X,d) by $S_Y(z,r)$, $S_X(z,r)$ respectively.

It is simple to verify that for $Y \subset X$, $z \in Y$, we have

$$S_Y(z,r) = Y \cap S_X(z,r).$$

Proposition 12 Let (X,d) be a metric space, $Z \subset Y \subset X$ and Z be open in (X,d). Then Z is open in (Y,d_Y).

Proof Let $z \in Z$. Then there is an $\varepsilon > 0$ such that $S_X(z,\varepsilon) \subset Z$. Thus $Y \cap S_X(z,\varepsilon) \subset Z$; hence $S_Y(z,\varepsilon) \subset Z$ and so z is an interior point of Z regarded as a subset of (Y,d_Y). Thus Z is open in (Y,d_Y). $\square$

Proposition 13 Let (X,d) be a metric space and $Z \subset Y \subset X$. Then Z is open in (Y,d_Y) if and only if there exists a set G open in (X,d) such that $Z = Y \cap G$.

Proof Assume that Z is open in (Y,d_Y). Then for each z in Z, there is an $\varepsilon > 0$, where ε depends on z, such that $S_Y(z,\varepsilon) \subset Z$. Define

$$G = \bigcup_{z \in Z} S_X(z,\varepsilon).$$

Then G is open in (X,d). Hence

$$Y \cap G = Y \cap \left\{ \bigcup_{z \in Z} S_X(z,\varepsilon) \right\} = \bigcup_{z \in Z} \{Y \cap S_X(z,\varepsilon)\} = \bigcup_{z \in Z} S_Y(z,\varepsilon).$$

But

$$\bigcup_{z \in Z} S_Y(z,\varepsilon) = Z.$$

Thus there is a set G open in (X,d) such that $Z = Y \cap G$.

Conversely, assume that there is a set G which is open in (X,d) and is such that $Z = Y \cap G$. If $z \in Z$, then $z \in G$, so there is an $\varepsilon > 0$ such that $S_X(z,\varepsilon) \subset G$. Thus $Y \cap S_X(z,\varepsilon) \subset Y \cap G = Z$, hence $S_Y(z,\varepsilon) \subset Z$; this means that z is an interior point of Z in (Y,d_Y). We therefore have that Z is open in (Y,d_Y). □

Corollary Let (X,d) be a metric space and $Z \subset Y \subset X$. Then Z is closed in (Y,d_Y) if and only if there exists a set F closed in (X,d) such that $Z = Y \cap F$.

Remark Let (X,d) be a metric space and $Z \subset Y \subset X$. Let $\overline{Z}^Y$ and $\overline{Z}^X$ denote the closures of Z in (Y,d_Y) and (X,d), respectively. It is simple to verify that

$$\overline{Z}^Y = Y \cap \overline{Z}^X.$$

Definition Let (X,d) be a metric space and $Z \subset Y \subset X$. We say that Z is *open relative to* Y if $Z = Y \cap G$ for some subset G of X open in (X,d).

Z is open relative to Y if and only if to each $z \in Z$ there is associated an $r > 0$ such that $y \in Z$ whenever $d(z,y) < r$ and $y \in Y$.

If $Z = (a,b)$, $Y = R^1$ and $X = R^2$ (where R^1 and R^2 are endowed with the usual Euclidean metric), then Z is open relative to Y without being an open subset of R^2; in other words, an open interval of the real line is not an open subset of the Cartesian plane.

The notion of *relative closedness* can be defined similarly.

Definition Let (X,d) be a metric space and let $(x_n)_{n=1}^{\infty}$ be an infinite sequence in X. Then a point p in X is called an *accumulation point* (or *cluster point*) *of the sequence* if there exists a subsequence $(x_{n_k})_{k=1}^{\infty}$ of $(x_n)_{n=1}^{\infty}$ which converges to p as $k \to \infty$.

Note that it is not required that all points of the subsequence be distinct.

Proposition 14 Let (X,d) be a metric space and $(x_n)_{n=1}^{\infty}$ be an infinite sequence in X; let p belong to X. Then p is an accumulation point of the sequence $(x_n)_{n=1}^{\infty}$ if and only if, given any $\varepsilon > 0$ and any integer m, there exists $n > m$ such that $d(x_n, p) < \varepsilon$.

Proof Let p be an accumulation point of the sequence $(x_n)_{n=1}^{\infty}$. Then there exists a subsequence $(x_{n_k})_{k=1}^{\infty}$ which converges to p as $k \to \infty$; so, given any $\varepsilon > 0$, there exists j such that $d(x_{n_k}, p) < \varepsilon$ for all $k \geq j$. Thus taking any $j' \geq j$ such that $n_{j'} > m$ and letting $n = n_{j'}$, we see that, given any $\varepsilon > 0$ and any integer m, there exists $n > m$ such that $d(x_n, p) < \varepsilon$.

Conversely, given any $\varepsilon > 0$ and any integer m, there exists $n > m$ such that $d(x_n, p) < \varepsilon$; we have to show that p is an accumulation point of the sequence. We define a subsequence of $(x_n)_{n=1}^{\infty}$ as follows. Let n_1 be the smallest integer such that $d(x_{n_1}, p) < 1$. Define n_2 to be the smallest integer such that $n_2 > n_1$ and $d(x_{n_2}, p) < \frac{1}{2}$. Define n_3 to be the smallest integer such that $n_3 > n_2$ and $d(x_{n_3}, p) < 1/3$, and so forth. The subsequence so obtained evidently converges to p; thus p is an accumulation point of $(x_n)_{n=1}^{\infty}$. □

Definition Let (X,d) be a metric space and $Y \subset X$. Then Y is said to be *everywhere dense* [in (X,d)] if $\overline{Y} = X$.

The proof of the following result is left as an exercise to the reader.

Proposition 15 Let (X,d) be a metric space and $Y \subset X$. Then the following statements are equivalent:

(i) Y is everywhere dense [in (X,d)];

(ii) for each x in X, there exists an infinite sequence $(x_n)_{n=1}^{\infty}$ of points of Y such that $x_n \to x$ as $n \to \infty$;

(iii) for each x in X, given any $\varepsilon > 0$, there exists x' in Y such that $d(x,x') < \varepsilon$;

(iv) if F is closed in (X,d) and $Y \subset F \subset X$, then $F = X$;

(v) if G is open in (X,d) and $Y \cap G = \emptyset$, then $G = \emptyset$;

(vi) Y intersects every open spere in (X,d).

Remarks The set of all rational points is everywhere dense in the real line R^1. The set of all real numbers of the form $\alpha x + y$, where α is a given irrational number and x and y are integers, is also everywhere dense in R^1 (see page 49 of the author's book "Real Analysis," American Elsevier Publishing Company, New York, 1973).

Proposition 16 If Y is a nonempty subset of a complete metric space (X,d), then (Y,d_Y) is complete if and only if Y is closed in (X,d).

Proof Let Y be closed in (X,d) and $(x_n)_{n=1}^{\infty}$ be Cauchy in (Y,d_Y). Then $(x_n)_{n=1}^{\infty}$ is also Cauchy in (X,d) and there exists x_0 in X such that $x_n \to x_0$. By Proposition 7, $x_0 \in \overline{Y}$ so $x_0 \in Y$. Hence (Y,d_Y) is complete.

Conversely suppose that (Y,d_Y) is complete and let x be any limit point of Y. Then there exists a sequence $(x_n)_{n=1}^{\infty}$ in Y which converges to x. Thus $(x_n)_{n=1}^{\infty}$ is a Cauchy sequence in (Y,d_Y) and so its limit x belongs to Y. Hence Y is closed. □

Remark In the second half of the proof of Proposition 16 we did not use that (X,d) is complete; hence we have the following result.

Proposition 17 If Y is a nonempty subset of a metric space (X,d) and if (Y,d_Y) is complete, then Y is closed in (X,d).

Proposition 18 If $(x_n)_{n=1}^{\infty}$ is a Cauchy sequence in a metric space (X,d) and if $(x_n)_{n=1}^{\infty}$ has an accumulation point p in X, then the sequence converges (to p).

Proof Given any $\varepsilon > 0$, there exists n_0 such that $d(x_m,x_n) < \frac{1}{2}\varepsilon$ for all m, $n > n_0$. Since p is an accumulation point of the sequence, by Proposition 14, there exists $n' > n_0$ such that $d(x_{n'},p) < \frac{1}{2}\varepsilon$. Thus, for any $n > n_0$,

$$d(x_n,p) \le d(x_n,x_{n'}) + d(x_{n'},p) < \varepsilon.$$

Therefore $(x_n)_{n=1}^{\infty}$ converges to p. □

Proposition 19 Every Cauchy sequence in a metric space can be contained in a sphere.

Proof Let $(x_n)_{n=1}^{\infty}$ be a Cauchy sequence in (X,d). Then there is an n_0 such that $d(x_m,x_n) < 1$ for all m, $n \ge n_0$, hence $x_n \in S(x_{n_0},1)$ for all $n \ge n_0$. There remain $x_1, x_2, \ldots, x_{n_0-1}$ which may not be in this sphere. We therefore define another sphere, center x_{n_0}, but with a suitably larger radius. Namely, let

$$r > \max\{1, d(x_1,x_{n_0}), d(x_2,x_{n_0}), \ldots, d(x_{n_0-1},x_{n_0})\}.$$

Then $x_n \in S(x_{n_0},r)$ for all n. It follows that the Cauchy sequence $(x_n)_{n=1}^{\infty}$ is contained in a sphere of finite radius. □

Corollary Every convergent sequence in a metric space can be contained in a sphere.

Proof Note that every convergent sequence is Cauchy; the corollary therefore follows from Proposition 19. □

Definition Let (X,d) be a metric space and $A \subset X$. We call the number

$$\sup\{d(x,y): x, y \in A\}$$

the *diameter of* A and denote it by $d(A)$.

Proposition 20 Let (X,d) be a metric space. Then (X,d) is complete if and only if whenever $(A_n)_{n=1}^{\infty}$ is a nested sequence of nonempty closed subsets of (X,d), that is,

$$A_1 \supset A_2 \supset \ldots \quad \text{such that} \quad \lim_{n\to\infty} d(A_n) = 0,$$

we have

$$\bigcap_{n=1}^{\infty} A_n = \{x\}$$

for some $x \in X$.

Proof Assume that $(A_n)_{n=1}^{\infty}$ is a nested sequence of nonempty closed subsets of (X,d) such that

$$\lim_{n\to\infty} d(A_n) = 0.$$

For each natural number n let $x_n \in A_n$. Then $m \geq n$ implies

$$d(x_m, x_n) \leq d(a_n) \to 0,$$

so $(x_n)_{n=1}^{\infty}$ is a Cauchy sequence. Let

$$x = \lim_{n\to\infty} x_n.$$

For each m, $x_n \in A_m$ for all large n, and A_m is closed, so $x \in A_m$. Thus

$$x \in \bigcap_{n=1}^{\infty} A_n.$$

If

$$y \in \bigcap_{n=1}^{\infty} A_n,$$

then $d(x,y) \leq d(A_n)$ for every n. Therefore $d(x,y) = 0$. Hence

$$\bigcap_{n=1}^{\infty} A_n = \{x\}.$$

To show the converse, suppose that (X,d) has the nested closed set property. Let $(x_n)_{n=1}^{\infty}$ be a Cauchy sequence in (X,d). For each natural number n, let

$$A_n = \overline{\{x_m : m \geq n\}}.$$

Then $(A_n)_{n=1}^{\infty}$ is a nested sequence of closed sets and since $(x_n)_{n=1}^{\infty}$ is a Cauchy sequence,

$$\lim_{n \to \infty} d(A_n) = 0.$$

Let

$$\bigcap_{n=1}^{\infty} A_n = \{x\}.$$

If $\varepsilon > 0$, then there is a natural number n_0 such that $d(A_{n_0}) < \varepsilon$. But $x \in A_{n_0}$ and thus $n \geq n_0$ implies $d(x_n,x) < \varepsilon$. □

Definition Let X be a nonempty set and $Y \subset X$. Suppose that

$$\mathcal{C} = \{C_t : t \in T\}$$

is a collection of subsets of X such that

$$Y \subset \bigcup_{t \in T} C_t.$$

Then $\mathcal{C}$ is said to be a *covering of* Y. If the indexing set T is finite, then $\mathcal{C}$ is called a *finite covering of* Y. If there exists a subset T' of T such that

$$\mathcal{C}' = \{C_t : t \subset T']$$

is also a covering of Y, then $\mathcal{C}'$ is called a *subcovering of* Y.

Definition If (X,d) is a metric space and $\mathcal{C} = \{C_t : t \in T\}$ is a covering of X such that C_t is open for all t in T, then $\mathcal{C}$ is called an *open covering of* (X,d).

Definition A metric space (X,d) is said to be *compact* if every open covering of (X,d) contains a finite subcovering of (X,d).

Definition Let (X,d) be a metric space and ε be an arbitrary positive number. Then a set $Y \subset X$ is called an *ε-net of* (X,d) if, given any $x \in X$, there is at least one point y in Y such that $d(y,x) < \varepsilon$. If the set Y is finite, then Y is said to be a *finite ε-net* of (X,d).

Definition A subset Y of a metric space (X,d) is said to be *bounded in* (X,d) if there is a sphere S such that $Y \subset S$. A metric space (X,d) is said to be *bounded* if the set $\{d(x,y): x, y \in X\}$ of real numbers is bounded. A metric space (X,d) is said to be *totally bounded* if given any $\varepsilon > 0$, there exists a finite ε-net.

Proposition 21 A totally bounded metric space (X,d) is bounded.

Proof If (X,d) is totally bounded, then there exists a finite 1-net, say $\{x_1, \ldots, x_n\}$; hence

$$X = \bigcup_{j=1}^{n} S(x_j,1).$$

But the union of a finite number of bounded sets is bounded. □

Proposition 22 Let (X,d) be a compact metric space. Then (X,d) is totally bounded.

Proof Note that if Y is an ε-net of (X,d), then

$$\bigcup_{y \in Y} S(y,\varepsilon) = X.$$

Thus (X,d) is totally bounded if and only if, given any $\varepsilon > 0$, every covering of X by open spheres of radius ε contains a finite subcovering. □

Proposition 23 Let (X,d) be a compact metric space. Then (X,d) is complete.

Proof Suppose, if possible, that (X,d) is a compact metric space which is not complete and let $(x_n)_{n=1}^{\infty}$ be a Cauchy sequence in (X,d) not having a limit in X.

Let $y \in X$; since $(x_n)_{n=1}^{\infty}$ does not converge to y, there is an $\varepsilon_0 > 0$ such that

$$d(x_n,y) \geq 2\varepsilon_0 \tag{4}$$

for infinitely many values of the index n. Since $(x_n)_{n=1}^{\infty}$ is Cauchy, there exists k such that $d(x_m,x_n) < \varepsilon_0$ for all $m, n > k$. Pick a value of n, say n_0, for which (4) is satisfied and such that $n_0 > k$. Then it follows that $d(x_m,y) > \varepsilon_0$ for all $m > k$ because

$$d(x_{n_0},y) \leq d(x_{n_0},x_m) + d(x_m,y).$$

Thus $S(y,\varepsilon_0)$ contains x_n for only a finite number of values of n.

In the same manner we can associate with each point $y \in X$ a sphere $S(y,\varepsilon_0)$, where ε_0 will depend on y. The collection of all such spheres forms an open covering of (X,d), which must contain, by assumption, a finite subcovering. Each sphere contains x_n for only a finite number of values of n, so the finite subcovering, and therefore X, must contain x_n for only a finite number of values of n; but this is impossible. □

Remark The statement "$S(y,\varepsilon_0)$ contains x_n for only a finite number of values of n" implies "$S(y,\varepsilon_0)$ contains only a finite number of terms of the sequence $(x_n)_{n=1}^{\infty}$," but the implication in the other direction is false.

Proposition 24 Let (X,d) be a complete and totally bounded metric space. Then (X,d) is compact.

Proof Suppose, if impossible, that (X,d) is complete and totally bounded, but is not compact. Then there exists an open covering $\mathcal{C} = \{C_t: t \in T\}$ of X which contains no finite subcovering of X. The proof proceeds in three steps.

1. Since (X,d) is totally bounded, it is bounded; hence for some real number r and some x_0 in X, $X = S(x_0,r)$ because $X \subset S(x_0,r)$. Let $\varepsilon_n = r/2^n$.

(X,d) being totally bounded, it can be covered by a finite number of spheres of radius ε_1. By assumption, at least one of these spheres, say $S(x_1,\varepsilon_1)$, cannot be covered by a finite number of the sets C_t. Again using that (X,d) is totally bounded, it follows that $S(x_1,\varepsilon_1)$ can be covered by a finite number of spheres of radius ε_2. In a similar manner, there exists one of these spheres, say $S(x_2,\varepsilon_2)$, such that

$$S(x_2,\varepsilon_2) \cap S(x_1,\varepsilon_1) \neq \emptyset,$$

and which cannot be covered by a finite number of the sets C_t.

Proceeding in this way, we define a sequence $(x_n)_{n=1}^{\infty}$ such that

$$S(x_n,\varepsilon_n) \cap S(x_{n-1},\varepsilon_{n-1}) \neq \emptyset,$$

and for which $S(x_n,\varepsilon_n)$ cannot be covered by a finite number of the sets C_t.

2. We now show that $(x_n)_{n=1}^{\infty}$ is Cauchy. Let x' be any point common to $S(x_{n+1},\varepsilon_{n+1})$ and $S(x_n,\varepsilon_n)$. Then

$$d(x_{n+1},x_n) \le d(x_{n+1},x') + d(x',x_n) < \varepsilon_{n+1} + \varepsilon_n < 2\varepsilon_n;$$

thus, for all $p \ge 0$,

$$\begin{aligned} d(x_{n+p},x_n) &\le d(x_{n+p},x_{n+p-1}) + d(x_{n+p-1},x_{n+p-2}) \\ &\quad + \dots + d(x_{n+1},x_n) \\ &\le 2\varepsilon_{n+p-1} + 2\varepsilon_{n+p-2} + \dots + 2\varepsilon_n \\ &< r2^{-n+2} \end{aligned}$$

since $\varepsilon_n = r/2^n$. Thus $(x_n)_{n=1}^{\infty}$ is a Cauchy sequence and so has a limit, say x_0, in X because (X,d) is complete.

3. Since $x_0 \in X$, there exists t_0 in T such that $x_0 \in C_{t_0}$. Moreover, since C_{t_0} is open, there exists $\delta > 0$ such that

$$S(x_0,\delta) \subset C_{t_0}.$$

Now take n sufficiently large, with value n_0 (say), such that

$$d(x_{n_0},x_0) < \frac{1}{2}\delta \quad \text{and} \quad \varepsilon_{n_0} < \frac{1}{2}\delta.$$

Then, if $d(x,x_{n_0}) < \varepsilon_{n_0}$, we obtain that

$$d(x,x_0) \le d(x,x_{n_0}) + d(x_{n_0},x_0) < \delta$$

and so

$$S(x_{n_0},\varepsilon_{n_0}) \subset S(x_0,\delta) \subset C_{t_0}.$$

Thus $S(x_{n_0},\varepsilon_{n_0})$ is covered by a finite number of the sets C_t in $\mathcal{C}$.

We have therefore reached a contradiction; so (X,d) is compact. □

Proposition 25 Let (X,d) be a compact metric space. Then every sequence in (X,d) has at least one accumulation point in X.

Proof Assume that (X,d) is compact and let $(x_n)_{n=1}^{\infty}$ be a sequence in X; suppose, if possible, that the sequence has no accumulation point in X.

Let $x \in X$, so x is not an accumulation point of $(x_n)_{n=1}^{\infty}$. By Proposition 14, it is not true that, given any $\varepsilon > 0$ and any integer $m > 0$, there exists an integer $n > m$ such that $d(x_n,x) < \varepsilon$; in other words, there exists $\varepsilon_0 > 0$ and a positive integer m_0 such that there is no $n > m_0$ for which $d(x_n,x) < \varepsilon_0$. Hence $d(x_n,x) < \varepsilon_0$ for at most a finite number of values of n.

Such an open sphere $S(x,\varepsilon_0)$ can be defined for each x in X, so the set of all such spheres forms an open covering of X. But (X,d) is compact by hypothesis and so there must be a finite subcovering; each sphere of such a subcovering contains x_n for only a finite number of values of n, so their union (which equals X) contains x_n for only a finite number of values of n. But this is impossible. □

Proposition 26 Let (X,d) be a metric space in which every sequence has at least one accumulation point. Then (X,d) is compact.

Proof In view of Proposition 24 we only need to show that (X,d) is complete and totally bounded.

Let $(x_n)_{n=1}^{\infty}$ be any Cauchy sequence in (X,d); let p be an accumulation point of this sequence. Then, by Proposition 18, $(x_n)_{n=1}^{\infty}$ converges to p; hence (X,d) is complete.

To see that (X,d) must be totally bounded, suppose, if possible, that there exists at least one $\varepsilon > 0$, say ε_0, such that there is no finite covering of X by open spheres of radius ε_0. We define a sequence $(x_n)_{n=1}^{\infty}$ as follows. Let x_1 be an arbitrary point of X. Having defined $x_1, \ldots, x_n$, let x_{n+1} be any point which is not in

$$S(x_1,\varepsilon_0) \cup \ldots \cup S(x_n,\varepsilon_0);$$

this is possible since no finite number of open spheres of radius ε_0 can cover X.

This sequence has no accumulation point. Indeed, for all m, n with $m \neq n$,

$$d(x_m,x_n) > \varepsilon_0; \tag{5}$$

if the sequence had an accumulation point, say p, then there would be a subsequence converging to p. This subsequence would be Cauchy [which contradicts (5)]. Thus $(x_n)_{n=1}^{\infty}$ has no accumulation point and (X,d) is totally bounded. □

Proposition 27 Let (X,d) be a metric space. Then the following two statements are equivalent:

(i) every infinite set in (X,d) has at least one limit point in X;

(ii) every infinite sequence in (X,d) has at least one accumulation point in X.

Proof Assume (i) and let $(x_n)_{n=1}^{\infty}$ be a sequence in (X,d). Then, either at least one term recurs infinitely often in $(x_n)_{n=1}^{\infty}$, and so will be an accumulation point, or no term recurs infinitely often, so $(x_n)_{n=1}^{\infty}$ contains an infinite number of distinct terms. In the latter case let Y be the infinite set consisting of these terms, having a limit point p, say.

A subsequence of $(x_n)_{n=1}^{\infty}$ which converges to p can be obtained as follows. Having defined n_{k-1}, let n_k be the smallest integer such that

$$n_k > n_{k-1} \quad \text{and} \quad 0 < d(p, x_{n_k}) < \frac{1}{k}.$$

Then $(x_{n_k})_{k=1}^{\infty}$ converges to p.

Assume (ii) and let Y be any infinite subset of X. Let $(x_n)_{n=1}^{\infty}$ be any sequence of distinct points of Y. By hypothesis there exists a convergent subsequence (in Y) with a limit p, say. Thus, by Proposition 4, p is a limit point of Y. □

We now summarize the foregoing results.

Proposition 28 Let (X,d) be a metric space. Then the following statements are equivalent:

(i) (X,d) is compact;

(ii) (X,d) is complete and totally bounded;

(iii) every infinite set in X has at least one limit point;

(iv) every sequence in X has at least one accumulation point.

Definition A subset K of a metric space is said to be *compact* if every open covering of K contains a finite subcovering.

Proposition 29 We have:

(i) Compact subsets of a metric space are closed.

(ii) Closed subsets of compact sets are compact.

Proof

(i) Let K be a compact subset of (X,d); we shall show that the complement $K^c = X - K$ of K is an open subset of (X,d).

Suppose $p \in X$, $p \notin K$. If $q \in K$, let V_q and W_q be neighborhoods of p and q, respectively, of radius less then $\frac{1}{2}\, d(p,q)$. Since K is compact, there are finitely many points $q_1, \ldots, q_n$ in K such that

$$K \subset W_{q_1} \cup \ldots \cup W_{q_n} = W.$$

If $V = V_{q_1} \cap \ldots \cap V_{q_n}$, then V is a neighborhood of p which does not intersect W. Hence $V \subset K^c$ and so p is seen to be an interior point of K^c.

(ii) Let $F \subset K \subset X$, F closed in (X,d) and K compact. Let $\{V_t : t \in T\}$ be an open covering of F. If F^c is adjoined to $\{V_t : t \in T\}$, we get an open covering $\mathcal{C}$ of K. Since K is compact, there is a finite subcovering $\mathcal{C}'$ of $\mathcal{C}$ which covers K, and hence F. If F^c is a member of $\mathcal{C}'$, we may remove it from $\mathcal{C}'$ and still retain an open covering of F. We therefore have that a finite subcollection of $\{V_t : t \in T\}$ covers F. □

3. *Continuity*

Definition Let (X,d), (X',d') be metric spaces and $E \subset X$. Let f be a function mapping E into X' and p be a limit point of E. We write

$$\lim_{x \to p} f(x) = q$$

if there is a point $q \in X'$ with the following property: For every $\varepsilon > 0$ there exists $\delta > 0$ such that

$$d'[f(x),q] < \varepsilon$$

for all points $x \in E$ for which $0 < d(x,p) < \delta$.

Remark In the definition of limit we have that $p \in X$, but p need not be in E. In addition, even if $p \in E$, we may have that $f(p) \neq \lim_{x \to p} f(x)$.

Proposition 30 Let (X,d), (X',d'), f, and p be as in the foregoing definition. Then

$$\lim_{x \to p} f(x) = q$$

if and only if

$$\lim_{n \to \infty} f(p_n) = q$$

for every sequence $(p_n)_{n=1}^{\infty}$ in E such that $p_n \neq p$, $\lim_{n\to\infty} p_n = p$.

Proof Suppose that $\lim_{x\to p} f(x) = q$. Choose $(p_n)_{n=1}^{\infty}$ in E satisfying $p_n \neq p$, $\lim_{n\to\infty} p_n = p$. Let $\varepsilon > 0$ be given. Then there exists $\delta > 0$ such that $d'[f(x),q] < \varepsilon$ if $x \in E$ and $0 < d(x,p) < \delta$. Also, there exists n_0 such that $n > n_0$ implies $0 < d(p_n,p) < \delta$. Thus, for $n > n_0$, we have $d'[f(p_n),q] < \varepsilon$ which shows that $\lim_{n\to\infty} f(p_n) = q$.

Conversely, suppose that $\lim_{x\to p} f(x) = q$ is false. Then there is some $\varepsilon > 0$ such that for every $\delta > 0$ there exists a point $x \in E$ (depending on δ), for which $d'[f(x),q] \geq \varepsilon$ but $0 < d(x,p) < \delta$. Letting $\delta_n = 1/n$, for $n = 1, 2, \ldots$, we thus find a sequence $(p_n)_{n=1}^{\infty}$ in E such that $p_n \neq p$, $\lim_{n\to\infty} p_n = q$, for which $\lim_{n\to\infty} f(p_n) = q$ is false. □

Corollary If f has a limit at p, this limit is unique.

Definition Let (X,d), (X',d') be metric spaces, $E \subset X$, and $p \in E$. Assume that f maps E into X'. Then f is said to be *continuous at* p if for every $\varepsilon > 0$ there exists $\delta > 0$ such that

$$d'[f(x),f(p)] < \varepsilon$$

for all points x in E for which $d(x,p) < \delta$. If f is continuous at every point of E, then f is said to be *continuous on* E.

Remark The function f has to be defined at p in order to be continuous at p. If p is an isolated point of E, then every function f having E as its domain of definition is continuous at p.

The next result follows directly from the corresponding definitions.

Proposition 31 Let (X,d), (X',d') be metric spaces, $E \subset X$, and $p \in E$. Assume that f maps E into X' and that p is a limit point of E. Then f is continuous at p if and only if $\lim_{x \to p} f(x) = f(p)$.

Proposition 32 Let (X,d), (X',d'), (X'',d'') be metric spaces, $E \subset X$, f map E into X', g map the range of f, that is, $f(E)$, into X", and h be the mapping of E into X" defined by

$$h(x) = g[f(x)]$$

for $x \in E$. If f is continuous at a point $p \in E$ and g is continuous at the point $f(p)$, then h is continuous at p.

Proof Let $\varepsilon > 0$ be given. Since g is continuous at $f(p)$, there exists $\gamma > 0$ such that

$$d''\{g(y),g[f(p)]\} < \varepsilon \quad \text{if} \quad d'[y,f(p)] < \gamma \quad \text{and} \quad y \in f(E).$$

Since f is continuous at p, there exists $\delta > 0$ such that

$$d'[f(x),f(p)] < \gamma \quad \text{if} \quad d(x,p) < \delta \quad \text{and} \quad x \in E.$$

It follows that

$$d''[h(x),h(p)] = d''\{g[f(x)],g[f(p)]\} < \varepsilon \quad \text{if} \quad d(x,p) < \delta \quad \text{and} \quad x \in E.$$

Thus h is continuous at p. □

Proposition 33 A mapping f of a metric space (X,d) into a metric space (X',d') is continuous on (X,d) if and only if for every open set V' of (X',d') the set

$$f^{-1}(V') = \{x \in X: f(x) \in V'\}$$

is open in (X,d).

Proof Suppose f is continuous on (X,d) and V' is an open set in (X',d'). We have to verify that every point of $f^{-1}(V')$ is an interior point of $f^{-1}(V')$. Thus, let $p \in X$ and $f(p) \in V'$. Since V' is open, there exists $\varepsilon > 0$ such that $y \in V'$ if $d'[f(p),y] < \varepsilon$, and since f is continuous at p there exists $\delta > 0$ such that $d'[f(x), f(p)] < \varepsilon$ if $d(x,y) < \delta$. Thus $x \in f^{-1}(V')$ whenever $d(x,p) < \delta$.

Conversely, suppose $f^{-1}(V')$ is open in (X,d) for every open set V' in (X',d'). Fix $p \in X$ and $\varepsilon > 0$; let V" be the set of all $y \in X'$ such that $d'[y,f(p)] < \varepsilon$. Then V' is open; hence $f^{-1}(V')$ is open; hence there exists $\delta > 0$ such that $x \in f^{-1}(V')$ whenever $d(p,x) < \delta$. But if $x \in f^{-1}(V')$, then $f(x) \in V'$, so that $d'[f(x),f(p)] < \varepsilon$. □

Corollary Let f, (X,d), and (X',d') be as in Proposition 33. Then f is continuous on (X,d) if and only if for every closed set V' in (X',d'), the set $f^{-1}(V')$ is closed in (X,d).

Proposition 34 Suppose f is a continuous mapping of a compact metric space (X,d) into a metric space (X',d'). Then f(X) is compact.

Proof Let $\{V_t : t \in T\}$ be a covering of f(X). Since f is continuous, Proposition 33 shows that each of the sets $f^{-1}(V_t)$ is open. Moreover, $\{f^{-1}(V_t) : t \in T\}$ is easily seen to be a covering of X. Since X is compact, there are finitely many indices, say $t_1, \ldots, t_n$, such that

$$X = \bigcup_{k=1}^{n} f^{-1}(V_{t_k}) = f^{-1}\left(\bigcup_{k=1}^{n} V_{t_k}\right).$$

It follows that

$$f(X) = f\left[f^{-1}\left(\bigcup_{k=1}^{n} V_{t_k}\right)\right] \subset \bigcup_{k=1}^{n} V_{t_k}. \qquad \square$$

Corollary If f is a continuous mapping of a compact metric space (X,d) into the real line (R^1, d_e), then f(X) is closed and bounded. Hence there exist points p and q in X such that

$$f(p) = \sup\{f(x): x \in X\} \quad \text{and} \quad f(q) = \inf\{f(x): x \in X\}.$$

Proposition 35 Let f be a continuous, one-to-one mapping [that is, $f(x_1) \neq f(x_2)$ whenever $x_1 \neq x_2$] of a compact metric space (X,d) onto a metric space (X',d'); onto means that f(X) = X'. Then the inverse function f^{-1} defined on (X',d') by

$$f^{-1}[f(x)] = x,$$

for all $x \in X$, is a continuous mapping of (X',d') onto (X,d).

Proof Applying Proposition 33 to f^{-1} instead of f, we see that it is enough to show that f(V) is an open set in (X',d') for every open set V in (X,d). We take such a set V.

The complement $V^c = X - V$ of V is closed in (X,d), hence compact (by Proposition 29). Thus $f(V^c)$ is a compact subset of (X',d') for every open set V in (X,d) (by Proposition 34) and so is closed in (X',d') (by Proposition 29). Since f is one-to-one and onto, f(V) is the complement of $f(V^c)$. Hence f(V) is open. This completes the proof. □

Definition Let f be a mapping of (X,d) into (X',d'). We say that f is *uniformly continuous on* (X,d) if for every $\varepsilon > 0$ there exists $\delta > 0$ such that

$$d'[f(p),f(q)] < \varepsilon$$

for all $p, q \in X$ for which $d(x,y) < \delta$.

Proposition 36 Let f be a continuous mapping of a compact metric space (X,d) into a metric space (X',d'). Then f is uniformly continuous on (X,d).

Proof Let $\varepsilon > 0$ be given. Since f is continuous, we can associate with each point $p \in X$ a positive number $\lambda(p)$ such that

$$q \in X,\ d(p,q) < \lambda(p) \quad \text{implies} \quad d'[f(p),f(q)] < \frac{\varepsilon}{2}. \tag{6}$$

Let $J(p)$ be the set of all $q \in X$ for which

$$d(p,q) < \frac{1}{2}\lambda(p).$$

Since $p \in J(p)$, the collection of all sets $J(p)$ is an open covering of (X,d); and, since (X,d) is compact, there is a finite set of points, say $p_1, \ldots, p_n$, in X such that

$$X \subset J(p_1) \cup \ldots \cup J(p_n). \tag{7}$$

We put

$$\delta = \frac{1}{2}\min\{\lambda(p_1), \ldots, \lambda(p_n)\}.$$

Then $\delta > 0$.

Now let q and p be points of X such that $d(p,q) < \delta$. By (7), there is an integer m, $1 \le m \le n$, such that $p \in J(p_m)$; hence

$$d(p, p_m) < \frac{1}{2}\lambda(p_m)$$

and we also have

$$d(q,p_m) \le d(p,q) + d(p,p_m) < \delta + \frac{1}{2}\lambda(p_m) \le \lambda(p_m).$$

Therefore, by (6),

$$d'[f(p),f(q)] \le d'[f(p),f(p_m)] + d'[f(q),f(p_m)] < \varepsilon.$$ □

Definition Let (X,d) be a metric space and Y be a nonempty subset of X. If $x \in X$, we call

$$\inf\{d(x,y): y \in Y\}$$

the *distance of* x *from* Y and denote it by $d(x,Y)$.

Proposition 37 Let (X,d) be a metric space, Y a nonempty subset of X and $x \in X$. Then the mapping f taking (X,d) into (R^1,d_e) defined by

$$f(x) = d(x,y),$$

for all $x \in X$, is uniformly continuous on X.

Proof For any x_0 in X it must be shown that, given any $\varepsilon > 0$, there exists $\delta > 0$, where δ is independent of x_0, for which

$$|d(x,Y) - d(x_0,Y)| < \varepsilon \quad \text{for all } x \text{ such that } d(x,x_0) < \delta. \tag{8}$$

But for all y in Y

$$d(x,Y) \le d(x,y) \le d(x,x_0) + d(x_0,y);$$

hence

$$d(x,Y) \le d(x,x_0) + d(x_0,Y),$$

and so

$$d(x,Y) - d(x_0,Y) \le d(x,x_0).$$

In a similar way we obtain that

$$d(x_0,Y) - d(x,Y) \le d(x_0,x);$$

thus

$$|d(x,Y) - d(x_0,Y)| \le d(x,x_0).$$

Taking $\delta = \varepsilon$ (so that δ is independent of x_0), (8) is satisfied and therefore f is seen to be uniformly continuous on X. □

Proposition 38 Let (X,d) be a metric space, $x \in X$ and Y be a nonempty subset of X. Then $d(x,Y) = 0$ if and only if $x \in \overline{Y}$.

Proof By Proposition 6, $x \in \overline{Y}$ if and only if given any $\varepsilon > 0$,

$$S(x,\varepsilon) \cap Y \neq \emptyset,$$

that is, $d(x,Y) < \varepsilon$. Thus $x \in \overline{Y}$ if and only if $d(x,Y) = 0$. □

Definition Let f be a mapping of a nonempty set X into itself. If there exists x' in X such that $f(x') = x'$, then x' is said to be a *fixed point of* f.

Definition Let f be a mapping of a metric space (X,d) into itself. Then f is called a *contraction mapping* if there exists λ such that, for all x, y in X,

$$d(f(x),f(y)) \le \lambda\, d(x,y),$$

where $0 \le \lambda < 1$.

The following result is an immediate consequence of the foregoing definition.

Proposition 39 Let (X,d) be a metric space and let f be a mapping of (X,d) into itself. If f is a contraction mapping, then f is uniformly continuous on X.

Proposition 40 (Banach's Fixed-Point Theorem) Let (X,d) be a complete metric space and f be a contraction mapping in (X,d). Then f has exactly one fixed point.

Proof Let $f^{(1)}(x) = f(x)$ and

$$f^{(n)}(x) = f[f^{(n-1)}(x)] \quad \text{for} \quad n = 2, 3, \ldots.$$

We take any point x_0 in X and define the sequence $(x_n)_{n=1}^{\infty}$ in (X,d) by

$$x_n = f^{(n)}(x_0)$$

for $n = 1, 2, \ldots$; thus $x_n = f(x_{n-1})$.

We now show that this sequence $(x_n)_{n=1}^{\infty}$ is Cauchy. Indeed,

$$\begin{aligned} d(x_n,x_{n+1}) &= d[f(x_{n-1}),f(x_n)] \le \lambda\, d(x_{n-1},x_n) \\ &\le \lambda^2\, d(x_{n-2},x_{n-1}) \\ &\ldots\ldots\ldots\ldots\ldots\ldots \\ &\le \lambda^n\, d(x_0,x_1) = \lambda^n c, \end{aligned}$$

where $c = d(x_0,x_1)$. Let m and n be integers such that $m < n$; then

$$\begin{aligned} d(x_m,x_n) &\le d(x_m,x_{m+1}) + d(x_{m+1},x_{m+2}) + \ldots + d(x_{n-1},x_n) \\ &\le \lambda^m c + \lambda^{m+1} c + \ldots + \lambda^{n-1} c \\ &< \frac{\lambda^m c}{1 - \lambda}, \end{aligned}$$

since $0 \leq \lambda < 1$. Therefore it follows that $d(x_m, x_n) \to 0$ as $m \to \infty$ (so $n \to \infty$ also), and $(x_n)_{n=1}^{\infty}$ is a Cauchy sequence in (X,d).

Since (X,d) is complete, $(x_n)_{n=1}^{\infty}$ converges to some element x' of X. The point x' is a fixed point of x; letting $n \to \infty$ in $x_{n+1} = f(x_n)$ and using that f is continuous (by Proposition 39), it follows that $x' = f(x')$.

To verify that the fixed point of f is unique, let x'' be any other fixed point of f. Then

$$d(x',x'') = d[f(x'),f(x'')] \leq \lambda\, d(x',x'').$$

Since $0 \leq \lambda < 1$, it follows that $d(x',x'') = 0$, so $x' = x''$. □

Definition A set E in a metric space (X,d) is said to be *connected* if there do not exist two disjoint open subsets A and B of X such that A intersects E, B intersects E, and $E \subset A \cup B$.

Proposition 41 If f is a continuous mapping of a connected metric space (X,d) into a metric space (X',d'), then $f(X)$ is connected.

Proof If $f(X)$ is not connected, there are disjoint open sets V and W in (X',d'), both of which intersect $f(X)$, such that $f(X) \subset V \cup W$. Since f is continuous, the sets $f^{-1}(V)$ and $f^{-1}(W)$ are open in (X,d); they are obviously disjoint and nonempty, and their union is X. But this means that X is not connected, a contradiction. □

Proposition 42 (Principle of Extension by Continuity) Let (X,d) and (X',d') be metric spaces, and let (X',d') be complete. Let A be an everywhere dense set in (X,d) and f be a uniformly continuous mapping of A into X'. Then f has a unique continuous extension F mapping X into X' with F being uniformly continuous on (X,d).

Proof If $x \in X$, there is an approximating sequence of points $(a_n)_{n=1}^{\infty}$ with $a_n \in A$ for $n = 1, 2, \ldots$ such that $\lim_{n\to\infty} a_n = x$. Since $(a_n)_{n=1}^{\infty}$ is convergent, it is Cauchy; since f is uniformly continuous, the sequence $[f(a_n)]_{n=1}^{\infty}$ is a Cauchy sequence in (X',d'). But (X',d') is complete and so there exists a point $F(x) \in X'$ with

$$\lim_{n \to \infty} f(a_n) = F(x).$$

To see that $F(x)$ is well defined, we verify that $F(x)$ depends only on x and not on the particular sequence $a_n \to x$. Let $(b_n)_{n=1}^{\infty}$ be another sequence in A with $\lim_{n \to \infty} b_n = x$. Then clearly

$$d(a_n,b_n) \to 0 \quad \text{as} \quad n \to \infty$$

and, since f is uniformly continuous,

$$d'[f(a_n),f(b_n)] \to 0 \quad \text{as} \quad n \to \infty$$

and so

$$\lim_{n \to \infty} f(b_n) = F(x).$$

It is now easy to see that $d(x,x') < \delta$ implies $d'[F(x),F(x')] < \varepsilon$ and from this the uniform continuity of F on (X,d) follows. The uniqueness of F is easy to establish and we leave its verification as an exercise to the reader. □

We conclude this section with an application of Proposition 40; the result we want to prove is a certain version of the *Implicit Function Theorem* and reads as follows:

Let S denote the strip $[a,b] \times R^1$, that is, the set of all ordered pairs (x,y) with $x \in [a,b]$ and $y \in R^1$, in the x, y plane. Let f be a real-valued continuous function on S satisfying

$$0 < m \leq \frac{f(x,y_2) - f(x,y_1)}{y_2 - y_1} \leq M \tag{9}$$

for all distinct y_1, y_2 in R^1. We shall show that there exists a unique real-valued continuous function g on $[a,b]$ such that

$$f[x,g(x)] = 0 \text{ for all } x \text{ in } [a,b].$$

Indeed, let C denote the set of all real-valued continuous functions on $[a,b]$. We associate with C the supremum metric d so that (C,d) is complete. We define a mapping F of (C,d) into itself by setting, for any $g \in C$,

$$\{F(g)\}(x) = g(x) - \frac{2}{M + m} f[x,g(x)].$$

For any g, $h \in C$, we have

$$\{F(g)\}(x) - \{F(h)\}(x) = g(x) - h(x) - \frac{2}{M + m}\{f[x,g(x)] - f[x,h(x)]\}$$

and so, by (9),

$$-\frac{M - m}{M + m} \le 1 - \frac{2}{M + m}\,\frac{f(x,y_2) - f(x,y_1)}{y_2 - y_1} \le \frac{M - m}{M + m}.$$

Thus

$$\begin{aligned} d[F(g),F(h)] &= \sup\{|[F(g)](x) - [F(h)](x)| : x \in [a,b]\} \\ &\le \sup\{|g(x) - h(x)| : x \in [a,b]\}\frac{M - m}{M + m} \\ &= \frac{M - m}{M + m} d(g,h) \end{aligned}$$

and F is seen to be a contraction mapping on (C,d). Hence there exists a unique element g of C such that $F(g) = g$, that is, such that $f[x,g(x)] = 0$ for all $x \in [a,b]$.

4. *Baire's Category Theorem*

Definition Let (X,d) be a metric space. A set $Y \subset X$ is called *nowhere dense in* (X,d) if $\overline{Y}$ contains no interior points. A set $F \subset X$ is said to be of *first category* if F is a countable union of nowhere dense sets. Subsets of X which are not of first category are said to be of *second category*.

Proposition 43 If Y is a nowhere dense subset of a metric space (X,d), then $X - \overline{Y}$ is everywhere dense in (X,d).

Proof Suppose that $X - \overline{Y}$ is not dense. Then there is a point $p \in X$ and an open sphere S such that $p \in S$ and $S \cap (X - \overline{Y}) = \emptyset$. Thus $p \in S \subset \overline{Y}$ and so p is an interior point of the closure of Y, but this is a contradiction. □

Proposition 44 If G is an open subset of a metric space (X,d) and Y is nowhere dense in (X,d), then there is an open ε-sphere $S(p,\varepsilon)$ such that $S(p,\varepsilon) \subset G$ and $S(p,\varepsilon) \cap Y = \emptyset$.

Proof Let $H = G \cap (X - \overline{Y})$. Then $H \subset G$ and $H \cap Y = \emptyset$. Since G and $X - \overline{Y}$ are open, H is open; since G is open and $X - \overline{Y}$ is dense, H is nonempty. Hence there is an $\varepsilon > 0$ such that $S(p,\varepsilon) \subset H$; moreover, $S(p,\varepsilon) \subset G$ and $S(p,\varepsilon) \cap Y = \emptyset$. □

Proposition 45 (Baire's Category Theorem) A complete metric space is of second category (as a subset of itself).

Proof We assume the contrary, i.e., we suppose that X is a complete metric space and

$$X = \bigcup_{n=1}^{\infty} M_n ,$$

where M_n are nowhere dense sets in X. Let S_0 be an open sphere of radius 1. Since M_1 is nowhere dense, there exists in S_0 an open sphere S_1 of radius $r_1 < \frac{1}{2}$ such that its closure $\overline{S}_1$ is disjoint from M_1. Since M_2 is nowhere dense there exists in S_1 an open sphere S_2 of radius $r_2 < 1/2^2$ such that its closure $\overline{S}_2$ is disjoint from M_2. Continuing this process, we obtain a sequence of nested closed sets

$$\overline{S}_1 \supset \overline{S}_2 \supset \overline{S}_3 \supset \cdots$$

and Proposition 20 is applicable. Now let x_0 be the point which is common to all closed sets $\overline{S}_n$, where $n = 1, 2, \ldots$. By our construction x_0 will not belong to any of the sets M_n; but this contradicts the assumption that

$$X = \bigcup_{n=1}^{\infty} M_n$$

and the proof is finished. □

As an application of Proposition 45 we prove the *existence of continuous nowhere differentiable functions*. We let X be the set of all real-valued continuous functions f on the interval [-1,1] satisfying f(-1) = f(1) and we endow X with the metric

$$d(f,g) = \sup\{|f(s) - g(s)|: s \in [-1,1]\}.$$

Since the uniform limit of continuous functions is itself continuous, it follows that (X,d) is a complete metric space. We let A_n denote the set of those elements of X for which there is some point s at which

$$|f(s + h) - f(s)| \le |h|n$$

for every h satisfying both $-1 \le s + h \le 1$ and $|h| \le 1/n$. If f is differentiable at some point s, then it belongs to A_n for a sufficiently high value of n. Thus it is enough to show that

$$\bigcup_{n=1}^{\infty} A_n$$

is a proper subset of X. Moreover, by Proposition 45 the space (X,d) is of second category so it will be sufficient to demonstrate that each A_n is nowhere dense in (X,d).

We first prove that A_n is a closed set by showing that its complement $X - A_n$ is open. This can be done by using a compactness argument: If $f \notin A_n$, then for any $s \in [-1,1]$ there is an h_s such that

$$|h_s| \le 1/n \quad \text{and} \quad |f(s + h_s) - f(s)| > |h_s|n.$$

Due to the strict inequality and the continuity of f we thus have

$$|f(\sigma + h_s) - f(\sigma)| > |h_s|n + 2\varepsilon_s \tag{10}$$

for some $\varepsilon_s > 0$ and for every σ in a suitable neighborhood J_s of s. Let $s_1, \ldots, s_m$ be chosen such that the corresponding neighborhoods J_{s_k} cover [-1,1] and let ε be the smallest one among the ε_{s_k}'s. Now, if $g \in X$ is such that

$$\sup\{|f(s) - g(s)|: s \in [-1,1]\} < \varepsilon,$$

then by (10) we have

$$|g(\sigma + h_{s_k}) - g(\sigma)| > |h_{s_k}|n$$

for every $\sigma \in J_{s_k}$ proving that $g \notin A_n$ and A_n is closed.

We complete the proof by showing that A_n has no interior points and consequently A_n is nowhere dense: Given $f \in A_n$ and $\varepsilon > 0$, by the Weierstrass Approximation Theorem (see Proposition 15 of Chapter 4) there is a polynomial p such that

$$\sup\{|f(s) - p(s)|: s \in [-1,1]\} < \frac{\varepsilon}{2}.$$

The derivative of p is bounded on [-1,1] and so by the Mean-Value Theorem (see Proposition 6 of Chapter 3) there is a constant M such that

$$|p(s + h) - p(s)| \le |h|M$$

for every s and h satisfying $|h| \le 1/n$. Let q be a "sawtooth" function on [-1,1] such that

$$\sup\{|q(s)|: s \in [-1,1]\} < \frac{\varepsilon}{2}$$

and the slopes of the linear segments are greater than M + n. Then $p + q \notin A_n$ though p + q belongs to the sphere $S(f,\varepsilon)$. This completes the proof of the existence of a continuous function f on [-1,1] satisfying f(-1) = f(1) and such that f is nowhere differentiable. By continuing f periodically one obtains a bounded, continuous, and nowhere differentiable function on the real line.

Definition Let Y be a subset of a metric space (X,d). The set of all interior points of Y is called the *interior of* Y and is denoted by Y°.

For each y in Y° there exists $\varepsilon_Y > 0$ such that $S(y,\varepsilon_Y) \subset Y$. Then it is easily verified that

$$\bigcup_{y \in Y^\circ} S(y,\varepsilon_Y) = Y^\circ.$$

It follows that Y° is open. It is left to the reader to show that Y° is also equal to the union of all open sets contained in Y.

N.B.: It is easy to show that $Y^\circ = X - \overline{X - Y}$. Indeed, first note that $Y^\circ \cap (\overline{X - Y}) = \emptyset$. For if $y \in Y^\circ$, there exists $\varepsilon > 0$ such that $S(x,\varepsilon) \subset Y$ and so $S(x,\varepsilon) \cap (X - Y) = \emptyset$; thus $x \notin \overline{X - Y}$. Next suppose that x is any point of X. Then either there exists $\varepsilon > 0$ such that $S(x,\varepsilon) \subset Y$, when $x \in Y^\circ$, or for all $\varepsilon > 0$

$$S(x,\varepsilon) \cap (X - Y) \neq \emptyset, \quad \text{when} \quad x \in \overline{X - Y}.$$

Using the formula $Y^\circ = X - \overline{X - Y}$, it is easy to see that the converse of Proposition 43 holds as well, that is, if $X - \overline{Y}$ is everywhere dense in (X,d), then Y is a nowhere dense subset of (X,d).

We now take up an important application of Proposition 45.

Proposition 46 (Uniform Boundedness Principle) Let F be any collection of continuous, real-valued functions on a complete metric space (X,d) such that for each $x \in X$, $|f(x)| \leq M_x$ for all $f \in F$. Then there exists a constant M and a nonempty open set U in (X,d) such that $|f(x)| \leq M$ for each $x \in U$ and each $f \in F$.

Proof Let $E_n = \{x \in X\colon |f(x)| \leq n \text{ for each } f \in F\}$. It is easily established that E_n is closed. By hypothesis, we have for each $x \in X$, a natural number n_x such that $|f(x)| \leq n_x$ for all $f \in F$. This implies $x \in E_{n_x}$. Hence

$$X = \bigcup_{n=1}^{\infty} E_n.$$

But (X,d) is of second category, by Proposition 45. Hence there is a natural number n such that E_n fails to be nowhere dense. Thus the open set

$$U = \text{interior of the closure of } E_n$$

is nonempty. Since E_n is closed, $U = (E_n)^\circ$. Consequently $|f(x)| \leq n$ holds for every $f \in F$ and every $x \in U$. □

The following result is also due to Baire.

Proposition 47 Let $(f_n)_{n=1}^{\infty}$ be a sequence of real-valued continuous functions defined on a metric space (X,d). Suppose that a finite limit

$$\lim_{n \to \infty} f_n(t) = f(t)$$

exists at every point t of X. Then the set of points at which the function f is discontinuous constitutes a set of first category.

Proof Let $\varepsilon > 0$ and put

$$P_m(\varepsilon) = \{t \in X: |f(t) - f_m(t) \leq \varepsilon\}$$

and

$$G(\varepsilon) = \bigcup_{m=1}^{\infty} P_m^\circ(\varepsilon),$$

where $P_m^\circ(\varepsilon)$ denotes the interior of $P_m(\varepsilon)$. Then we can show that

$$C = \bigcap_{n=1}^{\infty} G(1/n)$$

coincides with the set of all points at which f is continuous.

Suppose f is continuous at t_0. We shall show that $t_0 \in C$. Since $\lim_{n \to \infty} f_n(t) = f(t)$, there exists an m such that

$$|f(t_0) - f_m(t_0)| \le \frac{\varepsilon}{3}.$$

By the continuity of f and f_m at t_0, there exists an open set U_{t_0} containing t_0 such that

$$|f(t) - f(t_0)| \le \frac{\varepsilon}{3} \quad \text{and} \quad |f_m(t) - f_m(t_0)| \le \frac{\varepsilon}{3}.$$

whenever $t \in U_{t_0}$. Thus $t \in U_{t_0}$ implies

$$\begin{aligned} |f(t) - f_m(t)| &\le |f(t) - f(t_0)| + |f(t_0) - f_m(t_0)| \\ &\quad + |f_m(t_0) - f_m(t)| \\ &\le \varepsilon, \end{aligned}$$

which proves that $t_0 \in P_m^\circ(\varepsilon)$ and so $t_0 \in G(\varepsilon)$. Since $\varepsilon > 0$ was arbitrary, we must have

$$t_0 \in \bigcap_{n=1}^{\infty} G(1/n).$$

Let, conversely,

$$t_0 \in \bigcap_{n=1}^{\infty} G(1/n).$$

Then, for any $\varepsilon > 0$, $t_0 \in G(\varepsilon/3)$ and so there exists an m such that $t_0 \in P_m^\circ(\varepsilon/3)$. Thus there is an open set U_{t_0} containing t_0 such that $t \in U_{t_0}$ implies

$$|f(t) - f_m(t)| \le \varepsilon/3.$$

Hence, by the continuity of f_m and the arbitrariness of $\varepsilon > 0$, f must be continuous at t_0.

After these observations, we put

$$F_m(\varepsilon) = \{t \in X: |f_m(t) - f_{m+k}(t)| \le \varepsilon \quad (k = 1, 2, \ldots)\}.$$

This is a closed set by the continuity of the f_n's. We have

$$X = \bigcup_{m=1}^{\infty} F_m(\varepsilon)$$

because $\lim_{n\to\infty} f_n(t) = f(t)$. Again by $\lim_{n\to\infty} f_n(t) = f(t)$, we have

$$F_m(\varepsilon) \subset P_m(\varepsilon).$$

Thus $F_m^\circ(\varepsilon) \subset P_m^\circ(\varepsilon)$ and so

$$\bigcup_{m=1}^{\infty} F_m^\circ(\varepsilon) \subset G(\varepsilon).$$

On the other hand, for any closed set F, $(F - F^\circ)$ is a nowhere dense closed set. Thus

$$X - \bigcup_{m=1}^{\infty} F_m^\circ(\varepsilon) = \bigcup_{m=1}^{\infty} [F_m(\varepsilon) - F_m^\circ(\varepsilon)]$$

is a set of first category. Thus its subset $[G(\varepsilon)]^c = X - G(\varepsilon)$ is also a set of first category. Therefore the set of all points of discontinuity of the function f, which is expressible as

$$X - \bigcap_{n=1}^{\infty} G(1/n) = \bigcup_{n=1}^{\infty} [G(1/n)]^c,$$

is a set of first category. □

Remarks Every subset of a set of first category is of first category; the union of a finite or countably infinite number of sets of first category is of first category. If M_2 is a set of second category and M_1 is a subset of M_2 and M_1 is of first category, then $M_2 - M_1$ is of second category. The empty set is nowhere dense, hence of first category; thus a set of second category cannot be empty.

5. *Tietze's Extension Theorem*

The proof of the following simple result is left as an exercise to the reader.

Proposition 48 Let (X,d), (X',d') be metric spaces and f, g be continuous mappings of (X,d) into (X',d').

(i) If $f(x) = g(x)$ for all x in A, then $f(x) = g(x)$ for all x in $\overline{A}$; thus if $B = \{x: f(x) = g(x)\}$, then B is closed.

(ii) If $X' = R^1$, d' is the Euclidean metric and $C = \{x: f(x) \geq g(x)\}$, then C is closed.

Proposition 49 Let (X,d) be a metric space and Y be a nonempty closed subset of (X,d); let f be a bounded real-valued continuous function defined on (Y,d_Y) for which there exists $M > 0$ such that

$$\inf\{f(x): x \in Y\} = -M \quad \text{and} \quad \sup\{f(x): x \in Y\} = M. \tag{11}$$

Then there exists a continuous real-valued function g defined on (X,d) such that

$$|g(x)| \leq \frac{1}{3} M$$

for all $x \in X$ and

$$|g(x)| < \frac{1}{3} M \quad \text{on} \quad X - Y, \qquad |f(x) - g(x)| \leq \frac{2}{3} M \quad \text{on} \quad Y.$$

Proof Let

$$A = \left\{x: f(x) \leq -\frac{1}{3} M\right\}, \quad B = \left\{x: f(x) \geq \frac{1}{3} M\right\};$$

then the sets A, B are nonempty, disjoint and [by (ii) of Proposition 48] closed in (Y,d_Y). Since Y is closed in (X,d), the sets A and B are closed in (X,d). We define a real-valued function g on (X,d) by

$$g(x) = \frac{1}{3} M \frac{d(x,A) - d(x,B)}{d(x,A) + d(x,B)}.$$

By Proposition 37, g is continuous.

Evidently $|g(x)| \leq (1/3)M$ for all x in X. Moreover $x \in X - Y$ implies that $x \notin A$ and $x \notin B$, so $|g(x)| < (1/3)M$. Suppose that $x \in Y$. If $x \in A$, then $g(x) = -(1/3)M$, $-M \leq f(x) \leq -(1/3)M$ and so

$$|f(x) - g(x)| \leq \frac{2}{3} M;$$

similarly for $x \in B$. If $x \in Y - (A \cup B)$, then $|g(x)| < (1/3)M$, $|f(x)| < (1/3)M$ and so

$$|f(x) - g(x)| < \frac{2}{3} M. \qquad \square$$

Proposition 50 (Tietze's Extension Theorem) Let (X,d) be a metric space and Y be a nonempty closed subset of (X,d); let f be a continuous real-valued function on (Y,d_Y). Then there exists a continuous extension F of f which is defined on all of (X,d), that is, $F = f$ on Y and F is continuous on all of X. Moreover, if f is bounded, say

$$m = \inf\{f(x): x \in Y\}, \quad M = \sup\{f(x): x \in Y\},$$

then, provided $m < M$, there is a continuous extension F of f such that $m < F(x) < M$ on $X - Y$.

Proof First we assume that f is bounded; without loss of generality we may stipulate that $m = -M$ (for otherwise a constant can be added to f to achieve this).

We define a sequence $(g_n)_{n=1}^{\infty}$ of continuous real-valued functions on (X,d) as follows. We write f_1 in place of f; by Proposition 49 there exists a continuous real-valued function g_1 on (X,d) such that $|g_1(x)| \le (1/3)M$ for all x and

$$|g_1(x)| < \frac{1}{3} M \quad \text{on } X - Y, \qquad |f_1(x) - g_1(x)| \le \frac{2}{3} M \quad \text{on } Y.$$

Let f_2 on Y be defined by $f_2(x) = f_1(x) - g_1(x)$; then there exists a continuous real-valued function g_2 on X such that

$$|g_2(x)| \le \frac{1}{3}\left(\frac{2}{3} M\right)$$

for all x and

$$|g_2(x)| < \frac{1}{3}\left(\frac{2}{3} M\right) \quad \text{on } X - Y, \qquad |f_2(x) - g_2(x)| \le \left(\frac{2}{3}\right)^2 M \quad \text{on } Y.$$

Similarly let f_3 on Y be defined by $f_3(x) = f_2(x) - g_2(x)$; then there exists a continuous real-valued function g_3 on X such that

$$|g_3(x)| < \frac{1}{3}\left(\frac{2}{3}\right)^2 M$$

for all x and

$$|g_3(x)| < \frac{1}{3}\left(\frac{2}{3}\right)^2 M \text{ on } X - Y, \qquad |f_3(x) - g_3(x)| \le \left(\frac{2}{3}\right)^2 M \text{ on } Y.$$

Continuing in this manner, we obtain for each positive integer n a continuous real-valued function g_n on (X,d) such that

$$|g_n(x)| \le \frac{1}{3}\left(\frac{2}{3}\right)^{n-1} M \quad \text{for all x in X,} \tag{12}$$

$$|g_n(x)| < \frac{1}{3}\left(\frac{2}{3}\right)^{n-1} M \quad \text{on } X - Y, \text{ and} \tag{13}$$

$$|f_n(x) - g_n(x)| \le \left(\frac{2}{3}\right)^n M \qquad \text{on } Y,$$

where

$$f_n = f_{n-1} - g_{n-1} = f_{n-2} - (g_{n-2} + g_{n-1}) = \cdots$$
$$= f - (g_1 + \cdots + g_{n-1});$$

it follows that

$$\left|f(x) - \sum_{j=1}^{n} g_j(x)\right| \le \left(\frac{2}{3}\right)^n M \quad \text{on } Y. \tag{14}$$

Note that, as n increases, $g_1 + \cdots + g_n$ becomes a better and better approximation for f on Y, and that $g_1 + \cdots + g_n$ is continuous on all of X.

Define F by

$$F(x) = \sum_{j=1}^{\infty} g_j(x) \tag{15}$$

for all x in X. By (12) and Proposition 9 of Chapter 4, the series (15) is uniformly convergent over X; but each g_j is continuous, so F is continuous.

Letting $n \to \infty$ in (14), it follows that F(x) = f(x) on Y. Moreover, we get from (13) that

$$\left|f(x)\right| \le \sum_{j=1}^{\infty} \left|g_j(x)\right| < M \sum_{j=1}^{\infty} \frac{2^{j-1}}{3^j} = M$$

for all x in X - Y and so $|F(x)| < M$ on X - Y.

Now suppose that f is not bounded on Y. Let h be any one-to-one continuous mapping of R^1 onto the open interval (-1,1); for example let h be defined by

$$h(x) = \left(\frac{2}{\pi}\right) \tan^{-1} x$$

for all x in R^1 (the real line). Then the function h(f) is continuous and bounded on X; hence there exists a continuous extension H of h, where H is defined on all of (X,d). Define F on (X,d) by

$$F = h^{-1}(H);$$

then F is continuous on (X,d) and F(x) = f(x) on Y. □

Remark We have tacitly assumed that the reader can formulate the notion of uniform convergence for a sequence of real-valued continuous functions on a metric space (X,d) and can readily generalize those results in Chapter 4 which are needed in the foregoing proof (such as the fact that the uniform limit of a sequence of continuous real-valued functions on a metric space is continuous as well as the validity of the Weierstrass M-Test in these slightly more general circumstances).

Exercises

1. Let $(x_n)_{n=1}^{\infty}$ be a sequence in a metric space (X,d) such that the subsequences $(x_{2n})_{n=1}^{\infty}$, $(x_{2n+1})_{n=1}^{\infty}$, and $(x_{3n})_{n=1}^{\infty}$ are all convergent. Show that the sequence converges.

2. Let $(x_n)_{n=1}^{\infty}$ and $(y_n)_{n=1}^{\infty}$ be Cauchy sequences in a metric space (X,d). Show that the sequence $[d(x_n,y_n)]_{n=1}^{\infty}$ converges.

3. Let $(x_n)_{n=1}^{\infty}$, $(y_n)_{n=1}^{\infty}$ be sequences in (X,d) such that $(x_n)_{n=1}^{\infty}$ is Cauchy and $d(x_n,y_n) \to 0$ as $n \to \infty$. Show that

(i) $(y_n)_{n=1}^{\infty}$ is Cauchy;

(ii) $(y_n)_{n=1}^{\infty}$ converges to a limit p if and only if $(x_n)_{n=1}^{\infty}$ also converges to p.

4. Let $X = R^n$, $x = (x_1, \ldots, x_n)$, $y = (y_1, \ldots, y_n)$,

$$d(x,y) = \left(\sum_{j=1}^{n} |x_j - y_j|^p\right)^{1/p},$$

where $p \geq 1$, and

$$d'(x,y) = \max\{|x_j - y_j|: 1 \leq j \leq n\}.$$

Show that

$$d(x,y) \leq \sqrt[p]{n}\, d'(x,y) \quad \text{and} \quad d'(x,y) \leq d(x,y).$$

5. As was noted in Example 3 of Section 1, metric convergence implies coordinatewise convergence in ℓ_p, that is, if

$$x^{(n)} = (x_1^{(n)}, x_2^{(n)}, \ldots) \quad \text{and} \quad x = (x_1, x_2, \ldots),$$

then

$$d(x^{(n)}, x) = \left(\sum_{j=1}^{\infty} |x_j^{(n)} - x_j|^p\right)^{1/p} \to 0$$

implies $x_j^{(n)} \to x_j$ for $j = 1, 2, \ldots$. However, the converse is not true; for example, consider the sequence

$$x_j^{(n)} = x_j + \delta_{jn},$$

where

$$\delta_{jn} = 1 \quad \text{if} \quad j = n,$$

$$= 0 \quad \text{if} \quad j \neq n.$$

6. Let f be a mapping of (X,d) into (X',d'). Show that f is continuous over X if and only if $f(\overline{Y}) \subset \overline{f(Y)}$ for each set $Y \subset X$.

7. Let (X,d), (X',d') be two metric spaces and f be a uniformly continuous mapping of (X,d) into (X',d'). If $(x_n)_{n=1}^{\infty}$ is a Cauchy

sequence in (X,d), show that $[f(x_n)]_{n=1}^{\infty}$ is a Cauchy sequence in (X',d'). If, furthermore, f is one-to-one and onto and f^{-1} is also uniformly continuous, prove that (X,d) is complete if and only if (X',d') is complete.

8. Let f and g be continuous mappings of a metric space (X,d) into a metric space (X',d'), and let Y be an everywhere dense subset of (X,d). Prove that $f(Y)$ is dense in $f(X)$. If $g(x) = f(x)$ for all $x \in Y$, prove that $g(x) = f(x)$ for all $x \in X$. (In other words, a continuous mapping is determined by its values on a dense subset of its domain of definition).

9. Let A and B be disjoint nonempty closed sets in a metric space (X,d) and define

$$f(x) = \frac{d(x,A)}{d(x,A) + d(x,B)} .$$

Show that f is a continuous function on (X,d) whose range lies in $[0,1]$, that $f(x) = 0$ precisely when $x \in A$ and $f(x) = 1$ precisely when $x \in B$. Setting

$$V = f^{-1}\{[0,1/2)\}, \quad W = f^{-1}\{(1/2,1]\},$$

show that V and W are open and disjoint, and that $A \subset V$, $B \subset W$.

Hint See Exercise 14 at the end of Chapter 2.

10. Let f be a continuous real-valued function on the interval $[0,1]$. Let f_1 be any integral of f, that is,

$$f_1'(x) = f(x)$$

for all $x \in [0,1]$, f_2 any integral of f_1, and so on. If some f_k vanishes identically, so does f; we have only to differentiate f_k repeatedly.

Show that if for each $x \in [0,1]$ there is an integer k, possibly differing from one x to another, such that $f_k(x) = 0$, then f vanishes identically.

Solution Let E_k be the set of points x for which $f_k(x) = 0$; then our hypothesis says that every x in $[0,1]$ is in some E_k. By Baire's Category Theorem (see Proposition 45), not every E_k is nowhere dense. Hence there is some k for which the closure of E_k fills an interval I_k. For this particular k, since f_k is continuous and vanishes on E_k, we must have $f_k(x) = 0$ for every x in I_k. If I_k is not all of [0,1], we repeat this argument with any remaining part of [0,1] and so on. In this way we have $f(x) = 0$ for all points x of an everywhere dense set in [0,1]; and since f is continuous, it follows that $f(x) = 0$ for every x in [0,1].

Thus if $f \neq 0$, then no matter how the integrals f_k are selected, there must be some x (in fact, an everywhere dense set) such that $f_k(x) \neq 0$ for every k. □

11. Consider a continuous real-valued function f on [0,1]. If f has an n-th derivative that is identically zero, it is easily seen (by repeated application of Proposition 6 of Chapter 3) that f coincides with a polynomial (of degree at most n - 1). The following statement generalizes this in the spirit of Exercise 10.

Let f have derivatives of all orders on [0,1], and suppose that at each point some derivative of f is zero; that is, for each x there is an integer n(x) such that $f^{[n(x)]}(x) = 0$. Show that f then coincides on [0,1] with some polynomial.

N.B.: The reader interested the foregoing result is referred to the paper by E. Corominas and F. Sunyer Balaguer entitled "Condiciones para que una función infinitamente derivable sea un polinomo," *Revista Mat. Hisp.-Amer.* (4), 14 (1954), 26-43.

12. Show that the set Q of all rational numbers is of first category in R^1; hence the set of all irrationals is of second category (showing incidentally the existence of irrational numbers).

13. Prove that Cantor's (ternary) set is nowhere dense.

14. The reader is reminded that Propositions 5, 7, 15, 31, 39, and 48 need proofs.

15. Let $X = R^n$, $x = (x_1, \ldots, x_n)$, $y = (y_1, \ldots, y_n)$, and

$$d(x,y) = \left(\sum_{j=1}^{n} (x_j - y_j)^2 \right)^{1/2}.$$

Show that

(a) (X,d) is complete;

(b) a subset Y of X is compact if and only if it is closed and bounded.

16. Let C be the set of all real-valued continuous functions on a compact interval I. For $f, g \in C$, let

$$d(f,g) = \sup\{|f(x) - g(x)|: x \in I\}.$$

Show that a subset Y of C is uniformly bounded on I if and only if Y is a bounded subset of (C,d).

17. Let I be a compact interval and S be a set of real-valued continuous functions on I. Then S is said to be *equicontinuous on* I if, given any $\varepsilon > 0$, there exists a $\delta > 0$ such that

$$|f(x_1) - f(x_2)| < \varepsilon$$

for all f in S and all x_1, $x_2 \in I$ for which $|x_1 - x_2| < \delta$.

Let (C,d) be as in Exercise 16. Prove that a subset S of C is compact if and only if S is a closed subset of (C,d) and is both uniformly bounded and equicontinuous on I.

Hint See Section 5 of Chapter 4.

18. A metric space is said to be *separable* if it contains an everywhere dense subset which is countable.

Let (C,d) be as in Exercise 16; show that (C,d) is separable.

Hint See Exercise 30 at the end of Chapter 4.

19. Let (X,d) and (X',d') be metric spaces. A mapping f of X into X' is said to be *bounded* if $f(X)$ is a bounded subset of (X',d'). Let $C(X,X')$ denote the set of all bounded continuous mappings of (X,d) into (X',d') and let

$$D(f,g) = \sup\{d'(f(x), g(x)): x \in X\}.$$

for $f, g \in C(X,X')$.

Show that if (X',d') is complete, then $C(X,X')$ equipped with the distance function D is a complete metric space.

N.B.: Note the special case where (X',d') is the real line R^1.

20. Let (X,d) be a metric space, and fix a point p in X. Assign to every $u \in X$ the real-valued function f_u given by

$$f_u(x) = d(u,x) - d(p,x)$$

for all $x \in X$. Show that the mapping taking u to f_u is a distance-preserving mapping (also called an *isometry*) of (X,d) into $C(X,R^1)$.

N.B.: This shows that we can embed any metric space (X,d) in $C(X,R^1)$, the space of bounded real-valued continuous functions on X and therefore that every metric space has a completion. [Let (X,d) and (X',d') be metric spaces. We say that (X',d') is a *completion of* (X,d) if (X',d') is complete and (X,d) is isometric to a dense subset of (X',d'). A metric space (X,d) is said to be *isometric to* a metric space (X',d') if there exists a one-to-one mapping of X onto X' which preserves distance, that is, for all $p, q \in X$, $d(p,q) = d'[f(p),f(q)]$.]

21. The fact that a continuous function may fail to have a derivative at any point was considered in Proposition 12 of Chapter 4; in Section 4 of this chapter we saw that "most" continuous functions have this property. Equally surprising is the fact that a function may be everywhere oscillating and still have a finite derivative at every point.

Theorem There exists a real-valued, everywhere differentiable function H on R^1 such that H is monotone on no subinterval of R^1 and H' is bounded.

We divide the proof into seven steps.

I. Let r and s be real numbers.

(i) If $r > s > 0$, then $(r - s)/(r^2 - s^2) < 2/r$.

(ii) If $r > 1$ and $s > 1$, then $(r + s - 2)/(r^2 + s^2 - 2) < 2/s$.

Proof Assertion (i) is obvious. Inequality (ii) is equivalent to

$$(r - s)^2 + (r - 1)(s - 1) + r^2 + r + 3s > 5.$$

But this too is obvious when $r > 1$ and $s > 1$. □

II. Let $v(x) = (1 + |x|)^{-\frac{1}{2}}$ for $x \in R^1$. Then

$$\frac{1}{b - a} \int_a^b v(x)\, dx < 4 \min\{v(a), v(b)\}$$

whenever a and b are distinct real numbers.

Proof We may suppose that $a < b$. In case $0 \le a$, we have by step I that

$$\frac{1}{b - a} \int_a^b v(x)\, dx = \frac{2(\sqrt{1 + b} - \sqrt{1 + a})}{(1 + b) - (1 + a)}$$

$$< \frac{4}{\sqrt{1 + b}} = 4 \min\{v(a), v(b)\}.$$

Since $v(-x) = v(x)$, the case $b \le 0$ needs no special consideration. We therefore suppose that $a < 0 < b$. Then step I yields

$$\frac{1}{b - a} \int_a^b v(x)\, dx = \frac{2(\sqrt{1 + b} + \sqrt{1 - a} - 2)}{(1 + b) + (1 - a) - 2}$$

$$< 4 \min\{v(a), v(b)\}.$$ □

III. If v is as in step II and w is any function of the form

$$w(x) = \sum_{j=1}^{n} c_j\, v[\lambda_j(x - \alpha_j)],$$

where $c_1, \ldots, c_n$ and $\lambda_1, \ldots, \lambda_n$ are positive real numbers and $\alpha_1, \ldots, \alpha_n$ are any real numbers, then

$$\frac{1}{b - a} \int_a^b w(x)\, dx < 4\{\min w(a), w(b)\}.$$

Proof This follows directly from step II and the fact that

$$\frac{1}{b - a} \int_a^b v[\lambda(x - \alpha)]\, dx$$

$$= \frac{1}{\lambda(b - \alpha) - \lambda(a - \alpha)} \int_{\lambda(a - \alpha)}^{\lambda(b - \alpha)} v(t)\, dt. \qquad \square$$

IV. Let $(w_n)_{n=1}^{\infty}$ be any sequence of functions as in step III. For $x \in R^1$ and each n define

$$W_n(x) = \int_0^x w_n(t)\, dt.$$

Let $\Sigma_{n=1}^{\infty} w_n(a) = s < \infty$ for some $a \in R^1$. Then the series

$$F(x) = \sum_{n=1}^{\infty} W_n(x)$$

converges uniformly on every bounded subset of R^1, the function F is differentiable at a, and $F'(a) = s$. In particular, if

$$\sum_{n=1}^{\infty} w_n(t) = f(t) < \infty$$

for all $t \in R^1$, then F is differentiable everywhere on R^1 and $F' = f$.

Proof Let $b \in R^1$ satisfy $b \geq |a|$. Then, using step III, $-b \leq x \leq b$ implies

$$|W_n(x)| \leq \left| \int_0^a w_n(t)\, dt \right| + \left| \int_a^x w_n(t)\, dt \right|$$

$$\leq 4|a|\, w_n(a) + 4|x - a|\, w_n(a)$$

$$\leq 12b\, w_n(a).$$

Thus, uniform convergence on [-b, b] follows from Proposition 9 of Chapter 4.

To show that $F'(a) = s$, let $\varepsilon > 0$ be given. Choose n_0 such that

$$10 \sum_{n=n_0+1}^{\infty} w_n(a) < \varepsilon.$$

Since each w_n is continuous at a, there exists some $\delta > 0$ such that

$$\left| \frac{1}{h} \int_a^{a+h} w_n(t)\, dt - w_n(a) \right| < \frac{\varepsilon}{2n_0}$$

whenever $0 < |h| < \delta$ and $1 \le n \le n_0$. Therefore, using step III again, $0 < |h| < \delta$ implies that

$$\left| \frac{F(a+h) - F(a)}{h} - s \right| = \left| \sum_{n=1}^{\infty} \left[\frac{1}{h} \int_a^{a+h} w_n(t)\, dt - w_n(a) \right] \right|$$

$$\le \sum_{n=1}^{n_0} \left| \frac{1}{h} \int_a^{a+h} w_n(t)\, dt - w_n(a) \right|$$

$$+ \sum_{n=n_0+1}^{\infty} \left[\frac{1}{h} \int_a^{a+h} w_n(t)\, dt + w_n(a) \right]$$

$$< \frac{\varepsilon}{2} + \sum_{n=n_0+1}^{\infty} 5\, w_n(a) < \varepsilon. \qquad \square$$

V. Let $I_1, \ldots, I_n$ be disjoint open intervals, let α_j be the midpoint of I_j, and let ε and $y_1, \ldots, y_n$ be positive real numbers. Then there is a function w as in step III such that for each j

(i) $w(\alpha_j) > y_j$,

(ii) $w(x) < y_j + \varepsilon$ if $x \in I_j$,

(iii) $w(x) < \varepsilon$ if $x \notin I_1 \cup \cdots \cup I_n$.

Proof Choose $c_j = y_j + \varepsilon/2$ and define $v_j(x)$ to be $v_j(x) = c_j\, v(\lambda_j(x - \alpha_j))$, where λ_j is chosen so large that $v_j(x) < \varepsilon/2n$ if $x \notin I_j$ (one needs only to check this inequality at an endpoint of I_j). Take

$$w = v_1 + \cdots + v_n.$$

Since the I_j's are disjoint and because v_j takes its largest value c_j at α_j, properties (i), (ii), and (iii) are clear. □

VI. Let $\{\alpha_j\}_{j=1}^{\infty}$ and $\{\beta_j\}_{j=1}^{\infty}$ be disjoint countable subsets of R^1. Then there exists a real valued, everywhere differentiable function F on R^1 satisfying $F'(\alpha_j) = 1$, $F'(\beta_j) < 1$ for all j, and $0 < F'(x) \le 1$ for all x.

Proof We obtain F as in step IV by first constructing $F' = f = \Sigma_{n=1}^{\infty} w_n$, or, more precisely, the partial sums $f_n = \Sigma_{k=1}^{n} w_k$, in such a way that

$$f_n(\alpha_j) > 1 - \frac{1}{n} \qquad (1 \le j \le n), \tag{A_n}$$

$$f_n(x) < 1 - \frac{1}{n+1} \qquad (x \in R^1), \tag{B_n}$$

$$w_n(\beta_j) < \frac{1}{2n \cdot 2^n} \qquad (1 \le j \le n). \tag{C_n}$$

Supposing that this were done we would have

$$F'(\alpha_j) = \lim_{n\to\infty} f_n(\alpha_j) = 1,$$

$$0 < F'(x) = \lim_{n\to\infty} f_n(x) \le 1,$$

and, picking $n > j$,

$$F'(\beta_j) = f_{n-1}(\beta_j) + \sum_{k=n}^{\infty} w_k(\beta_j) < 1 - \frac{1}{n} + \sum_{k=n}^{\infty} \frac{1}{2k \cdot 2^k}$$

$$< 1 - \frac{1}{n} + \frac{1}{2n} \cdot 1 = 1 - \frac{1}{2n} < 1$$

and thus we would have the desired F.

We proceed inductively. First choose an open interval I with midpoint α_1 such that $\beta_1 \notin I$. Then apply step V with $\varepsilon = y_1 = 1/4$ to obtain $f_1 = w_1$ that satisfies (A_1), (B_1), and (C_1).

Suppose that $n > 1$ and that f_{n-1} and w_{n-1} have been chosen which satisfy (A_{n-1}), (B_{n-1}), and (C_{n-1}). Select disjoint open intervals $I_1, \ldots, I_n$ such that, for each $j \in \{1, \ldots, n\}$, α_j is the midpoint of I_j, $I_j \cap \{\beta_1, \ldots, \beta_n\} = \emptyset$ and

$$f_{n-1}(x) < f_{n-1}(\alpha_j) + \delta$$

for $x \in I_j$, where

$$\delta = \frac{1}{n(n+1)} - \frac{1}{2n \cdot 2^n} > 0.$$

We now apply step V with

$$\varepsilon = \frac{1}{2n \cdot 2^n} \quad \text{and} \quad y_j = 1 - \frac{1}{n} - f_{n-1}(\alpha_j) \quad \text{when} \quad 1 \le j \le n,$$

to obtain w_n. It is clear that (C_n) is satisfied. Also

$$f_n(\alpha_j) = f_{n-1}(\alpha_j) + w_n(\alpha_j) > f_{n-1}(\alpha_j) + y_j = 1 - \frac{1}{n}$$

when $1 \le j \le n$, and so (A_n) is satisfied. To check (C_n), observe that if $x \in I_j$, then

$$f_n(x) = f_{n-1}(x) + w_n(x) < f_{n-1}(\alpha_j) + \delta + y_j + \varepsilon$$
$$= 1 - \frac{1}{n} + \frac{1}{n(n+1)} = 1 - \frac{1}{n+1};$$

while if $x \notin \cup_{j=1}^{n} I_j$, then

$$f_n(x) = f_{n-1}(x) + w_n(x) < 1 - \frac{1}{n} + \varepsilon < 1 - \frac{1}{n+1}. \qquad \square$$

VII. *Proof of the Theorem:* Let $\{\alpha_j\}_{j=1}^{\infty}$ and $\{\beta_j\}_{j=1}^{\infty}$ be disjoint dense subsets of R^1. Apply step VI to obtain everywhere differentiable functions F and G on R^1 such that

$$F'(\alpha_j) = G'(\beta_j) = 1, \quad G'(\alpha_j) < 1, \quad F'(\beta_j) < 1,$$

$$0 < F'(x) \le 1, \quad 0 < G'(x) \le 1$$

for all j and x. Now put

$$H = F - G.$$

Then

$$H'(\alpha_j) > 0, \quad H'(\beta_j) < 0, \quad -1 < H'(x) < 1$$

for all j and x. Since $\{\alpha_j\}_{j=1}^{\infty}$ and $\{\beta_j\}_{j=1}^{\infty}$ are both dense in R^1, H cannot be monotone on an interval. □

Remarks The foregoing construction is due to Y. Katznelson and K. Stromberg. For other examples see A. Denjoy, "Sur les fonctions dérivées sommables", *Bull. Soc. Math. France* 43 (1915), 161 - 248.

22. Let H be any function as in the theorem of Exercise 21. Show that H has a local maximum and a local minimum in every interval of R^1.

23. Let f be a real-valued function defined on the real line R^1. Let J be any bounded open interval in R^1. We define

$$\omega(f;J) = \sup_{x \in J} f(x) - \inf_{x \in J} f(x)$$

and call $\omega(f;J)$ the *oscillation of* f *over* J. If $a \in R^1$, we define

$$\omega(f;a) = \inf \omega(f;J),$$

where the infimum is taken over all bounded open intervals J containing the point a.

One can easily see that if f is continuous at $a \in R^1$, then $\omega(f;a) = 0$ and if f is not continuous at $a \in R^1$, then $\omega(f;a) > 0$.

Suppose that $r > 0$ and let E_r be the set of all $a \in R^1$ such that $\omega(f;a) \ge 1/r$. Show that E_r is a closed set.

Solution Let x be any limit point of E_r. We have to show that $x \in E_r$, that is, we must show that $\omega(f;x) \ge 1/r$. To do this, it is enough to verify that if J is a bounded open interval containing x, then $\omega(f;J) \ge 1/r$, since $\omega(f;x)$ is the infimum of such $\omega(f;J)$.

Since the open interval J contains the limit point x of E_r, J must contain a point y of E_r. But then $\omega(f;J) \geq \omega(f;y) \geq 1/r$ and we have what we set out to do. □

24. A subset of the real line R^1 is said to be of *type* F_σ if it is a countable union of closed sets.

Let f be as in Exercise 23. Show that the set D of points of R^1 at which f is discontinuous is of type F_σ.

Solution If $x \in D$, then $(f;x) > 0$. For some natural number n we must have $\omega(f;x) \geq 1/n$. This shows that

$$D \subset \bigcup_{n=1}^{\infty} E_{1/n}$$

when $E_{1/n}$ is defined as in Exercise 23. Conversely, if

$$x \in \bigcup_{n=1}^{\infty} E_{1/n},$$

then $\omega(f;x) > 0$ and so $x \in D$. Hence

$$D = \bigcup_{n=1}^{\infty} E_{1/n}.$$

But each $E_{1/n}$ is closed by Exercise 23 and therefore D is a countable union of closed sets. □

25. Show that there is no real-valued function f on R^1 which is continuous at each rational point but discontinuous at each irrational point by verifying the following claims:

(i) The set of all irrational points of R^1 is of second category.

Solution The set of rational points of R^1 is of first category since it is the countable union of one-point sets and any one-point set of R^1 is closed and nowhere dense. If the set of irrational points of R^1 were of first category, then R^1 itself would have to be of first category as the union of two sets of first category; but R^1 is a complete metric space and is therefore of second category and we have reached a contradiction. □

(ii) The set of all irrational points of R^1 is not of type F_σ.

Solution Let A denote the set of all irrational points. If A were of type F_σ, then

$$A = \bigcup_{n=1}^{\infty} F_n ,$$

where each F_n is closed. But each F_n contains only irrational points. Hence F_n contains no nonempty open interval. Thus each F_n is closed and nowhere dense. This implies that A is of first category, contradicting part (i) above. □

(iii) Let f map R^1 into R^1 and let D denote the set of points at which f is not continuous. Then D is of type F_σ.

Hint See Exercise 24.

N.B.: Recall that at the end of Chapter 3 we considered an example of a function which was continuous at each irrational point but discontinuous at each rational point.

26. A real-valued function f on a metric space (X,d) is called *upper semicontinuous at* x_0 if for any $\varepsilon > 0$ there exists a $\delta(\varepsilon,x_0) > 0$ such that $f(x) \leq f(x_0) + \varepsilon$ whenever $d(x,x_0) \leq \delta(\varepsilon,x_0)$. If f is upper semicontinuous at every point of (X,d), it is said to be *upper semicontinuous on* (X,d).

Lower semicontinuity is analogously defined; that is, in place of $f(x) \leq f(x_0) + \varepsilon$, we put $f(x_0) - \varepsilon \leq f(x)$.

Clearly, f is continuous if and only if it is both upper semicontinuous and lower semicontinuous. Also, if f is lower semicontinuous, then -f is upper semicontinuous, and conversely.

Show that if f is upper semicontinuous on a compact metric space (X,d), then f is bounded above. Moreover, if $A = \sup\{f(x): x \in X\}$, then $f(p_0) = A$ for some $p_0 \in X$.

Solution By the upper semicontinuity, there is an open sphere S around every point $x \in X$ such that $x' \in S$ implies $f(x') \leq f(x) + \varepsilon$. The compactness of (X,d) gives a finite number of these spheres $S_1, \ldots, S_n$ covering X, and the number

$$\max\{f(x_1), \ldots, f(x_n)\} + \varepsilon$$

is then an upper bound for f. Now let $(x_n)_{n=1}^{\infty}$ be a sequence in (X,d) such that $\lim_{n\to\infty} f(x_n) = A$. By the compactness of (X,d) there is a convergent subsequence $(p_k)_{k=1}^{\infty}$ of $(x_n)_{n=1}^{\infty}$; let p_0 be the limit of this subsequence. If $f(p_0) < A$, then letting $\varepsilon = \frac{1}{2}[A - f(p_0)]$, there is a $\delta(\varepsilon,p_0)$ such that

$$f(p) \le f(p_0) + \varepsilon = \frac{1}{2}[A + f(p_0)] \quad \text{for} \quad d(p,p_0) \le \delta(\varepsilon,p_0).$$

But this is in contradiction to the fact that $\lim_{k\to\infty} f(p_k) = A$, and we conclude that $f(p_0) < A$ is impossible. Hence $f(p_0) = A$. □

27. Show that if f is lower semicontinuous on a compact metric space (X,d), then f is bounded below and $f(p_0) = \inf\{f(x): x \in X\}$ for some $p_0 \in X$.

28. Let f be an upper semicontinuous function on a compact metric space (X,d). Then there exists a sequence $(f_n)_{n=1}^{\infty}$ of real-valued continuous functions on (X,d) such that, for n = 1, 2, ..., $f_{n+1}(x) \le f_n(x)$ and $\lim_{n\to\infty} f_n(x) = f(x)$ for all $x \in X$.

Solution By Exercise 26, f is bounded above so that the function

$$\sup_{x_1 \in X} \{f(x_1) - n\, d(x,x_1)\}$$

is finite valued. To see that it is continuous, we observe that

$$f_n(x) \ge f(x_1) - n\, d(x_1,x)\}$$
$$\ge f(x_1) - n[d(x_1,x') + d(x',x)];$$

therefore

$$f_n(x) \ge \sup_{x_1 \in X} \{f(x_1) - n\, d(x_1,x')\} - n\, d(x',x)$$

or

$$f_n(x) \ge f_n(x') - n\, d(x',x).$$

Hence $f_n(x') - f_n(x) \le n\, d(x',x)$, and since we can interchange x' and x, we obtain

$$|f_n(x') - f_n(x)| \le n\, d(x',x),$$

which shows the continuity of f_n. The inequality

$$f(x_1) - n\,d(x,x') \geq f(x_1) - (n+1)\,d(x,x')$$

yields the monotonicity of the sequence $(f_n)_{n=1}^{\infty}$, that is,

$$f_1(x) \geq f_2(x) \geq \cdots$$

while

$$f_n(x) \geq f(x) - n\,d(x,x) = f(x)$$

gives $\lim_{n\to\infty} f_n(x) \geq f(x)$.

Now for a given x_0 we determine $\delta(\varepsilon,x_0)$ so that $d(x,x_0) \leq \delta(\varepsilon,x_0)$ implies $f(x) \leq f(x_0) + \varepsilon$; hence

$$f(x) - n\,d(x,x_0) \leq f(x_0) + \varepsilon$$

for such values of x, and all n. For $d(x,x_0) > \delta(\varepsilon,x_0)$, we have

$$f(x) - n\,d(x,x_0) \leq f(x) - n\,\delta(\varepsilon,x_0) \leq A - n\,\delta(\varepsilon,x_0),$$

where $A = \sup\{f(x): x \in X\}$. It follows that, for all sufficiently large n, we have

$$f(x) - n\,d(x,x_0) \leq f(x_0) + \varepsilon$$

and

$$f_n(x_0) \leq f(x_0) + \varepsilon.$$

Hence

$$\lim_{n\to\infty} f_n(x_0) \leq f(x_0) + \varepsilon$$

for all ε, and therefore

$$\lim_{n\to\infty} f_n(x_0) \leq f(x_0).$$

Combining this inequality with the one previously derived, we get

$$\lim_{n\to\infty} f_n(x_0) = f(x_0)$$

for all $x_0 \in X$. □

29. Show that if f is a real-valued function on a metric space

(X,d), and if there is a sequence $(f_n)_{n=1}^{\infty}$ of continuous functions such that, for $n = 1,2, \ldots, f_{n+1}(x) \le f_n(x)$ and $\lim_{n\to\infty} f_n(x) = f(x)$ pointwise, then f is upper semicontinuous.

An analogous result holds for increasing sequences.

30. *Baire's Insertion Theorem:* Let g and h be two real-valued functions defined on a compact metric space (X,d), g being lower semicontinuous and h being upper semicontinuous on (X,d), and suppose that $g(x) \ge h(x)$ for all $x \in X$. Show that there exists a function f, continuous on (X,d) and such that

$$g(x) \ge f(x) \ge h(x)$$

for all $x \in X$.

Solution By Exercise 28, there exist sequences $(g_n)_{n=1}^{\infty}$ and $(h_n)_{n=1}^{\infty}$ of continuous functions such that

$$g_1 \le g_2 \le g_3 \le \ldots, \; g_n \to g \text{ pointwise},$$

$$h_1 \ge h_2 \ge h_3 \ge \ldots, \; h_n \to h \text{ pointwise}.$$

We introduce the function f^+ by setting

$$f^+(x) = f(x), \quad \text{if } f(x) \ge 0,$$

$$= 0, \quad \text{if } f(x) < 0.$$

We have

$$h_1 - g_1 \ge h_1 - g_2 \ge h_2 - g_2 \ge h_2 - g_3 \ge \cdots$$

and consequently

$$(h_1 - g_1)^+ \ge (h_1 - g_2)^+ \ge (h_2 - g_2)^+ \ge (h_2 - g_3)^+ \ge \cdots.$$

The latter sequence of functions converges pointwise to $(h - g)^+$, that is, to 0, because $(h - g)(x) \le 0$ for all $x \in X$. Consider the series

$$g_1 + (h_1 - g_1)^+ - (h_1 - g_2)^+ + (h_2 - g_2)^+ - (h_2 - g_3)^+ + \cdots.$$

Beginning with its second term, this series alternates and is thus convergent (by Proposition 70 of Chapter 1); denote its limit by f. The terms of this series are continuous functions, and so are its partial sums. Now, the partial sums of odd orders form an increasing sequence, and those of even orders a decreasing sequence, both converging pointwise to the function f; thus, by Exercise 29, f is simultaneously lower semicontinuous and upper semicontinuous, thus it is continuous.

We have to show that $g(x) \geq f(x) \geq h(x)$. At those points x where $g(x) = h(x)$, we have $h_n(x) \geq g_m(x)$, thus

$$(h_n - g_m)^+(x) = (h_n - g_m)(x)$$

(for all n and m), consequently

$$f = g_1 + (h_1 - g_1) - (h_1 - g_2) + (h_2 - g_2) - (h_2 - g_3) + \cdots$$
$$= \lim_{n\to\infty} g_n = \lim_{n\to\infty} h_n = g = h.$$

At those points x where $g(x) > h(x)$, all terms of the series

$$g_1 + (h_1 - g_1)^+ - (h_1 - g_2)^+ + (h_2 - g_2)^+ - (h_2 - g_3)^+ + \cdots \tag{E.1}$$

from some term on are equal to 0, this term depending on the point considered. If, for a point x, the first term of (E.1) which vanishes is $(h_n - g_{n+1})^+$, then

$$f = g_1 + (h_1 - g_1) - (h_1 - g_2) + \cdots + (h_{n-1} - g_n) = g_n$$

and

$$g \geq g_n \geq h_n \geq h;$$

and if the first term of (E.1) which vanishes is $(h_n - g_{n+1})^+$, then

$$f = g_1 + (h_1 - g_1) - (h_1 - g_2) + \cdots + (h_n - g_n) = h_n$$

and

$$g \geq g_{n+1} \geq h_n \geq h,$$

thus in each case we have $g \geq f \geq h$. □

31. A subset D of the real line R^1 is said to have *measure zero* if, for any $\varepsilon > 0$, there is a countable set $(I_k)_{k=1}^{\infty}$ of intervals such that

$$\bigcup_{k=1}^{\infty} I_k \supset D \quad \text{and} \quad \sum_{k=1}^{\infty} |I_k| < \varepsilon,$$

where $|I_k|$ is the length of the interval I_k.

Clearly, any countable set $\{t_n : n = 1, 2, \ldots\}$ has measure zero, because the point t_n can be enclosed in an interval of length $\varepsilon/2^n$, and $\Sigma_{n=1}^{\infty} \varepsilon/2^n = \varepsilon$.

Show that if $E = \cup_{n=1}^{\infty} E_n$, where each E_n has measure zero, then E has measure zero.

Solution Given $\varepsilon > 0$, the set E_1 can be enclosed in the union $\cup_{k=1}^{\infty} I_{1,k}$, where $\Sigma_{k=1}^{\infty} |I_{1,k}| \le \varepsilon/2$, and in general, the set E_m can be enclosed in $\cup_{k=1}^{\infty} I_{m,k}$, where $\Sigma_{k=1}^{\infty} |I_{m,k}| \le \varepsilon/2^m$. The intervals $I_{m,k}$ with m, k = 1, 2, ... then satisfy

$$\bigcup_{m,k=1}^{\infty} I_{m,k} \supset \bigcup_{m=1}^{\infty} E_m = E, \quad \text{and}$$

$$\sum_{m,k=1}^{\infty} |I_{m,k}| \le \frac{\varepsilon}{2} + \frac{\varepsilon}{2^2} + \cdots + \frac{\varepsilon}{2^m} + \cdots = \varepsilon. \qquad \square$$

32. *Lebesgue's Characterization of the Riemann Integral:* A bounded function f on [a,b] is Riemann integrable if and only if the set D of discontinuities of f has measure zero.

Solution Let $E_n = \{t: \omega(t) \ge 1/n\}$, where $\omega(t) = \omega(f;t)$ with $\omega(f;t)$ as in Exercise 23. Clearly, f is continuous at t if and only if $\omega(t) = 0$. The set E_n is closed, and the set D of points where f is discontinuous is $\cup_{n=1}^{\infty} E_n$.

Now assume that f is integrable, so that for any $\varepsilon > 0$ there is a partition P_ε such that $U(f,P_\varepsilon) - L(f,P_\varepsilon) \le \varepsilon$. If

$$P_\varepsilon = \{a = t_0 < t_1 < \cdots < t_{n-1} < t_n = b\},$$

then the set of intervals

$$I_1 = (t_0, t_1), \quad \ldots, \quad I_n = (t_{n-1}, t_n)$$

can be split into two groups, where the intervals in the first group have nonempty intersection with E_m, and those of the second group do not meet E_m (m fixed). Then

$$U(f,P_\varepsilon) - L(f,P_\varepsilon) = \Sigma'(M_k - m_k)(t_k - t_{k-1}) + \Sigma''(M_k - m_k)(t_k - t_{k-1}) \le \varepsilon,$$

where the prime (resp. double prime) indicates summation over intervals of the first group (resp. second group). On the primed intervals, $M_k - m_k \ge 1/m$, so that

$$\frac{1}{m}\Sigma'(t_k - t_{k-1}) \le \Sigma'(M_k - m_k)(t_k - t_{k-1}) \le \varepsilon$$

and $\Sigma'(t_k - t_{k-1}) \le m\varepsilon$. Thus E_m can be enclosed in finitely many intervals of length less than or equal to $m\varepsilon$, and since ε is arbitrary, we see that E_m has measure zero. It now follows from Exercise 31 that D has measure zero.

Conversely, assume that $D = \cup_{n=1}^\infty E_n$ has measure zero and, hence, that E_n has measure zero for all n. Choose K so that $1/K < \varepsilon$. The set $E_K^c = [a,b] - E_K$ is open; therefore

$$E_K^c = \bigcup_{k=1}^\infty I_k,$$

where the intervals $(I_k)_{k=1}^\infty$ are open and disjoint. We now show that $\Sigma_{k=1}^\infty |I_k| = b - a$.

Suppose, on the contrary, that $\Sigma_{k=1}^\infty |I_k| = \lambda < b - a$. Since E_K has measure zero, we can find $(J_k')_{k=1}^\infty$ such that

$$\bigcup_{k=1}^\infty J_k' \supset E_K \quad \text{and} \quad \sum_{k=1}^\infty |J_k'| < \frac{b - a - \lambda}{2}.$$

The intervals $(J_k')_{k=1}^\infty$ and $(I_k)_{k=1}^\infty$ together cover [a,b], and hence

$$b - a \le \sum_{k=1}^\infty |J_k'| + \sum_{k=1}^\infty |I_k| < \frac{b - a - \lambda}{2} + \lambda = \frac{b - a + \lambda}{2} < b - a.$$

This contradiction shows that $\lambda = b - a$. We now see that, for some n,

$$\sum_{k=1}^{n} |I_k| > b - a - \frac{\varepsilon}{2}$$

and we can find closed intervals $J_1, \ldots, J_n$ so that for $j = 1, \ldots, n$,

$$I_j \subset J_j \quad \text{and} \quad \sum_{j=1}^{n} |J_j| > b - a - \varepsilon.$$

At each point t_0 of J_1 we have $\omega(t_0) < 1/K < \varepsilon$, and there is $\delta(t_0)$ such that, on the interval $[t_0 - \delta(t_0), t_0 + \delta(t_0)]$, the supremum of $f(t)$ minus the infimum of $f(t)$ is less than ε. By the compactness of J_1, finitely many such intervals cover J_1. We can use the endpoints of these intervals to form a partition P_1 of J_1, and similarly, we form partitions $P_2, \ldots, P_n$ of $J_2, \ldots, J_n$, thus obtaining a partition P of $[a,b]$.

Calculating $U(f,P) - L(f,P)$, we have

$$\begin{aligned} U(f,P) - L(f,P) &= \Sigma\, (M_k - m_k)(t_k - t_{k-1}) \\ &= \Big[\sum_{P_1} (M_k - m_k)(t_k - t_{k-1}) + \cdots \\ &\quad + \sum_{P_n} (M_k - m_k)(t_k - t_{k-1})\Big] \\ &\quad + \Sigma'\, (m_k - m_k)(t_k - t_{k-1}) \\ &\le \sum_{k=1}^{n} |J_k| + \varepsilon 2M \le \varepsilon(2M + b - a), \end{aligned}$$

where $M = \sup\{|f(t)| : t \in [a,b]\}$. It follows that f is Riemann integrable over $[a,b]$. □

33. Using Exercise 32, we show that there exists a function which is differentiable at every point of an interval and whose derivative is a bounded function, but this derivative is not Riemann integrable.

The following construction is due to V. Volterra. We remove from the middle of the closed interval $[0,1]$ the open interval of length $1/4$. From the middle of the two remaining closed intervals we remove open intervals of lenght 1/16. In the n-th step we remove from the middle of each of the 2^{n-1} closed intervals which remain after the (n - 1)-th step an open interval of length 4^{-n}. Continuing this process indefinitely we remove from the interval [0,1] a system of open intervals of total length

$$\frac{1}{4} + 2\,\frac{1}{4^2} + \cdots + 2^{n-1}\,\frac{1}{4^n} + \cdots = \frac{1}{2}.$$

The points which remain form a closed set E; E is clearly not of measure zero. Let δ_n denote the length of each of the closed intervals which remain after the n-th step. From the construction it is clear that $\delta_{n+1} < \delta_n/2$; hence $\delta_n \to 0$ as $n \to \infty$. This implies that no subinterval of [0,1], no matter how small, can be entirely contained in the set E.

Now we define the function f. At the points x of the set E we put $f(x) = 0$. If (α,β) is one of the removed open intervals we define

$$f(x) = (x - \alpha)^2 \sin \frac{1}{x - \alpha}$$

immediately to the right of α, and

$$f(x) = (x - \beta)^2 \sin \frac{1}{\beta - x}$$

immediately to the left of β, until we reach the maximum points $\overline{\alpha}$, $\overline{\beta}$

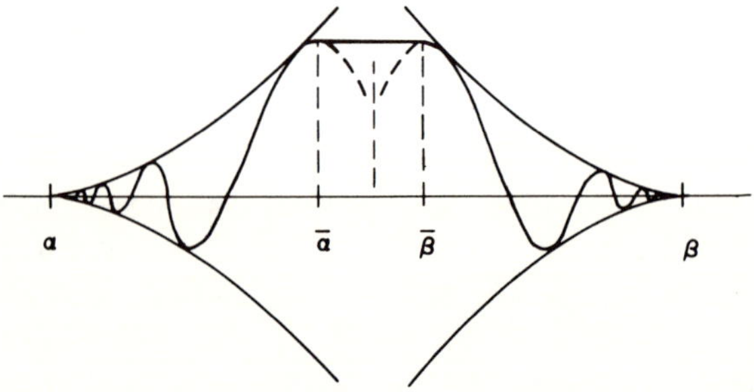

Figure 5 - 1

nearest the middle of (α,β); in the interval $(\bar{\alpha},\bar{\beta})$ we set f(x) equal to this maximum value (see Fig. 5 - 1).

We have thus defined the function f on the entire interval [0,1] and it is a continuous function.

It is clear that f is differentiable in each of the removed intervals (α,β), even at the points $\bar{\alpha}$, $\bar{\beta}$ where it has the derivative 0. For x near enough to α $(x > \alpha)$ we have

$$f'(x) = 2(x - \alpha) \sin \frac{1}{x - \alpha} - \cos \frac{1}{x - \alpha} ;$$

if $x \to \alpha$ the first term of the right-hand side converges to 0 while the second term oscillates between the values +1 and -1. The situation is analogous in a left-hand neighborhood of β.

$f'(x)$ exists at the point α, β and even at every point $x_0 \in E$, and we have

$$f'(x_0) = 0. \tag{E.2}$$

To show this, we assume first that $x > x_0$. If $x \in E$, then

$$f(x) - f(x_0) = 0 - 0 = 0,$$

and if x is contained in some of the removed intervals (α,β) then

$$|f(x) - f(x_0)| = |f(x)| \le (x - \alpha)^2 \le (x - x_0)^2;$$

thus in each case

$$\left|\frac{f(x) - f(x_0)}{x - x_0}\right| \le |x - x_0|.$$

We get the same result for $x < x_0$. Letting $x \to x_0$ this implies (E.2).

Thus $f'(x)$ exists everywhere. But at the points of E it is not continuous. Indeed, if $x_0 \in E$, then in every neighborhood of x_0 there is a point, and therefore also an endpoint, of one of the removed intervals, and we know that at such an endpoint the oscillation of the function f is equal to 2. But the set E is not of measure zero and so f' cannot be Riemann integrable according to the result in Exercise 32. □

34. Let f be a continuous and increasing function on an interval [a,b]. We desire to show that f has a finite derivative at every point of [a,b] with the possible exception of the points of a set of measure zero. The method of proof that we are going to employ is due to F. Riesz.

The function f is differentiable at x if, and only if, all four Dini derivatives have the same finite value there, and then

$$f'(x) = D^+f(x) = D_+f(x) = D^-f(x) = D_-f(x).$$

As a direct consequence of the definition of Dini derivatives we have

$$-\infty \le D_-f(x) \le D^-f(x), \quad -\infty \le D_+f(x) \le D^+f(x) \le \infty;$$

in the case of an increasing function the Dini derivatives are clearly nonnegative.

To prove our claim it will be enough to verify that for any increasing function we have, except for a set of measure zero,

(i) $D^+f(x) < \infty$,

(ii) $D^+f(x) \le D_-f(x)$.

In fact, applying (ii) to the function $-f(-x)$, it follows that we also have, except for a set of measure zero,

$$D^-f(x) \le D_+f(x),$$

and combining this with (i) and (ii) we obtain, except for a set of measure zero,

$$0 \le D^+f(x) \le D_-f(x) \le D^-f(x) \le D_+f(x) \le D^+f(x) \le \infty;$$

hence the equality sign holds, except for a set of measure zero, which was to be proved.

Assertion (i) means that the set E_∞ of points x for which $D^+f(x) = \infty$ is of measure zero. This set E_∞ is included in the set E_C of points x for which $D^+f(x) > C$, where C denotes a real number chosen as large as we wish. But the relation $D^+f(x) > C$ implies the existence of a $t > x$, such that

$$\frac{f(t) - f(x)}{t - x} > C,$$

that is,

$$g(t) > g(x),$$

where we set $g(x) = f(x) - Cx$. Hence the set E_C is included in the set E attached to the function g by Exercise 28 at the end of Chapter 2 and so it can be covered by a sequence of disjoint open intervals (a_k,b_k) for which

$$g(a_k) \le g(b_k),$$

that is,

$$f(b_k) - Cb_k \ge f(a_k) - Ca_k ;$$

hence

$$C(b_k - a_k) \le f(b_k) - f(a_k).$$

This yields, by addition,

$$C \sum_k (b_k - a_k) \le \sum_k [f(b_k) - f(a_k)] \le f(b) - f(a)$$

and shows that for sufficiently large C, the total length of the intervals (a_k,b_k) will be as small as we wish. That is, the set E_∞ is of measure zero.

Statement (ii) is proved by similar reasoning which is repeated alternately in two different forms. Let c and C be two positive numbers with $c < C$. We first show that the set E_{cC} of points x for which

$$D^+f(x) > C \quad \text{and} \quad D_-f(x) < c,$$

is of measure zero.

Consider first the condition $D_-f(x) < c$. Applying the result of Exercise 28 at the end of Chapter 2 to the function $g(x) = f(-x) + cx$ on the interval $[-b,-a]$, we get, for reasons similar to those just used, that the set E_{cC} can be covered by a sequence of disjoint open intervals (a_k,b_k), such that

$$f(b_k) - f(a_k) \le c(b_k - a_k). \tag{E.3}$$

Next we consider, inside each of the intervals (a_k, b_k), the points x where $D^+f(x) > C$; applying Exercise 28 at the end of Chapter 2 to the function $g(x) = f(x) - Cx$ on the interval (a_k, b_k), we see that these points can be covered by a sequence of disjoint intervals (a_{kj}, b_{kj}) such that

$$C(b_{kj} - a_{kj}) \le f(b_{kj}) - f(a_{kj});$$

hence

$$C \sum_j (b_{kj} - a_{kj}) \le \sum_j [f(b_{kj}) - f(a_{kj})] \le f(b_k) - f(a_k).$$

Taking into account (E.3), it follows that

$$C \sum_{k,j} (b_{kj} - a_{kj}) \le c \sum_k (b_k - a_k).$$

If $|\Sigma_1|$ and $|\Sigma_2|$ denote the total lengths of the systems $\Sigma_1 = \{(a_k, b_k)\}$ and $\Sigma_2 = \{(a_{kj}, b_{kj})\}$, respectively, it follows that

$$|\Sigma_2| \le \frac{c}{C} |\Sigma_1|.$$

Repeating the two steps alternately, we obtain a sequence Σ_1, Σ_2, ... of systems of intervals, each imbedded in the preceding, and we have, for $n = 1, 2, \ldots$,

$$|\Sigma_{2n}| \le \frac{c}{C} |\Sigma_{2n-1}| \le \frac{c}{C} |\Sigma_{2n-2}|.$$

It follows that

$$|\Sigma_{2n}| \le \left(\frac{c}{C}\right)^n |\Sigma_1| \to 0 \quad \text{as} \quad n \to \infty.$$

Thus the set E_{cC} can be covered by a sequence of intervals of total length as small as we wish and so E_{cC} is seen to be of measure zero.

Now we form the union E^* of all the sets E_{cC} corresponding to pairs c, C of positive rational numbers with $c < C$. As a union of countably many sets of measure zero, E^* itself is of measure zero (by Exercise 31). If at a point x we have $D_-f(x) < D^+f(x)$, we can wedge between $D_-f(x)$ and $D^+f(x)$ two rational numbers,

$$D_-f(x) < c < C < D^+f(x);$$

then x is a point of the set E_{cC} and consequently of E^*. Thus the points x where (ii) does not hold, form a set of measure zero. This completes the proof.

Remark It is of course possible to show that any monotonic function possesses a finite derivative at every point with the possible exception of the points of a set of measure zero. We leave to the reader the task of examining the details in Chapter 4 of the author's book "Real Analysis" (American Elsevier Publishing Company, New York, 1973).

35. Let E be a set of positive integers satisfying the following condition: There exists a positive number T such that any interval of length T on the positive part of the number line contains at least one element of the set E. We let

$$\chi_E(n) = 1 \quad \text{if} \quad n \in E,$$
$$= 0 \quad \text{if} \quad n \notin E$$

and put

$$A_k = \sum_{n=1}^{k-1} \chi_E(n)$$

and

$$\phi_k(n) = \frac{1}{A_k} \sum_{j=1}^{k-1} \chi_E(n + j)\, \chi_E(j).$$

Observe that $0 \le \phi_k(n) \le 1$. Thus, by Proposition 50 of Chapter 2, there exists a limit function ϕ_ω such that $0 \le \phi_\omega(n) \le 1$.

Prove the following results:

(a) If $\phi_\omega(n) > 0$, then n can be represented as the difference of two elements of the set E.

(b) For $n = 0, \pm 1, \pm 2, \ldots, \phi_\omega(n)$ is a positive definite sequence.

(c) By Exercise 22 at the end of Chapter 4, we can write

$$\phi_\omega(n) = \int_{-\pi}^{\pi} e^{int}\, dg(t), \quad (n = 0, \pm 1, \pm 2, \ldots),$$

where g is a nondecreasing function with total variation

$$\int_{-\pi}^{\pi} dg(t) = \phi_\omega(0) \le 1.$$

We decompose g into its monotonically increasing jump function g_d and its monotonically increasing continuous part g_c and define

$$\Psi(n) = \int_{-\pi}^{\pi} e^{int}\, dg_d(t) \quad \text{and} \quad \Phi(n) = \int_{-\pi}^{\pi} e^{int}\, dg_c(t).$$

Then $\Phi(n) \to 0$ as $n \to \infty$, $\Psi(n)$ is nonnegative, and

$$\lim_{K \to \infty} \frac{1}{K} \sum_{n=1}^{K-1} \Psi(n)$$

is strictly positive.

(d) Let

$$\Lambda_k(n) = \frac{1}{k} \sum_{j=1}^{k-1} \phi_\omega(n + j)\, \phi_\omega(j).$$

Then the limit

$$\Lambda(n) = \lim_{k \to \infty} \Lambda_k(n)$$

exists. Moreover, if $\Lambda(n) > 0$, then n is of the form

$$n = n_p + n_q - n_r - n_s,$$

where n_p, n_q, n_r, and n_s belong to the set E.

(f) For what n is $\Lambda(n) > 0$?

INDEX

C

D

R

S

T

U

V

W